“十二五”普通高等教育本科国家级规划教材

Probability Theory

An Elementary Course

LIN Zhengyan SU Zhonggen ZHANG Lixin

ZHEJIANG UNIVERSITY PRESS
浙江大学出版社
World Scientific

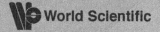

图书在版编目(CIP)数据

概率论＝Probability Theory：An Elementary Course：英文/林正炎，苏中根，张立新编著. —杭州：浙江大学出版社，2020.6(2022.7 重印)
ISBN 978-7-308-19879-0

Ⅰ. ①概… Ⅱ. ①林… ②苏… ③张… Ⅲ. ①概率论-高等学校-教材-英文 Ⅳ. ①O211

中国版本图书馆 CIP 数据核字(2019)第 286387 号

概率论

林正炎　苏中根　张立新 编著

责任编辑　李　晨
责任校对　虞雪芬　宁　檬
封面设计　春天书装
出版发行　浙江大学出版社
(杭州市天目山路 148 号　邮政编码 310007)
(网址：http://www.zjupress.com)
排　　版　杭州朝曦图文设计有限公司
印　　刷　浙江新华数码印务有限公司
开　　本　787mm×1092mm　1/16
印　　张　16.75
字　　数　503 千
版 印 次　2020 年 6 月第 1 版　2022 年 7 月第 4 次印刷
书　　号　ISBN 978-7-308-19879-0
定　　价　50.00 元

Preface to the Updated Edition

The first edition of the book was published in 2001, and the revised edition was published in 2008. It has been used by a number of universities. However, as we said in the preface to the revised edition, a textbook must be revised repeatedly in order to improve day by day. Taking advantage of the opportunity that the World Scientific intends to publish the book jointly with Zhejiang University Press, recently, we have made some major revisions to the book, for example, adding some distribution functions with important application backgrounds and adding some descriptions and examples of essential attributes that reflect probability and mathematical expectation; moreover, the conditional probability and conditional distribution are described in more detail, the proofs of some important theorems in the limit theory have been added. We have also revised some minor mistakes and inappropriate points in the original book.

We are entering the era of big data, and probability theory is getting more and more attention, because it is the main theoretical basis of data analysis.

This book can be used as a textbook or teaching reference book for mathematics, statistics and related majors in universities. It can also be used as a supporting material for our Chinese version of the textbook *Probability Theory* (《概率论》). For professionals who want to learn the principles and methods of data processing, it can also be used as an introductory material.

Lin Zhengyan, Su Zhonggen, and Zhang Lixin
January 2019
Yuquan Campus, Zhejiang University

Preface to the Revised Edition

Since the publishing of this book in 2001 (Second Printing in 2003), it has not only served as a required course teaching material for all majors from the Department of Mathematics in Zhejiang University, but has also been used by other colleges and universities, obtaining satisfactory effects. However, a good textbook can only be improved through a large amount of real teaching experiences, and unceasing and repeated revisions. Along with developments and changes on science in teaching environment, such a process shall be forever ongoing without end.

When time comes for repubilcation, the authors make revisions to the original edition according to practices and experiences of teaching in the past years. The authors correct typos and make suitable modification of the contents, and supplement materials on how to use the MATLAB software to calculate common probability values and to simulate the central limit theorems.

The authors wish to extend sincere thanks to Professor Lixin Zhang and Dr. Caiya Zhang for their valuable suggestions and support. The authors also wish to thank the Department of Mathematics in Zhejiang University for care and support toward the revision and the republication of this book.

Although much revision has been made to the original edition, there still certainly exist shortcomings and mistakes. Therefore, the authors earnestly request the readers to be liberal and forthcoming with any criticism and instruction.

Authors
April 2008
Zhejiang University

Preface to the First Edition

The subject matter of probability theory is the mathematical analysis of random events, that is, of those empirical phenomena which do not have deterministic regularity but possess some statistical regularity.

Probability theory, as a discipline, originated in the middle of the seventeenth century with Pascal, Fermat, and Huygens, although special calculations of probabilities in games of chance had been made earlier. The true history of probability theory begins with the work of James Bernoulli, *Ars Conjectandi* (*The Art of Guessing*) published in 1713, in which he proved quite rigorously the first limit theorem of probability theory, the law of large numbers; of de Moivre, *Miscellanea Analytica Supplementum* (*The Analytic Method*), in which the central limit theorem was stated and proved for the first time (for symmetric Bernoulli trials). The modern period in the development of probability theory begins with the axiomatization. Kolmogorov's book *Foundations of the Theory of Probability* appeared in 1933. He presented the axiomatic theory which has become generally accepted and not only is applicable to all the classical branches of probability theory, but also provides a firm foundation for the development of new branches that have arisen from questions in the sciences.

Indeed, in the last century the theory of probability has grown from a minor isolated theme into a broad and intensive discipline interacting with many other branches of mathematics. At the same time it is playing a central role in the mathematization of various applied science such as statistics, operation research, biology, economics, and psychology, to name a few. With the passage of time, probability theory and its applications have won a place in the college curriculum as a mathematical discipline essential to many fields of study. The elements of the theory are now given at different levels, sometimes even before calculus. The present textbook is intended for a course at about the sophomore level. It presuppose no prior acquaintance with the subject and the first chapter can be read largely without the benefit of calculus. The next two chapters require a working knowledge of infinite series, calculus and related topics. The fourth chapter is devoted to problems about weak convergence of probability distributions and the method of characteristic functions for proving limit theorems. The same part also discusses properties with probability one for sequences and sums of independent random variables.

Each chapter ends with Supplements and Remarks, which deal with more special or elaborate material and may be skipped, but a little browsing in them is recommended. A certain amount of flexibility and choice is left to the instructor who can best judge what is right for his or her class. There are over two hundred exercises throughout the four chapters. Many are routine and easy, and some are intended to give additional information about the circle of ideas that is under discussion.

A few words are necessary about the motivation for the textbook. In recent years, the Ministry of Education advocates bilingual teaching in colleges and universities with the aim to enable students to read foreign books as soon as possible and better join international academic activities. A major problem encountered in bilingual teaching is with no doubt how to choose a good textbook. It is really a difficult work to find a foreign probability textbook that not only emphasizes probability theory but also pays attention to applications and is consistent with our current teaching systems. Taking this into account, the authors make a daring attempt to write themselves an English textbook. There are certainly many errors and mistakes in English writing. The authors hope to gain experiences through teaching practice and to hear suggestions and comments from the readers as well.

The author of any elementary textbook owes of course a large debt to innumerable predecessors. The list of references only includes those books that have something to do with the present book. This book is based on a probability course that the authors have given roughly in the last decade. The Chinese textbook published earlier is recommended to the readers as a reference. The authors are grateful to all the students in those courses for the feedback and insight they provided. The writing of this book is supported by Zhejiang University Press and the Department of Mathematics of Zhejiang University. Finally, it is a pleasure to thank Mr. Yang Xiaoming for what he has done for the publication of this book.

Lin Zhengyan and Su Zhonggen
July 2005
Xixi Campus, Zhejiang University

Contents

Chapter 1 Events and Probabilities

1.1 Random phenomena and statistical regularity ······ 1
- 1.1.1 Random phenomena ······ 1
- 1.1.2 The statistical definition of probability ······ 2

1.2 Classical probability models ······ 3
- 1.2.1 Sample points and sample spaces ······ 3
- 1.2.2 Discrete probability models ······ 4
- 1.2.3 Geometric probability models ······ 10

1.3 The axiomatic definition of probability ······ 12
- 1.3.1 Events ······ 12
- 1.3.2 Probability space ······ 14
- 1.3.3 Continuity of probability measure ······ 21

1.4 Conditional probability and independent events ······ 22
- 1.4.1 Conditional probability ······ 22
- 1.4.2 Total probability formula and Bayes' rule ······ 26
- 1.4.3 Independent events ······ 34

Chapter 2 Random Variables and Distribution Functions

2.1 Discrete random variables ······ 51
- 2.1.1 The concept of random variables ······ 51
- 2.1.2 Discrete random variables ······ 52

2.2 Distribution functions and continuous random variables ······ 61
- 2.2.1 Distribution functions ······ 61
- 2.2.2 Continuous random variables and density functions ······ 65
- 2.2.3 Typical continuous random variables ······ 67

2.3 Random vectors ······ 74
- 2.3.1 Discrete random vectors ······ 74
- 2.3.2 Joint distribution functions ······ 77
- 2.3.3 Continuous random vectors ······ 78

2.4 Independence of random variables 84
2.5 Conditional distribution 86
2.5.1 Discrete case 86
2.5.2 Continuous case 87
2.5.3 The general case 90
2.5.4 The conditional probability given a random variable 91
2.6 Functions of random variables 92
2.6.1 Functions of discrete random variables 93
2.6.2 Functions of continuous random variables 95
2.6.3 Functions of continuous random vectors 97
2.6.4 Transforms of random vectors 101
2.6.5 Important distributions in statistics 105

Chapter 3 Numerical Characteristics and Characteristic Functions
3.1 Mathematical expectations 128
3.1.1 Expectations of discrete random variables 128
3.1.2 Expectations of continuous random variables 130
3.1.3 General definition 132
3.1.4 Expectations of functions of random variables 133
3.1.5 Basic properties of expectations 136
3.1.6 Conditional expectation 139
3.2 Variances, covariances and correlation coefficients 144
3.2.1 Variances 144
3.2.2 Covariances 148
3.2.3 Correlation coefficients 149
3.2.4 Moments 152
3.3 Characteristic functions 154
3.3.1 Definitions 154
3.3.2 Properties 156
3.3.3 Inverse formula and uniqueness theorem 159
3.3.4 Additivity of distribution functions 163
3.3.5 Multivariate characteristic functions 163
3.4 Multivariate normal distributions 164
3.4.1 Density functions and characteristic functions 165
3.4.2 Properties 166

Chapter 4 Probability Limit Theorems

4.1 Convergence in distribution and central limit theorems ······ 188

4.1.1 Weak convergence of distribution functions ······ 188

4.1.2 Central limit theorems ······ 194

4.2 Convergence in probability and weak law of large numbers ······ 199

4.2.1 Convergence in probability ······ 199

4.2.2 Weak law of large numbers ······ 204

4.3 Almost sure convergence and strong laws of large numbers ······ 208

4.3.1 Almost sure convergence ······ 208

4.3.2 Strong laws of large numbers ······ 209

Bibliography ······ 237

Appendix A Distribution of Typical Random Variables

A.1 Distribution of Typical Random Variables ······ 239

A.2 Distributions of Typical Random Variables ······ 240

Appendix B Tables

B.1 Table of Binomial Probabilities ······ 244

B.2 Table of Random Digits ······ 247

B.3 Table of Poisson Probabilities ······ 250

B.4 Table of Standard Normal Distribution Function ······ 252

B.5 Table of χ^2 Distribution ······ 254

B.6 Table of t Distribution ······ 256

Chapter 1

Events and Probabilities

1.1 Random phenomena and statistical regularity

1.1.1 Random phenomena

There are normally two classes of phenomena in nature and human society. One is deterministic phenomenon, which is bound to take place under a certain condition. For instance, the distance a free-fall object travels within t minutes is $gt^2/2$; water boils at 100 degrees centigrade under a standard atmosphere pressure. Furthermore, deterministic phenomena include two categories of events, certain events (denoted by Ω) and impossible events (denoted by $\varnothing$). There are a good many certain events in nature and human society.

The other class is random phenomenon, in which the events are not sure to take place. For example, it will rain on next January 1 in Hangzhou; 850 out of 1 000 seeds one plants will germinate; the distance between the target and the projectile which fires is less than 15 meters.

If we repeatedly observe a random phenomenon under a certain condition, the outcomes may be different from place to place. This is also true in random experiments. In particular, all possible outcomes are known, but the outcome is not predicted with certainty in each individual experiment. Nevertheless, the outcomes of a great deal of repeated experiments under a certain condition turn out to have some regularity — statistical regularity. For example, the distances between the target and projectiles may differ from time to time, but as a common knowledge, people know that the closer the distance is, the better the shooter's score is. Probability theory is a branch of mathematics that focuses on the study of various random phenomena. Universal existence of random phenomena determines the importance of this subject. Sometimes a deterministic phenomenon has to be dealt with as a random one when it is more deeply studied. To illustrate, consider the object's free-fall motion mentioned above. When atmosphere resistance, atmosphere flow and some extra factors are taken into account, the distance a free-fall object travels within t minutes is no longer sure to be $gt^2/2$.

Each possible outcome of a random experiment is termed a random event, or simply an event for short. As an outcome of the experiment, event A possibly happens. The possibility that A comes up may be large or small. The quantitative description of such a possibility is just the probability of the event we will discuss below.

1.1.2 The statistical definition of probability

If event A occurs n times in N repeated experiments under a certain condition, then frequency of A occurring in N experiments is defined as

$$F_N(A) = \frac{n}{N}.$$

In general, frequency is related to the times of doing experiment, and the frequency of N experiments may be different from experimenter to experimenter even though N is fixed. When N is large enough, however, the frequency turns out to have a kind of stability, i.e., the values of $F_N(A)$ show fluctuations which become progressively weaker as N increases, until ultimately $F_N(A)$ stabilizes to a constant. Thus it is very natural to use this limit constant, called probability, to estimate the possibility that A occurs in a single experiment and it is commonly denoted by $P(A)$. Such a definition of probability is said to be a statistical one.

Example 1 Obviously, a head or a tail will come up exclusively when a coin is tossed. Let $A = \{\text{head comes up}\}$. If the coin is fair, the frequency of head falling should be close to 50%. Many mathematicians in history have carried out this experiment. Table 1-1 shows two well-known examples.

Table 1-1

Experimenter	Number of tosses	Times of head	Frequency
Buffon	4 040	2 048	0.5069
Pearson	12 000	6 019	0.5016
Pearson	24 000	12 012	0.5005

Naturally, we take $P(A) = 1/2$ if the coin is fair.

Example 2 It is very helpful for coding and decoding of ciphers to study the use frequency of each English letter. Many statistics show that letter E is most frequently used, the use frequency of which is about 0.105; in the next places are letters T and O; and letters J, Q and Z are least frequently used, the use frequencies of which are roughly 0.001.

In everyday life and practice there are many situations in which people use frequency to estimate probability. For instance, the germination ratio of a set of seeds, the hitting-the-target rate, and the rejection rate of certain products.

In the above we have not given the accurate meaning of stability of frequencies. Nevertheless we will be able to understand it profoundly from argument on the famous large number law in probability theory in Chapter 4.

Although the statistical definition of probability does not enable us to grasp the accurate meaning of probability that an event comes up, it provides us with a method to estimate probability. The relationship between frequency and probability is just like that between the observation value of length of an object and the length itself. Length, which exists objectively, is one of the intrinsic characteristics of an object, while the observation value is to some extent an approximation of it. Similarly, probability is one of the intrinsic characteristics of a random event, while frequency is to some degree an approximation of it.

Note that in order to use the statistical definition of probability, all experiments must be conducted under basically identical condition and independently of each other, and the number of times of experiments is supposed to be large enough. The following three properties of frequency follow immediately from the statistical definition.

(1) Non-negativity: $F_N(A) \geqslant 0$;

(2) Normalization: $F_N(\Omega) = N/N = 1$, where Ω is a certain event;

(3) Additivity: Suppose that A and B will never come up simultaneously and that $A + B$ stands for the event that A or B comes up, then $F_N(A+B) = F_N(A) + F_N(B)$.

Property (3) can be generalized to the case of a finite number of events.

From the above properties of the frequeney and the statistical definition of the probability, we know that the probability also has the properties of non-negativity, normalization and additivity.

1.2 Classical probability models

1.2.1 Sample points and sample spaces

When a dice is thrown the number that comes up is unpredictable, though it must be one of 1,2,…, 6. We take six possible outcomes as a set of $\{1,2,\cdots,6\}$.

Each elementary outcome of a random experiment is termed a sample point, usually denoted by ω. The sample point is an event that cannot be decomposed into some combination of other events. The set of all sample points is termed sample space, usually denoted by Ω. Let $\omega_i = \{i \text{ comes up}\}$ in the proceeding example of throwing a dice, then $\Omega=\{\omega_1,\omega_2,\cdots,\omega_6\}$.

Sample point and sample space are two basic concepts in probability theory. The determination of a sample space depends on the problem under discussion. In other words, we need to choose an appropriate sample space according to the feature of the problem. Sample space is generally chosen before further study.

Example 3 A bag contains 10 balls, 3 of which are red, 3 white and 4 black. If a ball is drawn at random, then the sample space may be taken as $\Omega_1 = \{\text{a red ball, a white ball, a black ball}\}$. If we label these 10 balls with red balls by 1,2,3, white balls by 4,5,6 and black balls by 7, 8, 9, 10, and take a ball randomly, then the sample space may be taken as $\Omega_2 = \{\omega_1,\cdots,\omega_{10}\}$, where $\omega_i = \{\text{the } i\text{-th ball}\}$, $i = 1,2,\cdots,10$.

Example 4 In the above Example 3, if two balls are taken at a time, then each sample point is expressed as (i, j), where i and j are the numbers attached to balls and so the sample space is equal to $\{(1,2), (1,3), \cdots, (1,10), (2,3), \cdots, (2,10), \cdots, (9,10)\}$, which consists of $\binom{10}{2} = 45$ sample points in sum. Obviously, this sample space is 2-dimensional.

Note that the sample spaces shown in both Examples 3 and 4 are finite.

Example 5 The number of cosmic radials falling on a certain area within a unit time may be $0, 1, 2, \cdots$. This variable can take only finitely many values, but since it does not have a certain upper bound, it makes life easy assuming it takes infinitely many values. Thus we may take the sample space to be $\Omega = \{0, 1, 2, \cdots\}$, which contains an infinite number of sample points and all the elements of which can be arranged in a certain order (called countable infinity).

Example 6 The distance between the target and the projectile is a non-negative real number. In this case the sample space may be taken as $\Omega = [0, \alpha]$, a 1-dimensional continuous interval, where α is a positive real constant.

In practical applications, it is worthwhile studying at length how to construct a suitable sample space, as the problem may become very simple if the sample space is chosen correctly.

1.2.2 Discrete probability models

I The definition and calculation

Classical probability model, one of the simplest random experiment models, plays a basic role in probability calculation and has a wide range of applications in practice.

Classical probability model possesses the following two fundamental features.

(1) The sample space is finite, i. e., $\Omega = \{\omega_1, \omega_2, \cdots, \omega_n\}$, where ω_i, $i = 1, 2, \cdots, n$, are elementary events.

(2) Each elementary event comes up with equal possibility, i.e., their probabilities are identical.

There are many practical problems fulfilling or approximately fulfilling the above two conditions. Thus they can be treated as classical probability model. Now let us introduce the concept of classical probability based on equal possibility.

Definition 1 *If a random experiment possesses n elementary events of equal possibility and event A contains m of these elementary events, then the probability P(A) that event A comes up is defined as*

$$P(A) = \frac{m}{n} = \frac{\textit{number of sample points contained in } A}{\textit{total number of sample points in the sample space}}. \tag{1.1}$$

According to the definition, the probability in Example 3 that the ball randomly drawn from the bag is red equals 3/10.

We remark that when we try to check whether a problem can be treated as a classical probability model it is relatively easier to verify condition (1) than condition (2). For example, checking whether a coin is fair is equivalent to verifying condition (2). Once a problem can be treated as a classical probability model, we use formula (1.1) to calculate probabilities of the events under this problem. In calculations, combinations and permutations are frequently used. Some basic formulas about them are listed in Supplements and Remarks at the end of this chapter.

The following three properties of the probability follow immediately from the definition of the classical probability:

(1) Non-negativity: $P(A)\geqslant 0$;

(2) Normalization: $P(\Omega)=1$;

(3) Additivity: Suppose that the events A and B will never come up simultaneously, Then $P(A+B)=P(A)+P(B)$.

In particular, if A and B will never come up simultaneously, and at least one of A and B will come up, then $P(A)=1-P(B)$.

Ⅱ Examples

Example 7 There are n distinct balls and N distinct boxs ($n\leqslant N$). Suppose that each ball falls into a box with equal probability. Put these n balls into N boxs at random. Find the probabilities that

(1) each of n fixed boxs contains one ball;

(2) there are n boxs each containing one ball.

Solution We can treat this problem as a classical probability model again. Label these n balls by digits 1-n and regard each different way to place these balls as a sample point. The total number of sample points is N^n since each box can hold arbitrarily many balls.

(1) If we let $A=\{$each one of n fixed boxs contains one ball$\}$, then the number of sample points A contains is equal to permutations of n balls placed in n boxs. Therefore

$$P(A)=\frac{n!}{N^n}.$$

(2) If we let $B=\{$there are n boxs each containing one ball$\}$, then the number of sample points B contains is equal to permutations of selecting n boxs out of N boxs. Therefore

$$\begin{aligned}P(B)&=\frac{P_N^n}{N^n}=\frac{N(N-1)\cdots(N-n+1)}{N^n}\\&=\left(1-\frac{1}{N}\right)\left(1-\frac{2}{N}\right)\cdots\left(1-\frac{n-1}{N}\right).\end{aligned}$$

By noting $\log(1-x)=-x+O(x^2)$ as $x\to 0$, we have

$$\log\left(1-\frac{1}{N}\right)\left(1-\frac{2}{N}\right)\cdots\left(1-\frac{n-1}{N}\right)$$

$$=\sum_{k=1}^{n-1}\log\left(1-\frac{k}{N}\right)=-\sum_{k=1}^{n-1}\frac{k}{N}+O\left(\sum_{k=1}^{n-1}\frac{k^2}{N^2}\right)$$

$$=-\frac{n(n-1)}{2N}+O\left(\frac{n^3}{N^2}\right)$$

So, when N is much larger than n, we have the following approximation of the probability $P(B)$.

$$P(B)\approx\exp\left\{-\frac{n(n-1)}{2N}\right\}.$$

Example 8 (Birthday problem) This example is on the probability of having the same birthday and a special case of Example 7. If n people are in a room at present, what is the probability that at least two of them celebrate their birthday on the same day of the year? How large need n to be so that this probability is larger than 50%?

Solution We assume that each person celebrate his/her birthday on any one of 365 days with an equal probability. Let $\overline{A}$ be the event that no two of n people celebrate their birthday on the same day of the year, and A be the event that at least two of n people celebrate their birthday on the same day. By Example 7(2), the probability that we desire is as follows:

$$p_n=P(A)=1-P(\overline{A})=1-\frac{P_{365}^n}{365^n}=1-\frac{365!}{(365-n)!365^n}.$$

One can find that $p_{22}=0.4757\cdots$, $p_{23}=0.5073\cdots$. So when $n\geqslant 23$, the probability that at least two of n people celebrate their birthday on the same day of the year exceeds 50%.

When n is large, to compute the value of such probability is not an easy work because we need to compute the value of $n!$. The value of $n!$ grows very fast as n increases. For example, $10!=3\ 628\ 000$, $15!=1\ 307\ 674\ 368\ 700$, and $100!$ contains 158 digits. The amount of computation can be greatly reduced by the approximation formula as

$$p_n=1-\frac{P_{365}^n}{365^n}\approx 1-\exp\left\{-\frac{n(n-1)}{2\times 365}\right\}=\tilde{p}_n.$$

From Table 1-2, one can find that the approximation value and the exact probability are very close.

Table 1-2

n	20	30	40	50	60	70	80	22	23
p_n	0.411	0.706	0.891	0.970	0.994	0.9992	0.9999	0.4757	0.5073
$\tilde{p}_n$	0.406	0.696	0.882	0.965	0.992	0.9987	0.9998	0.4689	0.5000

Example 9 A bag contains a white balls and b black balls. These balls are drawn one by one randomly and without replacement. Find the probability that the k-th ball drawn is a white one.

Solution (1) We label these balls by digits $1,2,\cdots,a+b$ and arrange them in order as they are picked out one by one. Then each arrangement can be regarded as a sample point and the set of all sample points equals the set of permutations of $1,2,\cdots,a+b$. The event in the context is equivalent to one that ball k in an arrangement is a white one. Since each ball has been assigned a different digit, all black balls and white balls should be distinguishable. Thus, there are a ways to place a white ball in the k-th position in an arrangement. This means that the event comes up and there are $(a+b-1)!$ ways to place the other balls in the other positions. Therefore, the event under discussion contains $a(a+b-1)!$ sample points altogether and so its probability is

$$\frac{a}{(a+b)!}(a+b-1)! = \frac{a}{a+b}.$$

Solution (2) In this solution we do not label balls, all white balls are viewed as indistinguishable and so are black balls. Still, we arrange them in the order as they are taken out, but the arrangements are treated as just one arrangement, i. e., one sample point, as long as a white balls occupy a same positions. In this classical probability model, the sample space contains $\binom{a+b}{a}$ sample points in sum. The event considered here equals one that the k-th position is occupied by any one of a white balls and $a-1$ of other $a+b-1$ positions are occupied by $a-1$ white balls. Thus its probability is

$$\frac{\binom{a+b-1}{a-1}}{\binom{a+b}{a}} = \frac{a}{a+b}.$$

The essential difference between two solutions lies in the choice of sample spaces. In Solution (1), all balls are regarded as distinct and the position of each ball is concerned when we count the number of sample points that the event contains, so permutations are used. Instead, in Solution (2) balls with same color are viewed as indistinguishable, so only positions of white balls are concerned and thus combinations are used. Note that before calculating probabilities we should fix a sample space and find out both nominator and denominator in the right side of formula (1.1) in such a fixed sample space.

Since the probability we get in the preceding example has nothing to do with k, we conclude that the probability of gaining a white ball each time we pick a ball up, early or late, will never differ. In fact, it is equal to the proportion of white balls in all balls. Another example with same property is to draw lots. The chance of winning a prize in a lottery keeps the same no matter when he or she draws a lot, early or late.

Example 10 There are a defective products and b non-defective products and they are indistinguishable. If n ($n \leqslant a$) products are sampled from them, find the probability that the n products sampled contain k defective ones.

Solution Similarly to Example 9, products of the same level can be regarded as either distinct or indistinguishable. So there are two ways of solving this example. As shown in Example 9, the problem turns out to be simpler if we treat products of the same level as indistinguishable. If we take either way to get n products from $a+b$ products as a sample point, then the sample space contains $\binom{a+b}{n}$ sample points in all and the event under discussion has $\binom{a}{k}\binom{b}{n-k}$ sample points. Therefore

$$P = \frac{\binom{a}{k}\binom{b}{n-k}}{\binom{a+b}{n}}.$$

The model given in the example is commonly used in sampling inspection of products.

Example 11 One has two boxes of matches, each having n matches, in his pocket. Each time he wants to use match, he will randomly take out a box and draw one match from it. When he finds the box he takes out is empty, find the probability that the other box has just m matches.

Solution (1) In this solution only the first $2n+1-m$ drawings are considered. Since there are two ways to take a match out each time, there are 2^{2n+1-m} ways in sum to take $2n+1-m$ matches out. In order to count the number of ways in which the event under discussion will occur, we can consider two different situations. One is that the last match (namely the $(2n+1-m)$-th match) is drawn from the first box. The other is that the last match is from the second box. In the former case n of the first $2n-m$ match comes from the first box, and so there are $\binom{2n-m}{n}$ ways in sum. Similarly, in the latter case there are also $\binom{2n-m}{n}$ ways. Therefore, the desired probability is

$$\frac{2\binom{2n-m}{n}}{2^{2n+1-m}} = \frac{\binom{2n-m}{n}}{2^{2n-m}}.$$

Solution (2) Since each box contains only n matches, the event in the context must occur when he takes a box for a $2n+1$-th match. Obviously, he cannot get the $2n+1$-th match from his pocket. It is not hard to see that there are 2^{2n+1} ways in sum to try to take out $2n+1$ matches.

Now let us turn to the number of ways in which the fixed event will come up. Just as in Solution (1), we also consider two situations separately. First, suppose the first box is used up before the second one. In fact, when the $(n+r)$ -th match is drawn out, the first box may be already empty and so the second box had been taken out r times before. If the fixed event occurs, then the second box is taken out in all the next $n-m-r$ times and is taken out in the $(2n-m+1)$ -th time. At this point he finds the first box is empty. We assume that he puts the empty box into his pocket. All the remaining m matches certainly come from the second box, there are 2^m ways to take a box out. Therefore there are $\binom{n+r-1}{r}2^m$ ways in sum in which the event comes up provided the r is fixed. Thus there are $\sum_{r=0}^{n-m}\binom{n+r-1}{r}2^m$ ways for the event to come up. Similarly, there are also $\sum_{r=0}^{n-m}\binom{n+r-1}{r}2^m$ ways that the event occurs in the case when the second box is first found empty. So the probability asked is

$$2\frac{1}{2^{2n+1}}\sum_{r=0}^{n-m}\binom{n+r-1}{r}2^m=\frac{1}{2^{2n-m}}\sum_{r=0}^{n-m}\binom{n+r-1}{r}.$$

It is not difficult to verify the equivalence between probabilities obtained in Solutions (1) and (2). Hence, we have the following identity

$$\sum_{r=0}^{n-m}\binom{n+r-1}{r}=\binom{2n-m}{n}\quad\text{or}\quad\sum_{r=0}^{s}\binom{n+r-1}{r}=\binom{n+s}{n}.$$

Comparing these two solutions, we can find that Solution (2) is more specific but very complicated; while Solution (1) gives a concise answer by noting that at least $2n+1-m$ matches have been taken before the fixed event occurs.

Solution (3) The extension of the classical probability

In the classical probability model, the sample space $\Omega=\{\omega_1,\omega_2,\cdots,\omega_n\}$ is finite and each sample point appears with equal possibility. In general, if the sample space $\Omega=\{\omega_1,\omega_2,\cdots\}$ has many countable elements, and a sample ω_i appears with a possibility $p(\omega_i)$, where $p(\omega_i)\geqslant 0$, $\sum_{i=1}^{\infty}p(\omega_i)=1$. Then the probability of an event A is defined as

$$P(A)=\sum_{i:\omega_i\in A}p(\omega_i)$$

It is easily verified that this model has the following properties:

(1)Non-negativity: $P(A)\geqslant 0$;

(2)Normalization: $P(\Omega)=1$;

(3)Countable additivity: If no two of A_i, $i=1,2,\cdots$, will come up simultaneously (i. e., $A_i\cap A_j=\varnothing$, $i\neq j$), then

$$P(\sum_{i=1}^{\infty} A_i) = \sum_{i=1}^{\infty} P(A_i)$$

where $\sum_{i=1}^{\infty} A_i (\bigcup_{i=1}^{\infty} A_i)$ denotes the event that at least one of A_i, $i = 1,2,\cdots$, will come up.

1.2.3 Geometric probability models

In classical probability model, the total number of sample points is finite. Even though each point occurs with equal probability, classical probability model is not usable if the total number of sample points is infinite, especially uncountable. Nevertheless, the method shown in classical probability model can be extended to the case of infinitely many sample points.

It is obviously meaningless to count sample points if the sample space is a region Ω containing infinitely many points in $\mathbf{R}^n$. On this occasion, "equal possibility" can be interpreted as each sample point falls into two arbitrary regions with equal measure (length, area, volume, etc.) and the probability has nothing to do with the shape and the position of these two regions.

Motivated by the above viewpoint, the probability of the event $A_g = \{$a sample point falls into region $g \subset \Omega\}$ is defined as

$$P(A_g) = \frac{\text{measure of } g}{\text{measure of } \Omega}. \tag{1.2}$$

This is called the geometric probability. It can be easily verified that the geometric probability has the properties of the non-negativity, normalization and countable additivity.

Example 12 If a bus arrives at a stop on time every 5 minutes and it can hold arbitrarily many passengers. Find the probability that the time one needs to wait for a bus at the stop is less than 3 minutes.

Solution The time one arrives at the stop is between a and $a+5$ if the last bus left at time a. Let $\Omega = (a, a+5)$ and $A_g = \{$waiting time is less than 3 minutes$\}$. If A_g occurs, one has to get the stop at any time within $g = (a+2, a+5)$. Naturally we can assume that he gets the stop at any time with equal possibility. Thus according to the definition of the geometric probability, we have

$$P(A_g) = \frac{\text{length of } g}{\text{length of } \Omega} = \frac{3}{5}.$$

Example 13 (The arrangement problem) Two people make an appointment to meet at a park between 7 and 8 o'clock and the person who first arrives will keep waiting for the other for 20 minutes. Find the probability that they can meet.

Solution Since neither has claimed the exact time when he or she arrives, two numbers will decide a sample point. If we take 7 o'clock as the beginning time and assume that they arrive at x and y past 7 respectively, then the sample space is

$$\Omega = \{(x,y) \mid 0 \leqslant x \leqslant 60, 0 \leqslant y \leqslant 60\},$$

which can in turn be expressed as a square in a chart. Obviously, the necessary and sufficient condition for them to meet is that $|x-y| \leqslant 20$. Therefore the sample points such that the event $A = \{\text{they meet each other}\}$ happens constitute the shadow area in Figure 1-1. We denote this area by g, so we have

$$P(A) = \frac{\text{area of } g}{\text{area of } \Omega} = \frac{60^2 - (60-20)^2}{60^2} = \frac{5}{9}.$$

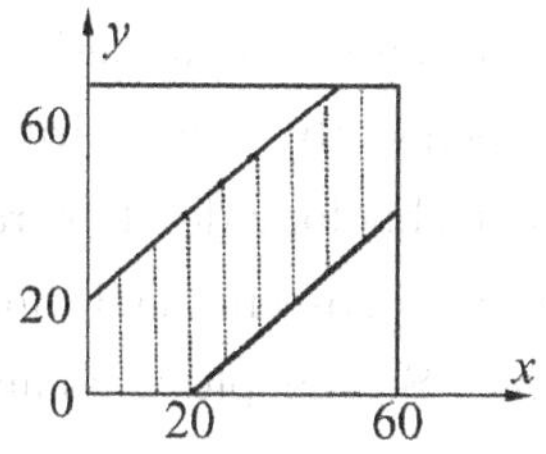

Figure 1-1

Example 14 (The problem of Buffon's needles) This is a charming problem that leads directly to some important (not all intuitive) conclusions. If a needle of length l is dropped at random on the middle of a horizontal surface ruled with parallel lines a distance $a > l$ apart, what is the probability that the needle will cross one of the lines?

Solution Consider the position the needle lies in, then each sample point is decided by two parameters, the distance x between needle's midpoint and the line closest to it and the angle φ between the needle and parallel lines (see Figure 1-2a).

So the sample space is

$$\Omega = \{(\varphi, x) \mid 0 \leqslant \varphi \leqslant \pi, 0 \leqslant x \leqslant \frac{a}{2}\},$$

a rectangle. The needle crosses one of parallel lines if and only if $x \leqslant \frac{l}{2}\sin\varphi$ (denote this condition by g) (see Figure 1-2b). Therefore the probability asked is

$$P = \frac{\text{area of } g}{\text{area of } \Omega} = \frac{\int_0^{\pi} \frac{l}{2}\sin\varphi \, d\varphi}{\pi a/2} = \frac{2l}{a\pi}.$$

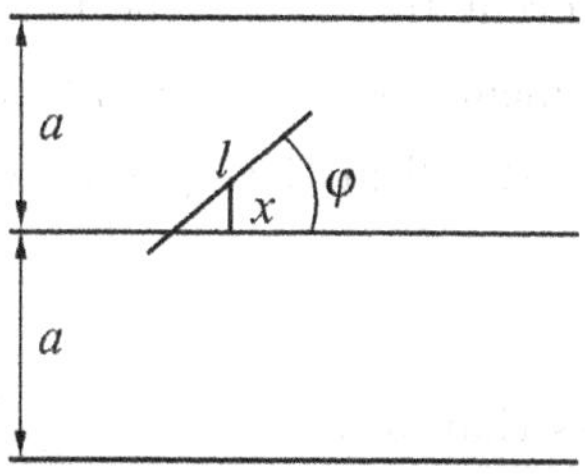

Figure 1-2a

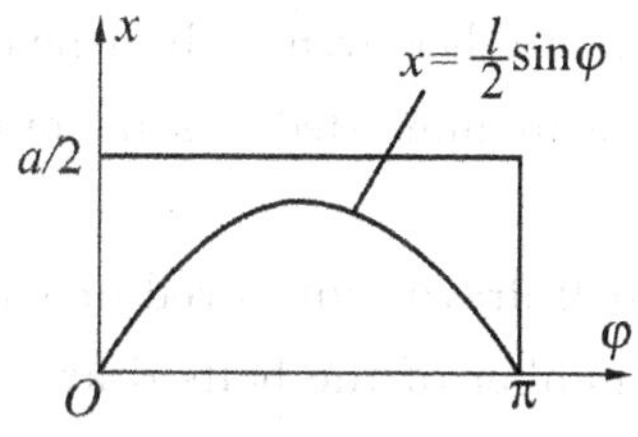

Figure 1-2b

Remark Since probability is approximated by frequency obtained from a large number of repeated independent experiments, one can get an approximate value of π by using the model above. The specific steps are as follows. Throw a needle N times, or throw N needles at a time, and then count the times (denoted by n) the needle crosses one of the lines or the number of needles crosses one of the lines. In this way we gain an approximate value of the probability, namely $P \approx n/N$. Note that $2l/a\pi \approx n/N$, we have $\pi \approx 2lN/an$, which is just an approximate value of π as both l and a are measured. Many people in history have conducted this experiment in order to get an approximate value of π. One of the best approximate values is $\pi \approx 3.1415929$, for which needle is repeatedly thrown 3 408 times.

In order to determine a parameter, an idea is to design an appropriate random experiment and to conduct it repeatedly and independently. The spirit of the Monte Carlo method is just based on this thinking. As electronic technology develops rapidly, it becomes easier and easier to conduct repeatedly a certain random experiment, and thus the Monte Carlo method becomes more and more efficient and applicable. See Supplements and Remarks of Chapter 4.

1.3 The axiomatic definition of probability

In the previous sections, we have studied the intuitive definition of probability from statistical viewpoint, and looked at classical and geometrical probability models with their applications. But these reflect just a part of the truth. We do not give a rigorous mathematical definition of probability yet. Although classical and geometric probability models can be used to characterize and explain a great deal of random phenomena, the "equal likelihood" hypothesis prevents us from studying other kinds of random experiments without such a property. Still, different understanding of "equal likelihood" may lead to different probability values of a certain event.

In this section, we are ready to set up a formal axiom system which permits rigorous proof of some important properties of probability. This is the way that most mathematical research actually proceeds from concrete to abstract rather than the other way around.

1.3.1 Events

A sample point (often called an outcome) is an event that cannot be broken down into some combination of other events. In a random experiment, besides elementary outcomes — sample points, we are interested in some other results. Recall in Example 3, we consider the following results:

$A = \{$the ball drawn out is red or white$\}$;

$B = \{$the number of the balls drawn out is less than 5$\}$;

$C = \{$the ball drawn out is not a red one$\}$,

These are all events.

If sample space is regarded as a universal set Ω, and sample points as elements of this set, then an event is defined as a certain subset of the sample space, a certain set composed of sample points. In the preceding example we have discussed, if Ω is taken as a sample space, then

$$A = \{\omega_1, \omega_2, \omega_3, \omega_4, \omega_5, \omega_6\};$$
$$B = \{\omega_1, \omega_2, \omega_3, \omega_4\};$$
$$C = \{\omega_4, \omega_5, \omega_6, \omega_7, \omega_8, \omega_9, \omega_{10}\}.$$

Generally speaking, events are denoted by capital letters A, B, C, $\cdots$. If a particular sample point ω appears in an experiment and $\omega \in A$, then we say that event A occurs. Naturally, sample space Ω can be considered as an event. Because there must be one sample point of sample space Ω occurring in each experiment, i. e., Ω takes place with certainty, and Ω is called inevitable event. Similarly, the empty set $\varnothing$ can also be regarded as an event and it will never come up in any experiment, therefore $\varnothing$ is called an impossible event.

To treat events as sets of sample points enables us to study events by the method of set theory, and in particular to study the relation and operations among events in a very similar way to set. Now let us start with the relation of events.

Let A and B be two events. If event A happens whenever event B happens, in other words, $\omega \in A$ if $\omega \in B$, then B is contained or included in A and is a subset of A, while A is a superset of B. We write this in one of the following two ways:

$$B \subset A,\ A \supset B.$$

Two events A and B are identical if A happens and if and only if B happens, and we write $A = B$.

The union of events A and B is the event that happens if at least one of A and B happens. We write $A \cup B$ for this.

The intersection (or product) of events A and B is the event that happens if both A and B happen. We write $A \cap B$ (or AB) for this.

The difference of A and B is the event that A happens but B does not happen. We write $A - B$ for this.

If it is impossible for both A and B to come up together, namely $A \cap B = \varnothing$, then A and B are said to be mutually exclusive. In this situation, $A \cup B$ is often written as $A + B$.

If it is impossible for both A and B to come up together and at least one of A and B will come up for certainty, namely, $A \cap B = \varnothing$ and $A \cup B = \Omega$, then the event B is said to be the complement of event A, denoted symbolically by $B = \overline{A}$ (or A^c). Of course, the event A is also said to be the complement of the event B.

Obviously, $A - B = A\overline{B}$.

The relation and operations among events possess properties similar to those for sets in set theory. For instance,

Commutative law

$$A \cup B = B \cup A, \quad AB = BA;$$

Associative law

$$(A \cup B) \cup C = A \cup (B \cup C), \quad (AB)C = A(BC);$$

Distributive law

$$(A \cup B) \cap C = AC \cup BC,$$
$$(A \cap B) \cup C = (A \cup C) \cap (B \cup C);$$

de Morgan's law

$$\overline{A \cup B} = \overline{A} \cap \overline{B}, \quad \overline{A \cap B} = \overline{A} \cup \overline{B}.$$

When a number of events and even infinitely countable events are dealt with, de Morgan's law still holds.

We should learn to translate operations of events in the terms of sets and learn to use operations of events to decompose complicated events into simple ones.

Example 15 Suppose that A, B, C are three events, then

(1) {both A and B come up while C does not} can be written as $AB\overline{C}$ or $AB - C$ or $AB - ABC$;

(2) { A, B and C all come up} can be written as ABC;

(3) {at least one of A, B and C comes up} can be written as $A \cup B \cup C$ or $A\overline{B}\,\overline{C} + \overline{A}B\overline{C} + \overline{A}\,\overline{B}C + AB\overline{C} + \overline{A}BC + A\overline{B}C + ABC$.

Example 16 A system is composed of components A, B and C with A and B in parallel connection and C series connected with the circuit composed by A and B, as shown in Figure 1-3.

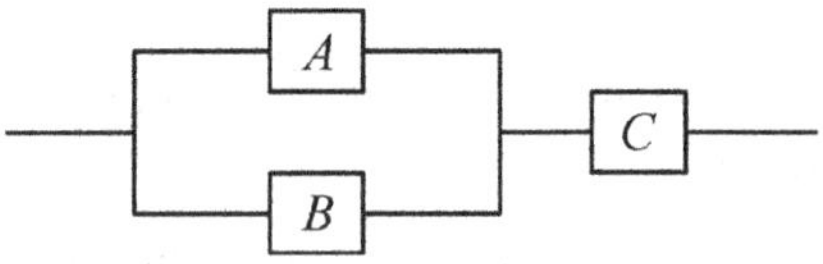

Figure 1-3

If A, B and C respectively denote the events that corresponding component works orderly, then event W = {the system works orderly} = {at least one of A and B works orderly and so does C} = $(A \cup B)C$ or $AC \cup BC$.

1.3.2 Probability space

A probability space contains three basic elements. The first element is the sample space Ω, the set of all sample points ω. The sample space should be decided in advance according to the problem background in the context.

The second is the σ-algebra $\mathscr{F}$ of events, a family of Ω subsets satisfying:

(1) $\Omega \in \mathscr{F}$;

(2) If $A \in \mathscr{F}$, then $\overline{A} \in \mathscr{F}$;

(3) If $A_1, A_2, \cdots, A_n, \cdots \in \mathscr{F}$, then $\bigcup_{n=1}^{\infty} A_n \in \mathscr{F}$.

$\mathscr{F}$ satisfying the above three hypotheses is termed a σ-algebra (or σ-field) in Ω and the elements of $\mathscr{F}$ (subsets of Ω) are called events.

The following properties about σ-algebra of events can be deduced from the above three hypotheses:

(4) $\varnothing \in \mathscr{F} (\varnothing = \overline{\Omega})$;

(5) If $A_1, A_2, \cdots, A_n, \cdots \in \mathscr{F}$, then $\bigcap_{n=1}^{\infty} A_n \in \mathscr{F}$. Indeed,

$$\bigcap_{n=1}^{\infty} A_n = \overline{\bigcup_{n=1}^{\infty} \overline{A_n}};$$

(6) If $A_1, A_2, \cdots, A_n \in \mathscr{F}$, then $\bigcup_{k=1}^{n} A_k \in \mathscr{F}$, $\bigcap_{k=1}^{n} A_k \in \mathscr{F}$.

We conclude from the above statement that inevitable event, the impossible event, complements, finite unions, finite intersections, countable unions, and countable intersections of events are all still events and thus the operations like complement, union and intersection in σ-algebra of events are all meaningful.

A sample space Ω may have many σ-algebras, the simplest one of which is $\mathscr{F}_1 = \{\varnothing, \Omega\}$, and the greatest is $\mathscr{F}_2 = \{\text{all subsets of } \Omega\}$. So when dealing with specific problems, we need to select a suitable σ-algebra of events so that we can easily solve them. The algebras of events widely used are as follows.

If Ω consists of finite or countable sample points, the family of all subsets of Ω is chosen to be $\mathscr{F}$, as in the classical probability model. It is not difficult to verify that $\mathscr{F}$ is a σ-algebra.

If $\Omega = \mathbf{R}^1$ (the space of all real numbers), the family of all intervals and sets of (finite or countable) unions, (finite or countable) intersections and complements of them is chosen to be $\mathscr{F}$, often denoted by $\mathscr{B}$, called one-dimensional Borel σ-algebra. Such a $\mathscr{B}$ is much bigger than the family of all intervals, and sets in $\mathscr{B}$ are called one-dimensional Borel sets.

If $\Omega = \mathbf{R}^n$ (the n-dimensional real number space), the collection of all n-dimensional rectangles and sets of (finite or countable) unions, (finite or countable) intersections and complements of them is chosen to be $\mathscr{F}$, often denoted by $\mathscr{B}^n$, called n-dimensional Borel σ-algebra.

Such a $\mathscr{F}$ is large enough to contain all sets of interest to us.

If we are only interested in certain sub-family $\mathfrak{G}$ of Ω, we can take the smallest σ-algebra containing $\mathfrak{G}$ as the algebra $\mathscr{F}$ of events. To show such a smallest σ-algebra really exists, we note the following two facts:

(i) There is at least one σ-algebra containing $\mathfrak{G}$, i. e., the above $\mathscr{F}_2$;

(ii) If there are many σ-algebras containing $\mathfrak{S}$, then the intersection of them is still a σ-algebra and is the smallest one.

In particular, if we are only interested in a subset A of Ω, then we can take the smallest σ-algebra $\mathscr{F}=\{\varnothing, A, \overline{A}, \Omega\}$.

The third element of a probability space is probability P. The definition of probability should be harmonic with the statistical intuition as well as classical and geometric probability models. So we define it as follows:

Probability P is a real function defined on $\mathscr{F}$: $A \in \mathscr{F} \longrightarrow P(A)$, satisfying

P_1 (non-negativity) $P(A) \geqslant 0$ for all $A \in \mathscr{F}$;

P_2 (normalization) $P(\Omega)=1$;

P_3 (countable additivity) If $A_1, \cdots, A_n, \cdots$ are mutually disjoint events, then

$$P(\sum_{n=1}^{\infty} A_n)=\sum_{n=1}^{\infty} P(A_n).$$

In terms of measure theory, probability is a normed measure defined on a σ-algebra.

The triple $(\Omega, \mathscr{F}, P)$ is called a probability space.

Example 17 We shall construct a probability measure P on an arbitrary σ-field $\mathscr{F}$ in an arbitrary non-empty Ω. Suppose that $\omega_0 \in \Omega$, define

$$P(A)=I_A(\omega_0)=\begin{cases}1, & \omega_0 \in A; \\ 0, & \text{otherwise}\end{cases} \tag{1.3}$$

for $A \in \mathscr{F}$.

P is clearly a discrete probability measure — a unit mass at ω_0.

Example 18 Let $\Omega=\{\omega_1, \omega_2, \cdots\}$, $\mathscr{F}=\{A: A \subset \Omega\}$. Suppose $p_i=P(\{\omega_i\})>0$ and $\sum_{i=1}^{\infty} p_i=1$. Define

$$P(A)=\sum_{i: \omega_i \in A} p_i.$$

Then $(\Omega, \mathscr{F}, P)$ is a probability space. In particular, if Ω is a finite set and p_is are equal, then $(\Omega, \mathscr{F}, P)$ is just the classical probability model.

Example 19 Let $\Omega=[0,1]$, and $\mathscr{F}=\beta_{([0,1])}$ be the Borel σ-field on $[0,1]$, m be the Lebesgue measure. Then $(\Omega, \mathscr{F}, m)$ is a probability space which just describes the geometric probability model on $[0, 1]$.

Remark Suppose that β^n is the Borel σ-field on the n-dimensional real space $\mathbf{R}^n$. Then there exists a unique function $\mathrm{m}: \beta^n \rightarrow \mathbf{R}_+$ satisfying

(i) For any mutually disjoint sets $A_1, A_2, \cdots \in \beta^n$,

$$m(\sum_{j=1}^{\infty} A_j)=\sum_{j=1}^{\infty} m(A_j);$$

(ii) For any real numbers $a_1<b_1, a_2<b_2, \cdots, a_n<b_n$,

$$m((a_1, b_1] \times \cdots \times (a_n, b_n])=(b_1-a_1) \cdots (b_n-a_n).$$

The function m is called the Lebesgue measure on $\mathbf{R}^n$.

The axiomatic definition of probability formulates basic properties that probability must have. But one rarely finds a probability in specific situations according to the definition. The significance of the axiomatic structure is to lay a solid foundation for rigorous probability theory.

Usually, when constructing a probability model for a specific problem, we find it not difficult to choose a sample space and a σ-algebra of events, but we need enough background knowledge related to the problem in order to assign a probability to each basic event. One of main tasks of the subject of probability theory is to study how to calculate probabilities of complicated events from those of simple events, while the probability model is always supposed to be given.

We have the following properties for probability from the above definition.

(1) $P(\varnothing) = 0$;

(2) If $A_iA_j = \varnothing$ for $i \neq j$, then $P(\sum_{i=1}^{n} A_i) = \sum_{i=1}^{n} P(A_i)$;

(3) $P(\overline{A}) = 1 - P(A)$;

(4) If $B \subset A$, then $P(A - B) = P(A) - P(B)$;

(5) $P(A \cup B) = P(A) + P(B) - P(AB)$.

Proof (1) $P(\Omega) = P(\Omega + \varnothing + \varnothing + \cdots) = P(\Omega) + P(\varnothing) + P(\varnothing) + \cdots$. Cancelling $P(\Omega)$ in both sides yields $P(\varnothing) = 0$.

(2) is called finite additivity, which is deduced from (1) and countable additivity (for $i > n$, take $A_i = \varnothing$). Combining P_2 and (2) yields (3).

In order to show (4), we only note that $A = B + (A - B)$ and $B \cap (A - B) = \varnothing$, and use (2).

Observe that (4) follows under condition $B \subset A$. In general, we have $A - B = A - AB$ and so

(6) $P(A - B) = P(A) - P(AB)$.

Proof (5) $A \cup B = A + (B - AB)$, $A \cap (B - AB) = \varnothing$, and $AB \subset B$. So $P(A \cup B) = P(A) + P(B - AB) = P(A) + P(B) - P(AB)$.

In the case that $AB = \varnothing$, (5) is equivalent to (2).

By an induction argument, (5) can be extended to the case of the union of arbitrarily many events.

(7) (Exclusion-inclusion)

$$\begin{aligned} &P(A_1 \cup A_2 \cup \cdots \cup A_n) \\ &= \sum_{i=1}^{n} P(A_i) - \sum_{1 \leqslant i < j \leqslant n} P(A_iA_j) + \cdots + (-1)^{n-1} P(A_1A_2\cdots A_n). \end{aligned} \tag{1.4}$$

(8) (Sub-additivity)

$$P(A_1 \cup A_2 \cup \cdots \cup A_n) \leqslant \sum_{i=1}^{n} P(A_i). \tag{1.5}$$

For $n=2$, (1.5) follows form Property (5). In general, the inequality can be shown by induction. Actually, (1.5) also holds for $n=\infty$.

$$P(\bigcup_{i=1}^{\infty} A_i) \leqslant \sum_{i=1}^{\infty} P(A_i). \tag{1.6}$$

In specific problems, we can decompose complex events into unions, intersections, differences and complements of relatively simpler events and then calculate the probabilities of them by formulae above.

Example 20 A bag contains n ($n \geqslant 3$) balls numbered 1, 2, ⋯, n respectively. Take three balls randomly; find the probability that at least one of ball 1 and ball 2 is taken.

Solution (1) Let $A_i = \{\text{ball } i \text{ is taken}\}$, $i = 1, 2$, and then the desired probability is

$$P(A_1 \cup A_2) = P(A_1) + P(A_2) - P(A_1 A_2)$$
$$= \frac{\binom{n-1}{2}}{\binom{n}{3}} + \frac{\binom{n-1}{2}}{\binom{n}{3}} - \frac{\binom{n-2}{1}}{\binom{n}{3}}.$$

Solution (2) The complement of $A_1 \cup A_2$ is $\overline{A_1 \cup A_2} = \overline{A}_1 \cap \overline{A}_2$. So

$$P(A_1 \cup A_2) = 1 - P(\overline{A}_1 \cap \overline{A}_2) = 1 - \frac{\binom{n-2}{3}}{\binom{n}{3}}.$$

The proof that the above two answers are identical is left to the reader.

Example 21 Suppose that a person types n letters, types the corresponding addresses on n envelopes, and then places the n letters in the n envelopes in a random manner. Find the probability P_n that at least one letter will be placed in the correct envelope.

Solution Let $A_i = \{\text{letter } i \text{ is placed in the correct envelope}\}$, $i = 1, 2, \cdots, n$, and then the required probability P_n is $P(\bigcup_{i=1}^{n} A_i)$

Note that

$$P(A_i) = \frac{1}{n},\ P(A_i A_j) = \frac{1}{n(n-1)}, 1 \leqslant i < j \leqslant n;$$

$$\cdots\cdots$$

$$P(A_1 A_2 \cdots A_n) = \frac{1}{n!}.$$

Hence

$$P_n = P(\bigcup_{i=1}^{n} A_i)$$

$$= \frac{n}{n} - \frac{\binom{n}{2}}{n(n-1)} + \frac{\binom{n}{3}}{n(n-1)(n-2)} - \cdots + (-1)^{n-1}\frac{1}{n!}$$

$$= 1 - \frac{1}{2!} + \frac{1}{3!} - \cdots + (-1)^{n-1}\frac{1}{n!}.$$

The Table 1-2 gives the values of P_n for $n = 5, 6, 7, 8, 9$ and ∞.

Table 1-2

n	5	6	7	9	∞
P_n	0.633333	0.631944	0.632143	0.632121	0.632121

This probability has the following interesting feature. As $n \to \infty$, the value of P_n approaches the limit $1 - 1/e$. It follows that for a large value of n, the probability P_n is approximately 0.63212.

We remark that P_n converges to the limit very rapidly. In fact, for $n = 7$ the exact value P_7 and the limiting value of P_n agree to four decimal places. Hence regardless of whether seven letters are placed at random in seven envelopes or seven million letters are placed at random in seven million envelopes. The probability that at least one letter will be placed in the correct envelope is approximately 0.6321.

The problem like this example is sometimes called the matching problem. The same problem can be expressed in various entertaining contexts. For example, suppose that all the cards in a deck of n different cards are placed in a row, and that the cards in another similar deck are then shuffled and placed in a row on the top of the cards in the original deck. It is desired to determine the probability that there will be at least one match between the corresponding cards from the two decks. As another example, we could suppose that the photographs of n famous film actors are paired in a random manner with n photographs of the same actors taken when they were babies. It could then be desired to determine the probability that the photograph of at least one actor will be paired correctly with this actor's own baby photograph.

Example 22 The complete graph G having n vertices is defined to be a set of n points (called vertices) in the plane and the $\binom{n}{2}$ lines (called edges) connecting each pair of vertices. For a fixed integer $k \leqslant n$, each group of k vertices and related edges in G constitute a complete sub-graph of G on k vertices. As there are totally $\binom{n}{k}$ such sub-graphs, we denote them by G_i, $i = 1, \cdots, \binom{n}{k}$. Suppose now that each edge in the complete graph G on n vertices is to be colored either red or blue. A question of interest is whether there is a way of coloring the edges of G so that none of the sub-graphs G_i has all its $\binom{n}{2}$ edges the same color.

Solution We will show, by a probabilistic argument, that if n is not too large, then the answer is yes. The argument runs as follows. Suppose that the edges are colored randomly, and each edge is equally likely to be colored either red or blue. That is, edge is red with probability 1/2. Let

$$E_i = \{\text{sub-graph } G_i \text{ has all its edges the same color}\}.$$

Then $\bigcup_i E_i$ is the event that at least one of the sub-graphs having all its $\binom{k}{2}$ edges the same color. Note

$$\begin{aligned} P(E_i) &= P(G_i \text{ has all its edges red}) + P(G_i \text{ has all its edges blue}) \\ &= 2\frac{1}{2^{k/2}} = \left(\frac{1}{2}\right)^{k(k-1)/2-1} \end{aligned}$$

So

$$P(\bigcup_i E_i) \leqslant \sum_i P(E_i) = \binom{n}{k}\left(\frac{1}{2}\right)^{k(k-1)/2-1}$$

Hence, if

$$\binom{n}{k} < 2^{k(k-1)/2-1},$$

then $P(\bigcup_i E_i) < 1$, which implies that $\bigcap_i \overline{E}_i \neq \varnothing$. Therefore, under the preceding condition on n and k, there is at least one way of coloring the edges of G for which none of the complete sub-graphs on k vertices has all of its edges the same color.

Example 23 (Random walk) We now present a model which describes the random walk of a particle.

Let the particle start at the origin, and after unit time let it take a unit step left with probability $1-p$ or right with probability p. Consequently after n steps the particle can have moved at most n units left or n units right. It is clear that each path π of the particle is completely specified by a set $(w_1, w_2, \cdots, w_n)$, where $w_i = 1$ if the particle moves right at the i-th step, $w_i = -1$ if it moves left. Let us assign to each path π weight $p(\pi) = p^{s(\pi)}(1-p)^{n-s(\pi)}$, where $s(\pi)$ is the number of 1's in the sequence $\pi = (w_1, w_2, \cdots, w_n)$, i. e., $s(\pi) = [(w_1 + \cdots + w_n) + n]/2$, and $0 \leqslant p \leqslant 1$.

Since $\sum_\pi p(\pi) = 1$, the set of probabilities $p(\pi)$ together with space Ω of paths $\pi = (w_1, w_2, \cdots, w_n)$ and its subsets define an acceptable probabilistic model of the motion of the particle for n steps.

Let us find the probability of the event A_k that after n steps the particle is at a point with ordinate k. The condition is satisfied by these paths π for which $s(\pi) - (n - s(\pi)) = k$, i. e.,

$$s(\pi) = \frac{n+k}{2}.$$

When $n+k$ is even, the number of such paths is $\binom{n}{\frac{n+k}{2}}$, and therefore

$$P(A_k) = \binom{n}{\frac{n+k}{2}} p^{(n+k)/2}(1-p)^{n-(n+k)/2}.$$

When $n+k$ is odd, $P(A_k) = 0$.

Consequently the binomial distribution $(P(A_{-n}), \cdots, P(A_0), \cdots, P(A_n))$ can be said to describe the probability distribution for the position of the particle after n steps.

Note that in the symmetric case ($p = 1/2$) the probability of each individual path is equal to 2^{-n}, and hence for n and k with the same parity,

$$P(A_k) = \binom{n}{\frac{n+k}{2}} 2^{-n}.$$

1.3.3 Continuity of probability measure

Suppose that (Ω, $\mathscr{F}$, P) is a probability space and A_1, A_2, $\cdots$, is a sequence of increasing events, i. e. , $A_1 \subset A_2 \subset \cdots \subset A_n \subset \cdots$. Let $A = \bigcup_{n=1}^{\infty} A_n$, termed as the limit of A_n. From the axiomatic definition of probability, we know that A is still an event. The following theorem gives the probability of A.

Theorem 1 *Suppose that* $A_1, A_2, \cdots$, *is a sequence of increasing events with* A *as its limit*, *then*

$$P(A) = \lim_{n\to\infty} P(A_n).$$

Proof Let $B_k = A_k - A_{k-1}, k = 2, 3, \cdots$, and then $A = A_1 \cup B_2 \cup B_3 \cup \cdots$, the union of a series of disjoint events. By the countable additivity of probability, we have

$$\begin{aligned} P(A) &= P(A_1) + P(B_2) + P(B_3) + \cdots \\ &= P(A_1) + \lim_{n\to\infty} \sum_{k=2}^{n} P(B_k). \end{aligned}$$

Also, $P(B_k) = P(A_k) - P(A_{k-1})$. Therefore we have

$$\begin{aligned} P(A) &= P(A_1) + \lim_{n\to\infty} \sum_{k=2}^{n} [P(A_k) - P(A_{k-1})] \\ &= \lim_{n\to\infty} P(A_n), \end{aligned}$$

which completes the proof.

By the above theorem, (1.6) can follow from (1.5) by letting $n\to\infty$.

If A_1, A_2, $\cdots$ is a increasing sequence of events and denote $\lim\limits_{n\to\infty} A_n = \bigcup\limits_{n=1}^{\infty} A_n$, then the above theorem tells us that

$$P(\lim_{n\to\infty} A_n) = \lim_{n\to\infty} P(A_n).$$

So we say that a probability has the continuity property.

Similarly, if $A_1, A_2, \cdots$ is a decreasing sequence of events and denote $\lim\limits_{n\to\infty} A_n = \bigcap\limits_{n=1}^{\infty} A_n$, then

$$P(\lim_{n\to\infty} A_n) = \lim_{n\to\infty} P(A_n).$$

To distinguish, the former is called upper-continuity and the latter is called lower-continuity. In general, for a sequence of events $A_1, A_2, \cdots$, we denote

$$\liminf_{n\to\infty} A_n = \bigcup_{n=1}^{\infty}\bigcap_{m=n}^{\infty} A_m \text{ and } \limsup_{n\to\infty} A_n = \bigcap_{n=1}^{\infty}\bigcup_{m=n}^{\infty} A_m .$$

It can be verified that $\liminf_{n\to\infty} A_n \subset \limsup_{n\to\infty} A_n$. If $\liminf_{n\to\infty} A_n = \limsup_{n\to\infty} A_n$, we say that the limit of A_n exits and denote the limit by $\lim_{n\to\infty} A_n$. If the limit $\lim_{n\to\infty} A_n$ exists, then we also have $P(\lim_{n\to\infty} A_n) = \lim_{n\to\infty} P(A_n)$. The proof for this is left to the reader.

By Property 2 and Theorem 1, the countable additivity of the probability implies the finite additivity and continuity. Conversely, if a non-negative function $Q:\mathscr{F}\to[0, 1]$ is finite additive and

$$A_n \searrow \emptyset \Rightarrow Q(A_n) \searrow 0, \tag{1.7}$$

then Q is countably additive. And so, the countable additivity is equivalent to the finite additivity and lower-continuity.

In fact, suppose that $A_1, A_n, \cdots \in \mathscr{F}$ are mutually exclusive events. Let $B_n = \sum_{j=n+1}^{\infty} A_j$. Then $A_1, \cdots, A_n, B_n$ are mutually exclusive events, and so

$$Q(\sum_{n=1}^{\infty} A_n) = Q(A_1 + \cdots + A_n + B_n) = \sum_{j=1}^{n} Q(A_j) + Q(B_n) ,$$

by the finite additivity. On the other hand, $B_n \searrow \bigcap_{j=1}^{\infty} B_j = \emptyset$. By (1.7), we have $Q(B_n)\searrow 0$. Hence

$$Q(\sum_{n=1}^{\infty} A_n) = \lim_{n\to\infty}\sum_{j=1}^{n} Q(A_j) + \lim_{n\to\infty} Q(B_n) = \sum_{j=1}^{\infty} Q(A_j) .$$

The proof of countable additivity is completed.

Example 24 Toss a fair coin infinitely many times independently, the probability that no head comes up is obviously 0. In what follows, we use the above continuity theorem to explain this fact rigorously. Let A_n be the event that at least one head comes up in the first n tosses, then $A_n \subset A_{n+1}$. Let $A = \bigcup_{n=1}^{\infty} A_n$, which indicates that a head will come up eventually. Then we have

$$P(A) = \lim_{n\to\infty} P(A_n) = \lim_{n\to\infty}\left[1 - \left(\frac{1}{2}\right)^n\right] = 1.$$

1.4 Conditional probability and independent events

1.4.1 Conditional probability

A random experiment is conducted under a certain condition and under this condition the event A occurs with some probability. With some additional condition, the probability of event A coming up may differ from that of event A occurring without any additional conditions. This additional condition can be described in terms of another event B.

Suppose we are given a probability space $(\Omega, \mathscr{F}, P)$ and we consider the probability that event A occurs. Of course, we cannot know the ultimate outcome of a random experiment for certainty. However, we may get some information about the experiment before the outcome appears. This information affects no doubt our judgment on the ultimate outcome. For example, if we throw a dice and get the information that it falls on an even number face, our judgment on the outcome surely differs from that without this information. Generally speaking, provided that some event B related to event A occurs, the probability that event A occurs will not be $P(A)$ any longer.

The probability that event A occurs provided that event B has occurred is called the conditional probability of event A given event B, denoted by $P(A \mid B)$. Conditional probability is one of basic tools in probability theory.

On some occasions, adding a condition is equivalent to reducing the original sample space and corresponding probabilities can be directly calculated in the reduced sample space.

Example 25 A box contains balls shown as in Table 1-3. Now draw a ball randomly and let $A = \{\text{the ball selected is blue}\}$ and $B = \{\text{the ball selected is a glass one}\}$. Obviously, this is a classical probability model, in which Ω have 16 sample points in sum and A consists of 11 points. Thus $P(A) = 11/16$.

Table 1-3

	Glass	Wooden	Total
Red	2	3	5
Blue	4	7	11
Total	6	10	16

Provided that the ball selected is glass, the probability that the ball is blue is just the conditional probability of A given B, written as $P(A \mid B)$. In this case, the total number of sample points equals the number of all glass balls, i. e. , the original sample space is reduced to a sample space of all glass balls. Moreover, the sample points that A contains under B are just blue glass balls. Therefore, $P(A \mid B) = 4/6 = 2/3$.

Generally speaking, in a classical probability model, conditional probability can be obtained by the method in preceding example. In the original sample space, conditional probability $P(A \mid B)$ is given by the following formula

$$P(A \mid B) = \frac{\text{number of sample points contained by } A \text{ given } B}{\text{number of sample points contained by } B}$$

$$= \frac{\dfrac{\text{number of sample points contained by } AB}{\text{total number of sample points}}}{\dfrac{\text{number of sample points contained by } B}{\text{total number of sample points}}}$$

$$= \frac{P(AB)}{P(B)},$$

as long as $P(B) \neq 0$. The above formula is also suitable for geometric probability model. Therefore we formulate a general definition of conditional probability as follows.

Definition 2 *If A, B are two events and $P(B) \neq 0$, then the conditional probability of A given B, written as $P(A \mid B)$, is defined as*

$$P(A \mid B) = \frac{P(AB)}{P(B)}. \tag{1.8}$$

The conditional probability can be used to express the probability of product of A and B through the following two equations

$$P(AB) = P(A \mid B)P(B),\ P(B) \neq 0, \tag{1.9}$$

and

$$P(AB) = P(B \mid A)P(A),\ P(A) \neq 0. \tag{1.10}$$

We call them multiplicative formulae.

From the definition of conditional probability, one easily knows that it has the same properties as the usual probability. In other words, conditional probability $P(\cdot \mid B)$ is nonnegative, countably additive and $P(\Omega \mid B) = 1$. Therefore, we can do the same operations as we do on the usual probability by adding a phrase like conditioning on event B.

The preceding multiplicative formulae on A and B can be generalized to the intersection of n events as follows

$$\begin{aligned} P(A_1A_2\cdots A_n) = P(A_1) \cdot P(A_2 \mid A_1) \cdot P(A_3 \mid A_1A_2) \\ \cdots P(A_n \mid A_1A_2\cdots A_{n-1}). \end{aligned} \tag{1.11}$$

In solving problems, conditional probability can be calculated according to the definition as well as in a reduced sample space like in the preceding example. Likewise, the probability of product of two events can be calculated by formulaes (1.9) and (1.10) as well as by a direct approach, which is simpler and better and is dependent on problems under discussion.

Example 26 There is a prizewinning ticket in n lottery tickets. These n lottery tickets are supposed to be sold to n different persons randomly.

(1) If the first $k-1$ customers do not get the prizewinning ticket, find the probability that the k-th customer gets the prizewinning ticket;

(2) Find the probability that the k-th customer gets the prizewinning ticket.

Solution We first remark that (1) is equivalent to finding the conditional probability given the condition that the first $k-1$ customers do not get the prizewinning ticket, and (2) is equivalent to finding a probability without any additional condition.

Let $A_i=\{$the i-th customer gets the prizewinning ticket$\}$. Then the condition in (1) is $\overline{A}_1\overline{A}_2\cdots\overline{A}_{k-1}$. If we consider the event A_k in the reduced sample space by $\overline{A}_1\overline{A}_2\cdots\overline{A}_{k-1}$, we can obtain by a direct application of classical probability model

$$P(A_k\mid\overline{A}_1\overline{A}_2\cdots\overline{A}_{k-1})=\frac{1}{n-k+1}.$$

As for (2), $A_k=\overline{A}_1\overline{A}_2\cdots\overline{A}_{k-1}A_k$ obviously holds. So by (1.8) we have

$$\begin{aligned}P(A_k)&=P(\overline{A}_1\overline{A}_2\cdots\overline{A}_{k-1}A_k)\\&=P(\overline{A}_1)P(\overline{A}_2\mid\overline{A}_1)P(\overline{A}_3\mid\overline{A}_1\overline{A}_2)\cdots P(A_k\mid\overline{A}_1\overline{A}_2\cdots\overline{A}_{k-1})\\&=\frac{n-1}{n}\cdot\frac{n-2}{n-1}\cdot\frac{n-3}{n-2}\cdots\frac{n-k+1}{n-k+2}\cdot\frac{1}{n-k+1}\\&=\frac{1}{n}.\end{aligned}$$

$P(A_k)=1/n$ implies that the probability of one customer getting the prizewinning ticket has nothing to do with when he or she buys a lottery ticket.

Example 27 In Example 21 of Subsection 1.3.2, we have found the probability that at least one letter will be placed in the correct envelope. Now find the probability that exactly k of the letters will be placed in the correct envelopes ($0\leqslant k\leqslant n$).

Solution Denote the probability that exactly k of the letters will be placed in the correct envelopes by $P_k^{(n)}$. By Example 21 of Subsection 1.3.2,

$$P_0^{(n)}=\frac{1}{2!}-\frac{1}{3!}+\frac{1}{4!}-\cdots+\frac{(-1)^n}{n!}.$$

Let A_i be the event that the k-th letter is placed in its correct envelopes. To obtain the probability that exactly k of the n letters will be placed in the correct envelopes, we first fix attention on a particular set of k letters, say, for example the i_1-th, the i_2-th, $\cdots$, the i_k-th letters. The probability that they will be placed in the correct envelopes is

$$\begin{aligned}&P(A_{i_1}\cdots A_{i_k}\overline{A}_{i_k+1}\cdots\overline{A}_{i_n})\\&=P(A_{i_1})P(A_{i_2}\mid A_{i_1})\cdots P(A_{i_k}\mid A_{i_1}\cdots A_{i_{k-1}})\cdot P(\overline{A}_{i_{k+1}}\cdots\overline{A}_{i_n}\mid A_{i_k}\cdots A_{i_k})\\&=\frac{1}{n}\frac{1}{n-1}\cdots\frac{1}{n-(k-1)}q_{n-k}=\frac{(n-k)!}{n!}q_{n-k},\end{aligned}$$

where q_{n-k} is the probability that, knowing these k letters are placed in the correct envelopes, none of the other $n-k$ letters will be placed in its correct envelope. So,

$$q_{n-k}=P_0^{(n-k)}=\frac{1}{2!}-\frac{1}{3!}+\frac{1}{4!}-\cdots+\frac{(-1)^{n-k}}{(n-k)!}.$$

Hence, as there are $\binom{n}{k}$ possible selections of a group of k letters, the desired probability is thus

$$\begin{aligned} P_k^{(n)} &= \sum_{i_1<\cdots<i_k} P(A_{i_1}\cdots A_{i_k}\overline{A}_{i_k+1}\cdots\overline{A}_{i_n}) \\ &= \binom{n}{k}\cdot\frac{(n-k)!}{n!}q_{n-k} \\ &= \frac{1}{k!}\frac{1}{2!}-\frac{1}{3!}+\frac{1}{4!}-\cdots+\frac{(-1)^{n-k}}{(n-k)!}, \end{aligned}$$

where for n large is approximately $e^{-1}\frac{1}{k!}$. The values $e^{-1}\frac{1}{k!}$, $k=0,1,\cdots$, are of some theoretical importance as they represent the values associated with the Poisson distribution. This is elaborated upon in the next chapter.

1.4.2 Total probability formula and Bayes' rule

To calculate probability of a complex event, we need to make full use of some of basic formulae mentioned above. Now let us introduce a basic concept of disjoint decomposition.

Definition 3 *Suppose that* $\{A_1,A_2,\cdots,A_n,\cdots\}$ *is a set of events satisfying*: (1) $A_i, i=1,2,\cdots$, *are mutually disjoint and* $P(A_i)>0$; (2) $\sum_{i=1}^{\infty}A_i=\Omega$. *Then* $\{A_1,A_2,\cdots,A_n,\cdots\}$ *is called a complete set of disjoint events in* Ω, *or a partition of* Ω.

Theorem 2 (*Total probability formula*) *If* $\{A_1,A_2,\cdots,A_n,\cdots\}$ *is a complete set of disjoint events, then for any event* B

$$P(B)=\sum_{i=1}^{\infty}P(A_i)P(B\mid A_i). \tag{1.12}$$

Proof Since $A_iB\subset A_i$, we have $(A_iB)\cap(A_jB)=\varnothing, i\neq j$. Therefore by the property of countable additivity, we have

$$\begin{aligned} P(B) = P(B\Omega) &= P(B\sum_{i=1}^{\infty}A_i) = P(\sum_{i=1}^{\infty}A_iB) \\ &= \sum_{i=1}^{\infty}P(A_iB) = \sum_{i=1}^{\infty}P(A_i)\cdot P(B\mid A_i). \end{aligned}$$

Here is a useful interpretation. Suppose that the event B may occur under a number of mutually exclusive circumstances (or causes). Then the formula shows how its total probability is computed from the probabilities of the various circumstances, and the corresponding conditional probabilities figured under the respective hypotheses.

Example 28 There are 5 table tennis, 3 new and 2 old. If two table tennis are drawn in succession without replacement, find the probability that the second is a new one.

Solution Let $A=\{$the first table tennis is new$\}$, $B=\{$the second is old$\}$. Obviously, the probability that the second is new depends on whether the first is a new one. In other words, B is related to the complete set $\{A,\overline{A}\}$ of disjoint events. It is easy to know that $P(A) = 3/5$, $P(\overline{A}) = 2/5$, $P(B \mid A) = 2/4$ and $P(B \mid \overline{A}) = 3/4$. Substituting these numerical values into formula (1.9) yields

$$P(B) = P(A) \cdot P(B \mid A) + P(\overline{A}) \cdot P(B \mid \overline{A}) = \frac{3}{5}.$$

Example 29 The following example is famous and illustrative, but somewhat artificial. Imagine a population of $N+1$ urns, each containing N red and white balls; the urn numbered k contains k red and $N-k$ white balls ($k = 0, 1, \cdots, N$). An urn is chosen at random and n random drawings are made from it, the ball drawn being replaced each time. Suppose that all n balls turn out to be red, find the probability that next drawing will also yield a red ball.

Solution Denote by A the event that all n balls turn out to be red, B the event that the next drawing will also yield a red ball. If the first choice falls on urn numbered k, then the probability of extracting in succession n red balls is $(k/N)^n$. Hence by (1.9),

$$P(A) = \frac{1^n + \cdots + N^n}{N^n(N+1)}. \tag{1.13}$$

The event AB means that $N+1$ drawings yield red balls, and therefore

$$P(AB) = \frac{1^{n+1} + \cdots + N^{n+1}}{N^{n+1}(N+1)}. \tag{1.14}$$

The required probability is $P(B \mid A) = P(AB)/P(A)$.

The sums in (1.13) and (1.14) can be considered as Riemann sums approximating integrals, so that when N is large

$$\frac{1}{N}\sum_{k=1}^{N}\left(\frac{k}{N}\right)^n \sim \int_0^1 x^n \mathrm{d}x = \frac{1}{n+1}. \tag{1.15}$$

We have therefore for large N approximately

$$P(B \mid A) \approx \frac{n+1}{n+2}. \tag{1.16}$$

This formula can be interpreted roughly as follows: if all compositions of an urn are equally probable, and if n trials yielded red balls, the probability of a red ball at the next trial is $(n+1)/(n+2)$. This is the so-called law of succession of Laplace.

Theorem 3 *(Bayes' rule) Under the assumption and notation of Theorem 2, we have*

$$P(A_i \mid B) = \frac{P(B \mid A_i)P(A_i)}{\sum_{i=1}^{\infty} P(B \mid A_i)P(A_i)} \tag{1.17}$$

provided $P(B) > 0$.

Proof The denominator above is equal to $P(B)$ by Theorem 2, so the equation may be multiplied out to read

$$P(B \mid A_i)P(A_i) = P(A_i \mid B)P(B).$$

This is true since both sides are equal to $P(A_iB)$.

This simple theorem with an easy proof is very famous under the name of Bayes' rule. It is supposed to yield an *inverse probability*, or probability of the *cause* A, on the basis of the observed "effect" B. Whereas $P(A_i)$ is the priori probability, $P(A_i \mid B)$ is the *posteriori probability* of the cause A_i. Numerous applications are made in all areas of natural phenomena and human behavior. For instance, if B is a "body" and the A_i's are the several suspects of the murder, then the theorem will help the jury or court to decide the whodunit (Jurisprudence was in fact a major field of early speculations on probability). If B is an earthquake and the A_i's are the different physical theories to explain it, then Bayes' rule will help the scientists to choose between them. In modern times Bayes lent his name to a school of statistics. For our discussion here let us merely comment that Bayes has certainly hit upon a remarkable turn-around for conditional probabilities, but the practical utility of his formula is limited by our usual lack of knowledge on various priori probabilities.

Example 30 Urn one contains 2 black and 3 red balls; urn two contains 3 black and red balls. We toss an unbiased coin to decide on the urn to draw from but we do not know which is which. Suppose that the first ball drawn is black and it is put back. Find the probability that the second ball drawn from the same urn is also black.

Solution Call these two urns U_1 and U_2 respectively; the probability that either one is chosen by coin-tossing is $1/2$:

$$P(U_1) = \frac{1}{2},\ P(U_2) = \frac{1}{2}.$$

Denote the event that the first ball is black by B_1, and that the second ball is black by B_2. We have by (1.17)

$$P(U_1 \mid B_1) = \frac{\frac{1}{2} \cdot \frac{2}{5}}{\frac{1}{2} \cdot \frac{2}{5} + \frac{1}{2} \cdot \frac{3}{5}} = \frac{2}{5};\ P(U_2 \mid B_1) = \frac{3}{5}.$$

Now we use (1.14) to compute the probability that the second ball is also black. Let $A_1 = \{B_1 \text{ is from } U_1\}$ and $A_2 = \{B_1 \text{ is from } U_2\}$. Then

$$P(A_1) = P(U_1 \mid B_1) = \frac{2}{5},\ P(A_2) = P(U_2 \mid B_1) = \frac{3}{5}.$$

On the other hand, it is obvious that

$$P(B_2 \mid A_1) = \frac{2}{5},\ P(B_2 \mid A_2) = \frac{3}{5}.$$

Hence we obtain the conditional probability

$$P(B_2 \mid B_1) = \frac{2}{5} \cdot \frac{2}{5} + \frac{3}{5} \cdot \frac{3}{5} = \frac{13}{25}.$$

Compare this with

$$P(B_2) = P(U_1)P(B_2 \mid U_1) + P(U_2)P(B_2 \mid U_2) = \frac{1}{2}.$$

Note that $P(B_2 \mid U_1) = P(B_1 \mid U_1)$. We see that the knowledge of the first ball drawn being black has strengthened the probability of drawing a second black ball, because it has increased the likelihood that we have picked the urn with more black balls. To proceed one more step, given that the first two balls drawn are both black and put back, what is the probability of drawing a third black ball from the same urn? We have in notation similar to the above:

$$P(U_1 \mid B_1B_2) = \frac{\frac{1}{2}\left(\frac{2}{5}\right)^2}{\frac{1}{2}\left(\frac{2}{5}\right)^2 + \frac{1}{2}\left(\frac{3}{5}\right)^2} = \frac{4}{13},$$

$$P(U_2 \mid B_1B_2) = \frac{9}{13},$$

$$P(B_3 \mid B_1B_2) = \frac{4}{13} \cdot \frac{2}{5} + \frac{9}{13} \cdot \frac{3}{5} = \frac{35}{65}.$$

This is greater than $13/25$, so a further strengthening has occurred. Now it is easy to see that we can extend the result to any number of drawings. Thus,

$$P(U_1 \mid B_1 \cdots B_n) = \frac{\frac{1}{2}\left(\frac{2}{5}\right)^n}{\frac{1}{2}\left(\frac{2}{5}\right)^n + \frac{1}{2}\left(\frac{3}{5}\right)^n} = \frac{1}{1+\left(\frac{3}{2}\right)^n},$$

where we have divided the denominator by the numerator in the middle term. It follows that as n becomes larger and larger, the conditional probability of U_1 becomes smaller and smaller; in fact it decreases to zero and consequently the conditional probability of U_2 increases to one in the limit. Thus we have

$$\lim_{n\to\infty} P(B_{n+1} \mid B_1 \cdots B_n) = \frac{3}{5} = P(B_1 \mid U_2).$$

This simple example has important implications on the empirical viewpoint of probability. Replace the two urns above by a coin which may be biased (as all real coins are). Assume that the probability p of heads is either $2/5$ or $3/5$ but we do not know which is the true value. The two possibilities are then two alternative hypotheses between which we must decide. If they both have the $1/2$ probability, then we are in the situation of the two urns. The outcome of each toss will affect our empirical estimate of the value p. Suppose for some reason we believe that $p = 2/5$. Then if the coin falls heads 10 times in a row, can we still maintain that $p = 2/5$ and give probability $(2/5)^{10}$ to this rare event? Or shall we concede that really $p = 3/5$ so that the same event will have probability $(3/5)^{10}$? This is very small but $(3/5)^{10}$ is still larger than the other. In certain problems of probability theory it is customary to consider the value of p as fixed and base the rest of our calculations on it. The query is why we have to maintain such a fixed stance in the face of damaging evidence given by observed outcomes. From the axiomatic point of view, a simple answer is this: our formulas are correct for each arbitrary value of p, but axioms of course do not tell us what this value is, nor even whether it makes sense to assign any value at all. In other words, mathematics appears to be a deductive science. The problem of evaluating, estimating or testing the value of p lies outside its eminent domain. Of course, it is of the utmost importance in practice, and statistics was invented to cope with this kind of problem. But it need not concern us too much here.

Example 31 (The gambler's ruin problem) Suppose that two gamblers A and B are playing a game against each other. Let p be a given number $(0 < p < 1)$, and suppose that on each play of the game, the probability that gambler A will win one dollar from gambler B is p and the probability that gambler B will win one dollar from gambler A is $q = 1 - p$. Suppose also that the initial fortune of gambler A is i dollars and the initial fortune of gambler B is $k - i$ dollars, where i and $k - i$ are given positive integers. Finally suppose that the gamblers continue playing the game until the fortune of one of them has been reduced to 0 dollar. Find the probability that the fortune of gambler A reaches k dollars before it reaches 0 dollar. Because one of the gamblers will have no money left at the end of the game, this problem is called the gambler's ruin problem.

Solution Let p_i denote the probability that the fortune of gambler A will reach k dollars before it reaches 0 dollar, given that his initial fortune is i dollars. If $i = 0$, then gambler A is ruined; and if $i = k$, then gambler A has won the game. Therefore, we shall assume that $p_0 = 0$ and $p_k = 1$. We shall determine the value of p_i for $i = 1, \cdots, k-1$.

Let $A_1 (B_1)$ denote the event that gambler A wins (resp. loses) one dollar on the first play of the game, and let W denote the event that the fortune of gambler A will reach k dollars before it reaches 0 dollar. Then

$$\begin{aligned} P(W) &= P(A_1)P(W \mid A_1) + P(B_1)P(W \mid B_1) \\ &= pP(W \mid A_1) + qP(W \mid B_1). \end{aligned} \tag{1.18}$$

Since the initial fortune of gambler A is i dollars $(i = 1, \cdots, k-1)$, then $P(W) = p_i$. Furthermore, if gambler A wins one dollar on the first play of the game, then his fortune becomes $i+1$ dollars and the probability $P(W \mid A_1)$ that his fortune will ultimately reach k dollars is therefore p_{i+1}. If A loses one dollar on the first play of the game, then his fortune becomes $i - 1$ dollars and the probability $P(W \mid B_1)$ that his fortune will ultimately reach k dollars is therefore p_{i-1}. Hence, by (1.12)

$$p_i = pp_{i+1} + qp_{i-1}. \tag{1.19}$$

Then, by some elementary algebra, we obtain

$$p_i - p_{i-1} = \frac{q}{p}(p_{i-1} - p_{i-2}) = \left(\frac{q}{p}\right)^{i-1} p_1, \ i = 2, \cdots, k.$$

By equating the sum of the left sides of these $k - 1$ equations with the sum of the right sides, we obtain the relation

$$1 - p_1 = p_1 \sum_{i=1}^{k-1} \left(\frac{q}{p}\right)^i. \qquad (1.20)$$

Suppose now that $p \neq q$. Then (1.20) can be rewritten in the form

$$1 - p_1 = p_1 \frac{(q/p)^k - q/p}{q/p - 1}.$$

Hence,

$$p_1 = \frac{q/p - 1}{(q/p)^k - 1}.$$

Each of the other values of p_i for $i = 2, \cdots, k-1$ can now be determined in turn from (1.20). In this way we obtain the following complete solution:

$$p_i = \frac{(q/p)^i - 1}{(q/p)^k - 1}, \text{ for } i = 1, \cdots, k-1.$$

Suppose that $p = q = 1/2$. Then $q/p = 1$, and it follows directly from (1.20) that $1 - p_1 = (k-1)p_1$, from which $p_1 = 1/k$. In turn, we obtain from equation (1.19) the following solution:

$$p_i = \frac{i}{k}, \text{ for } i = 1, \cdots, k-1.$$

Example 32 (Polya's urn scheme) An urn contains b black and r red balls. Random drawings are made. The ball drawn is always replaced, and in addition, c balls of the colors drawn are added to the urn. Here we are given conditional probabilities only. If the first ball is black, the (conditional) probability of a black ball at the second drawing is $(b+c)/(b+c+r)$. The absolute probability of the sequence black is therefore

$$\frac{b}{b+r} \cdot \frac{b+c}{b+c+r}.$$

If the first two drawings result in black, then the urn contains $b+r+2c$ balls among which $b+2c$ are black. The (conditional) probability of a black ball at the third trial becomes

$$\frac{b+2c}{b+2c+r}.$$

In this way we can calculate all probabilities. It is easily seen that any sequence of n drawings resulting in n_1 black and n_2 red balls ($n_1 + n_2 = n$) has the same probability as the event of extracting first n_1 black and then n_2 red balls, namely,

$$p_{n_1, n_2} = \frac{b(b+c)\cdots(b+n_1c-c)r(r+c)\cdots(r+n_2c-c)}{(b+r)(b+r+c)\cdots(b+r+nc-c)}.$$

This scheme is devised for the analysis of phenomena like contagious diseases, where the occurrence of certain events increases their future probabilities.

Remark $P(A_i)$ in equation (1.12) is the understanding of the possibility that A_i occurs without any information (without knowing whether B has occurred); while the conditional probability $P(A_i \mid B)$ in (1.17) is a new estimation of possibility that A_i occurs under a certain information (knowing that B has occurred).

If we regard B as result and A_i, $i = 1,2,\cdots$, as factors that B occurs, then the formula for total probability uses factors to estimate result; while Bayes' formula uses result to estimate factors. Conditioning on B, the possibility of each factor occurring can be obtained by Bayes' formula.

This is consistent with people's daily experience: some event that is unlikely to occur will occur with high likelihood because of another event, or vice versa.

Example 33 A doctor uses a blood test method to diagnose patients in order to see whether they suffer from liver cancer. Let C be the event that a patient suffers from liver cancer, and A the event that a patient is diagnosed suffering from liver cancer. Suppose

$$P(A \mid C) = 0.95,\ P(\overline{A} \mid \overline{C}) = 0.90,\ P(C) = 0.0004,$$

find the probability that one patient diagnosed suffering from liver cancer really gets liver cancer.

Solution According to Bayes' formula, we have

$$P(C \mid A) = \frac{P(C) \cdot P(A \mid C)}{P(C) \cdot P(A \mid C) + P(\overline{C}) \cdot P(A \mid \overline{C})}. \tag{1.21}$$

In addition, $P(\overline{C}) = 1 - P(C) = 0.9996$, $P(A \mid \overline{C}) = 1 - P(\overline{A} \mid \overline{C}) = 0.10$.

Substituting these numerical values into (1.21) yields

$$P(C \mid A) = 0.0038.$$

$P(A \mid C)$ is the probability that a patient suffering from liver cancer is diagnosed suffering from liver cancer and $P(\overline{A} \mid \overline{C}) = 0.90$ is the probability that a patient without liver cancer is diagnosed not suffering from liver cancer. Since $P(\overline{A} \mid \overline{C}) = 0.95$ and $P(\overline{A} \mid \overline{C}) = 0.90$, both of them being very big, this diagnostic method is rather reliable. However, the probability $P(C \mid A)$ that a patient diagnosed suffering from liver cancer really gets a liver cancer is small, just 0.0038. Without knowledge of probability theory, we might well think this diagnostic method is not efficient. In fact, the proportion of patients with liver cancer in all people is small ($P(C) = 0.0004$) and the diagnostic method is not completely accurate. So there is still a large number of healthy people diagnosed incorrectly suffering from liver cancer. On the other hand, since the ratio of patients with liver cancer is rather small, even though all patients with liver cancer are one hundred percent diagnosed, they account just for a small fraction of all the patients diagnosed suffering from liver cancer.

Example 34 A factory produces the same kind of product by four assembly lines. Their output account for 12%, 25%, 25%, 38% of the total, and the rates of defective product are 0.06, 0.05, 0.04, 0.03, respectively. Now a customer finds what he/she buys is defective and asks for compensation of 10,000 yuan. If the loss is paid by four lines together, then how much should each line compensate for resepctively?

Solution Let B be the event "a sample product is defective", and A_i the event "the sample product comes from the i-th assembly linc", $i=1, 2, 3, 4$. We have

$$P(B) = \sum_{i=1}^{4} P(B \mid A_i)P(A_i)$$
$$= 0.12 \times 0.06 + 0.25 \times 0.05 + 0.25 \times 0.04 + 0.38 \times 0.03$$
$$= 0.0411.$$

This means that the defective fraction of the product is 4.11%. Now a customer finds what he/she buys is defective. That is, B occurs. Then we shall find out which line it is made by. We need to calculate the conditional probability $P(A_i|B), i=1,2,3,4$. By the Bayesian formula, it follows

$$P(A_1 \mid B) = \frac{P(B \mid A_1)P(A_1)}{P(B)}$$
$$= \frac{0.12 \times 0.06}{0.0411} \approx 0.175.$$

Similarly,

$$P(A_2 \mid B) \approx 0.304, \quad P(A_3 \mid B) \approx 0.243, \quad P(A_4 \mid B) \approx 0.278.$$

Therefore, the assembly lines should give compensation of 1 750 yuan, 3 040 yuan, 2 430 yuan and 2 780 yuan respectively.

The above example is an application of Bayesian method in the decision making.

Example 35 Consider a medical partitioner dealing with a disease A by the following rule. If he is at least 85% certain that a patient has this disease, then he always recommends surgery, whereas if he is not quite as certain, then he recommends additional test that are expensive and sometime painful. Now, initially the doctor was only 60% certain that a patient named Jones had the disease A, so Jones was ordered the series B test, which always gives a positive result when the patient has the disease A and almost never does when he is healthy. The test result of Jones is positive, but at the same time it is told that Jones had another disease which would cause a positive result of the test B with probability 25%. Now what should the doctor do? Recommend more test or surgery?

Solution In order to decide whether or not to recommend surgery, the doctor should first compute his updated probability that Jones has the disease given that the B test result was positive. Let A denote the event that Jones has the disease A, and B the event of a positive B test result. So, $P(A)=0.6$, $P(B|A)=1$ and $P(B|\overline{A})=0.25$. By Bayes' Formula, the desired conditional probability $P(A|B)$ is obtained by

$$P(A \mid B) = \frac{P(AB)}{P(B)} = \frac{P(A)P(B \mid A)}{P(B \mid A)P(A) + P(B \mid \overline{A})P(\overline{A})}$$
$$= \frac{1 \times 0.6}{1 \times 0.6 + 0.25 \times 0.4} = 0.857.$$

Hence, as the doctor should now be over 85% certain that Jones has the disease A, he should reeommend surgery.

1.4.3 Independent events

Ⅰ Independence of two events

Suppose A and B are two events. That B occurs will usually cause some change of the probability that A occurs. If the probability that A occurs has nothing to do with the event B at all, i. e.,

$$P(A \mid B) = P(A), \tag{1.22}$$

then we have $P(AB) = P(B) \cdot P(A \mid B) = P(A) \cdot P(B)$. Conversely, if

$$P(AB) = P(A) \cdot P(B), \tag{1.23}$$

then

$$P(A \mid B) = \frac{P(AB)}{P(B)} = \frac{P(A) \cdot P(B)}{P(B)} = P(A).$$

When either equation (1. 22) or equation (1. 23) holds, A and B are said to be independent. Instead A and B are said to be dependent if they are not independent. Obviously, (1. 22) is equivalent to (1. 23) given that $P(B) > 0$. Generally, the definition of independence adopts equation (1. 23) since it is symmetric in terms of A and B and can be generalized to the n case of events. Moreover, (1. 23) does not require the condition that $P(B) > 0$. In fact, if $B = \varnothing$, then $P(B) = 0$ and thus $P(AB) = 0$. Therefore, (1. 23) holds whatever the event A is. In other words, any event is independent of event $\varnothing$. Similarly, any event is independent of inevitable event Ω.

Example 36 An urn contains a black balls and b white balls. If two balls are drawn in succession and we denote by A the event that the first ball drawn is black, B the event that the second ball drawn is black, are A and B independent of each other?

Solution We need to consider two different situations: (1) with replacement, (2) without replacement.

In case (1) the first ball drawn is put back in the urn, so the experiment condition of the second drawing is completely same as in the first drawing. Thus A and B are clearly independent of each other. In fact, using the total probability formula, we have

$$\begin{aligned} P(B) &= P(A)P(B \mid A) + P(\overline{A})P(B \mid \overline{A}) \\ &= \frac{a}{a+b} \cdot \frac{a}{a+b} + \frac{b}{a+b} \cdot \frac{a}{a+b} \\ &= \frac{a}{a+b} = P(B \mid A). \end{aligned}$$

For case (2), we have

$$P(B \mid A) = \frac{a-1}{a+b-1}, \quad P(B \mid \overline{A}) = \frac{a}{a+b-1}.$$

By the formula for total probability, we have

$$P(B) = \frac{a}{a+b} \cdot \frac{a-1}{a+b-1} + \frac{b}{a+b} \cdot \frac{a}{a+b-1}$$

$$= \frac{a}{a+b} \neq P(B \mid A),$$

which shows that A and B are not independent.

Example 37 Show that A and B are not independent if A and B are mutually disjoint and $P(A)P(B) \neq 0$.

Proof Since A and B are mutually disjoint, $P(AB) = 0 \neq P(A)P(B)$. Therefore A and B are not independent.

Example 38 Suppose A and B are two events independent of each other, show that so are A and $\overline{B}$, $\overline{A}$ and B, $\overline{A}$ and $\overline{B}$.

Proof Since A and B are independent, $P(AB) = P(A)P(B)$, then

$$\begin{aligned} P(A\overline{B}) &= P(A - AB) = P(A) - P(AB) \\ &= P(A) - P(A)P(B) = P(A)(1 - P(B)) \\ &= P(A)P(\overline{B}). \end{aligned}$$

This shows that A and $\overline{B}$ are independent. Replacing A by $\overline{A}$ in preceding equation yields that $\overline{A}$ and $\overline{B}$ are independent. A similar argument is applicable to $\overline{A}$ and B.

In many practical problems it is very hard to check whether two events are independent by (1.22) or (1.23). In these situations, we should make a judgment by taking the intuitive understanding of independence into account. For examples, malfunctions of two different components in a circuit system are often regarded as independent, while the temperature and rainfall in a certain region cannot be treated as independent.

Definition 4 *Two σ-algebras $\mathscr{F}_1$ and $\mathscr{F}_2$ are said to be independent with regard to P, if*

$$P(A_1 A_2) = P(A_1)P(A_2)$$

holds for any $A_1 \in \mathscr{F}_1$, and any $A_2 \in \mathscr{F}_2$.

In particular, let $\mathscr{F}_1 = \{A_1, \overline{A_1}, \varnothing, \Omega\}$ and $\mathscr{F}_2 = \{A_2, \overline{A_2}, \varnothing, \Omega\}$. Then $\mathscr{F}_1$ and $\mathscr{F}_2$ are independent if and only if A_1 and A_2 are independent.

Ⅱ Independence of several events

When n events are involved, besides the pair-wise independence, we consider the independence among them. Now let us start with the independence among three events.

Definition 5 *Events A, B and C are said to be independent if*

$$\left.\begin{aligned} P(AB) &= P(A) \cdot P(B) \\ P(AC) &= P(A) \cdot P(C) \\ P(BC) &= P(B) \cdot P(C) \end{aligned}\right\} \tag{1.24}$$

and

$$P(ABC) = P(A) \cdot P(B) \cdot P(C). \tag{1.25}$$

Equation (1.24) implies that A, B and C are pair-wise independent. So the independence of three events implies pair-wise independence. However, pair-wise independence does not imply the independence of three events since mutual independence requires both (1.24) and (1.25) hold simultaneously. Then does equation (1.24) imply equation (1.25)? The answer to this question is NO. See the following example.

Example 39 Consider a fair regular tetrahedron. The first side is red, the second side white, the third side black, and the fourth side consists of red, white and black. Now throw randomly this regular tetrahedron and let A, B and C denote the events that the bottom is red, white and black, respectively. Since two sides of this regular tetrahedron are red, $P(A)=\frac{1}{2}$; similarly $P(B)=P(C)=\frac{1}{2}$. Only when the face is the fourth side, the face contains three colors. So

$$P(AB)=P(AC)=P(BC)=P(ABC)=\frac{1}{4}.$$

Hence

$$P(AB)=P(A)\cdot P(B),$$
$$P(AC)=P(A)\cdot P(C),$$
$$P(BC)=P(B)\cdot P(C),$$

which shows that equation (1.24) holds, i.e., A, B and C are pairwise independent. However,

$$P(ABC)=\frac{1}{4}\neq P(A)\cdot P(B)\cdot P(C)=\frac{1}{8},$$

which shows that equation (1.25) does not hold.

Conversely, there are examples showing that (1.25) does not imply (1.24). Thus both (1.24) and (1.25) are necessary in the definition of mutual independence of A, B and C. Similarly, the mutual independence of n events is defined as follows.

Definition 6 *Suppose that $A_1, A_2, \cdots, A_n$ are n events. If for $1\leqslant i<j<k<\cdots\leqslant n$,*

$$\left.\begin{aligned} &P(A_iA_j)=P(A_i)P(A_j),\\ &P(A_iA_jA_k)=P(A_i)P(A_j)P(A_k),\\ &\cdots\cdots\\ &P(A_iA_j\cdots A_n)=P(A_i)P(A_j)\cdots P(A_n)\end{aligned}\right\} \tag{1.26}$$

hold, then $A_1, A_2, \cdots, A_n$ are said to be independent.

(1.26) contains $C_n^2+\cdots+C_n^n=2^n-n-1$ equations in sum and indicates that any k $(2\leqslant k<n)$ events of n independent events are independent.

Example 40 Suppose that $A_1, A_2, \cdots, A_n$ are independent, and $P(A_i)=p_i$, $i=1,2,\cdots,n$. Find the probabilities that

(1) none of them occurs;

(2) at least one of them occurs;

(3) only one of them occurs.

Solution First we represent the above events in form of unions, intersections and complements of $A_1, A_2, \cdots, A_n$. Then we use probability formulae to calculate the desired probabilities.

(1) {none of them occurs} $= \overline{A}_1 \overline{A}_2 \cdots \overline{A}_n$, as in Example 31, we can verify that $\overline{A}_1, \overline{A}_2, \cdots, \overline{A}_n$ are mutually independent. Therefore, we have

$$P(\overline{A}_1 \overline{A}_2 \cdots \overline{A}_n) = P(\overline{A}_1)P(\overline{A}_2)\cdots P(\overline{A}_n) = \prod_{i=1}^{n}(1-p_i).$$

(2) {at least one of them occurs} $= A_1 \cup A_2 \cup \cdots \cup A_n$, which is the complement of the event in (1). So

$$P(A_1 \cup A_2 \cup \cdots \cup A_n) = 1 - P(\overline{A}_1 \overline{A}_2 \cdots \overline{A}_n) = 1 - \prod_{i=1}^{n}(1-p_i).$$

Of course, we can also obtain this probability by the additive rule in Section 1. 3. However, using the additive rule here is much more complicated.

(3) {only one of them occurs} $= \overline{A}_1\overline{A}_2\cdots\overline{A}_{n-1}A_n + \overline{A}_1\overline{A}_2\cdots A_{n-1}\overline{A}_n + \cdots + A_1\overline{A}_2\cdots\overline{A}_n$.

The items in the right side are mutually disjoint and the events in each item are independent. Therefore, the desired probability is

$$\begin{aligned} & P\left(\sum_{k=1}^{n}\overline{A}_1\overline{A}_2\cdots\overline{A}_{k-1}A_k\overline{A}_{k+1}\cdots\overline{A}_n\right) \\ = & \sum_{k=1}^{n}P(\overline{A}_1\overline{A}_2\cdots\overline{A}_{k-1}A_k\overline{A}_{k+1}\cdots\overline{A}_n) \\ = & \sum_{k=1}^{n}P(\overline{A}_1)P(\overline{A}_2)\cdots P(\overline{A}_{k-1})P(A_k)P(\overline{A}_{k+1})\cdots P(\overline{A}_n) \\ = & \sum_{k=1}^{n}p_k\prod_{i=1,i\neq k}^{n}(1-p_i). \end{aligned}$$

Example 41 The probability that a system can work in order is called reliability. Two systems are both composed of electronic components A, B, C and D of the same kind as in Figure 1-4.

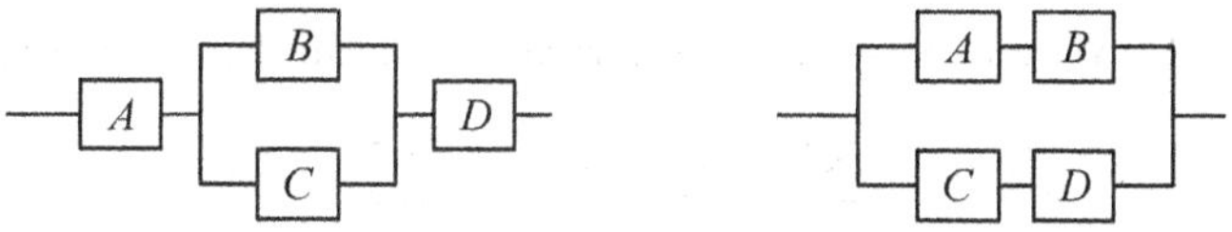

Figure 1-4

The reliability of each component is p, and find the reliability of both systems.

Solution Denote by R_1, R_2 the reliability of these two systems respectively and denote by A, B, C and D the events that corresponding components work in order respectively. We may and do assume that A, B, C and D are independent. Thus we have

$$\begin{aligned} R_1 &= P(A(B \cup C)D) = P(ABD \cup ACD) \\ &= P(ABD) + P(ACD) - P(ABCD) \\ &= P(A)P(B)P(D) + P(A)P(C)P(D) \\ &\quad - P(A)P(B)P(C)P(D) \\ &= p^3(2-p), \end{aligned}$$

$$\begin{aligned} R_2 &= P(AB \cup CD) = P(AB) + P(CD) - P(ABCD) \\ &= p^2(2-p^2). \end{aligned}$$

Obviously, $R_2 > R_1$.

Reliability theory is widely applied in system science, and so the study of reliability of system is of great significance.

Example 42 (Branching process) Consider a population consisting of individuals able to produce offspring of the same kind. Suppose that each individual will, by the end of its lifetime, have produced k new offspring with probability p_k, $k=0, 1, 2 \cdots$, independently of the number produced by any other individual. Assume that the number of individuals initially present (the 0-th generation) is 1. Show that when $m=\sum_{k=0}^{\infty} kp_k \leqslant 1$ and $p_1<1$, the population will eventually die out with the probability 1.

Proof Let A be the event that the population will eventually die out, and B_k be the event that the first generation has k individuals (i. e., the 0-th generation produces k new offspring). By the total probability formula, the desired probability is obtained as

$$q = P(A) = \sum_{k=0}^{\infty} P(A \mid B_k)P(B_k) = \sum_{k=0}^{\infty} P(A \mid B_k)p_k$$

Conditional on B_k, the population has k individuals, and the group of each individual and its offspring will eventually die out with the probability q. Since individuals produce their offspring independently, $P(A|B_k)=q^k$. Hence,

$$q = \sum_{k=0}^{\infty} q^k p_k.$$

That is that q is a solution of the equation $g(s)=s$, where $g(s) = \sum_{k=0}^{\infty} s^k p_k \, (0 \leqslant s \leqslant 1)$. Obviously, $g(1)=1$. When $0 \leqslant s < 1$, the derivative of the function $g(s)$ is as follows.

$$g'(s) = \sum_{k=1}^{\infty} s^{k-1} kp_k = p_1 + \sum_{k=2}^{\infty} s^{k-1} kp_k.$$

If $p_0 + p_1 < 1$, then we must have $p_k > 0$ for some $k \geqslant 2$. In such case,

$$g'(s) = p_1 + \sum_{k=2}^{\infty} s^{k-1} kp_k < p_1 + \sum_{k=2}^{\infty} kp_k = m \leqslant 1;$$

If $p_0 + p_1 = 1$, then

$$g'(s) = p_1 < 1.$$

So, in any case we have that $(g(s)-s)' < 0$, $0 \leqslant s < 1$. It follows that $g(s)-s$ is strictly decreasing in $[0, 1]$. So, $q=1$ is the unique solution of the equation $g(s)=s$. The proof is now completed.

Ⅲ The independence of experiments

The independence of experiments is closely related to that of events. Suppose E_1, E_2, …, E_n are n experiments, then each possible outcome of each experiment can be treated as an event. E_1, E_2, …, E_n are said to be independent if A_1, A_2, …, A_n are independent, where $A_1 \in E_1$, $A_2 \in E_2$, …, $A_n \in E_n$.

Suppose that Ω_i is a sample space of E_i. To describe these n experiments, we construct a compound experiment $E = (E_1, E_2, \cdots, E_n)$ with sample space $\Omega = \Omega_1 \times \Omega_2 \times \cdots \times \Omega_n$, which is the direct product of n sample spaces, and let sample points $\omega = (\omega^{(1)}, \cdots, \omega^{(n)})$, where $\omega^{(i)} \in \Omega_i$. In a compound sample space, event $A^{(i)}$ can be represented as $(\Omega_1, \cdots, A^{(i)}, \cdots, \Omega_n)$, which we still denote by $A^{(i)}$. Then the independence of E_1, E_2, …, E_n can be expressed in terms of

$$P(A^{(1)}A^{(2)}\cdots A^{(n)}) = P(A^{(1)})P(A^{(2)})\cdots P(A^{(n)}) \tag{1.27}$$

for all $A^{(i)}$ of E_i, $i = 1, 2, \cdots, n$.

Drawing balls n times with replacement is one example of n independent experiments with $\Omega = \Omega_1 = \cdots = \Omega_n$ and with equal probability of the same event in each experiment. Such n experiments are said to be n repeated independent experiments. Indeed, we had learned this concept when we discussed the statistical definition of probability. On the other hand, drawing balls n times without replacement is one example of n dependent experiments.

Ⅳ The Bernoulli model

A trial is called Bernoulli trial if there are only two possible outcomes for each trial. For instance, tossing a coin appears either a head or a tail; a circuit is either connected or disconnected; a piece of information we received is either right or wrong; a seed we plant will either germinate or not; a machine we observe is either working in order or not, and so forth. Sometimes a trial has many possible outcomes, but can still be regarded as a Bernoulli one, if we are only interested in whether an event A occurs or not. For example, sampling from a set of products normally has several results. But it can also be regarded as a Bernoulli trial if we are only interested in whether this product is up to standard or not. Let A denote "success" and $\overline{A}$ "failure" in a Bernoulli trial, and then the sample space is $\Omega = (\omega_1, \omega_2)$, where $\omega_1 = A$ and $\omega_2 = \overline{A}$, and σ-algebra is $\mathscr{F} = \{\varnothing, A, \overline{A}, \Omega\}$. Given $P(A) = p (0 < p < 1)$, $P(\overline{A}) = 1 - p$, we will get probabilities of all events in a Bernoulli trial.

Repeated independent Bernoulli trials are widely studied. We call this probability model the Bernoulli model. As seen from the end of last paragraph, its sample points are $\omega = (\omega^{(1)}, \cdots, \omega^{(n)})$, where $\omega^{(i)}$ is A or $\overline{A}$ and the total number of its sample points is 2^n. Although its sample space is finite, the Bernoulli model is not a classical probability model since the probabilities of its sample points are not necessarily equal.

Each sample point of a Bernoulli model is an elementary event, and either a part or all of them will constitute a compound event. Using operation rules for event and probability, we can calculate the probabilities of compound events.

Example 43 Consider a Bernoulli model of n repeated independent trials. Let $A_k = \{A$ occurs only in the first k trials$\}$, $B_k = \{A$ occurs exactly k times$\}$. Find (1) $P(A_k)$, (2) $P(B_k)$.

Solution (1) It is easy to see

$$A_k = \underbrace{AA\cdots A}_{k}\underbrace{\overline{A}\,\overline{A}\cdots\overline{A}}_{n-k}.$$

Since A occurs independently with probability p each trial, then $P(A_k) = p^k q^{n-k}$.

(2) Note that B_k limits the number of times A occurs, but does not require A to occur in certain trials. A may occur in the first k trials, or in the last k trials. Without confusion, we represent each elementary event in form of the product of k A's and $(n-k)$ $\overline{A}$'s. Then B_k is the union of all these elementary events, i. e.,

$$\begin{aligned} B_k = {} & \underbrace{AA\cdots A}_{k}\underbrace{\overline{A}\,\overline{A}\cdots\overline{A}}_{n-k} + A\overline{A}\,\underbrace{A\cdots A}_{k-1}\underbrace{\overline{A}\,\overline{A}\cdots\overline{A}}_{n-k-1} \\ & + \cdots + \underbrace{\overline{A}\,\overline{A}\cdots\overline{A}}_{n-k}\underbrace{AA\cdots A}_{k}. \end{aligned} \tag{1.28}$$

The right side of the above equation contains $\binom{n}{k}$ mutually disjoint summands (events) in sum, and each summand consists of independent events. Therefore, the probability of each summand is $[P(A)]^k[P(\overline{A})]^{n-k} = p^k q^{n-k}$, and using finite additivity yields $P(B_k) = \binom{n}{k} p^k q^{n-k}$. This is usually written

$$\begin{gathered} b(k;\ n,\ p) = C_n^k p^k q^{n-k} = \frac{n!}{k!(n-k)!} p^k q^{n-k}, \\ k = 0, 1, 2, \cdots, n, \end{gathered} \tag{1.29}$$

which appears in the expansion $(p+q)^n = \sum_{k=0}^{n} C_n^k p^k q^{n-k}$ with the total sum 1. So call $b(k;\ n,\ p)$ the binomial distribution. The probability obtained above is one of the most important in a Bernoulli model. In fact, probabilities of many other events can be derived from it.

Example 44 Consider n machines of the same type. Suppose the probability is p that each machine breaks down over a certain period. Find the probability that at least m machines work well during this period.

Solution Observe the probability that each machine works well during this period is $q = 1 - p$, and $\{$at least m machines work well during this period$\} = \sum_{k=m}^{n} \{$just k

machines work well during this period} and all the events in the sum are mutually disjoint. We have

$$\sum_{k=m}^{n} P\{k \text{ machines work well during this period}\} = \sum_{k=m}^{n} b(k;n,p).$$

Revisit Example 11 Each time he draws a match, the match is from either the first box or the second. Let $A=$ {the match is from the first box} and $\overline{A}=$ {the match is from the second box}, and consider a Bernoulli model of $2n-m+1$ repeated independent trials. If the first box is first found empty, then the first box is drawn n times and the second box is drawn $n-m$ times and the last drawing is from the first box. So, the probability is $\binom{2n-m}{n}(1/2)^n(1/2)^{n-m}1/2$. We can consider the case that the second box is first found empty is in a very similar way. Therefore, the probability desired is

$$2\binom{2n-m}{n}\left(\frac{1}{2}\right)^n\left(\frac{1}{2}\right)^{n-m}\frac{1}{2} = \binom{2n-m}{n}\left(\frac{1}{2}\right)^{2n-m}.$$

Remark A trial is no longer a Bernoulli one when it has more than two outcomes. However, it can be dealt with by a similar analytic method. Suppose a trial has outcomes $A_1, A_2, \cdots, A_k (k \geqslant 3)$, which constitute a complete set of disjoint sets and $P(A_i) = p_i$, $\sum_{i=1}^{k} p_i = 1$. Then the probabilities that $A_1, A_2, \cdots, A_k$ occurs $n_1, n_2, \cdots, n_k$ times in n repeated independent trials are

$$\frac{n!}{n_1! \cdots n_k!} p_1^{n_1} \cdots p_k^{n_k},$$

respectively.

(Hint: For clarity, assume first that A_1 occurs in the first n_1 trials, A_2 occurs in the second n_2 trials, $\cdots$, and A_k occurs in the last n_k trials. The probability of such an event is $p_1^{n_1}\cdots p_k^{n_k}$. Then, exchange the order of $A_1, A_2, \cdots, A_n$, and note that all the probabilities of the events in different orders are $p_1^{n_1}\cdots p_k^{n_k}$. Finally, there are $\binom{n}{n_1}\binom{n-n_1}{n_2}\cdots\binom{n_k}{n_k} = n!/(n_1!\cdots n_k!)$ different possible orders. Therefore, the desired probability is the sum of all the probabilities of different events in different orders.)

Supplements and Remarks

1. Probability theory originates from ancient games of chance. But the introduction of mathematical models of probability is generally attributed to French mathematicians Pascal (1623 - 1662) and Fermat (1601 - 1655). They discussed mathematical problems on games of throwing the dice in their correspondence, and accurately calculated the probabilities of some events by means of permutations, combinations and binomial coefficients. Meanwhile, they established the theory on permutations, combinations and binomial coefficient. Later on, the subject matter of probability theory became gradually

rich by the works due to Bernoulli (1654 - 1705), de Moivre (1667 - 1754), Bayes, Buffon, Legendre, Lagrange and others. By the age of Laplace (1749 - 1827), the frame of the so-called classical probability theory had been basically completed with the great treatise "Théorie Analytigue des Probabilitiés" (Laplace 1812) as a landmark. The statistical and empirical viewpoint on probability was mainly developed by Fisher and Von-Mises. The concept of sample space introduced by Von-Mises makes it possible to develop mathematical theory of probability on the basis of measure theory. In 1930's, Kolmogorov (1903 - 1987), a Soviet mathematician, put forward axiomatic system of probability theory. This does not only provide a firm logic foundation to the study of infinite sequences of random experiments and general stochastic processes, but also promotes greatly the development of mathematical statistics.

2. Permutations and combinations

Consider an experiment in which a card is selected and removed from a deck of n different cards, a second card is then selected and removed from the remaining $n-1$ cards, and finally a third card is selected from the remaining $n-2$ cards. A process of this kind is called sampling without replacement, since a card that is drawn is not replaced in the deck before the next card is selected. In this experiment, any one of the n cards could be selected first. Once this card has been removed, any one of the other $n-1$ cards could be selected second. Therefore, there are $n(n-1)$ possible outcomes for the first two selections. Finally for any given outcome of the first two selections, there are $n-2$ other cards that can be selected third. Therefore the total number of possible outcomes for all three selections is $n(n-1)(n-2)$. Thus, each outcome in the sample space Ω of this experiment will be some arrangement of three cards from the deck. Each different arrangement is called a permutation. The total number of possible permutations for the described experiment will be $n(n-1)(n-2)$.

This reasoning can be generalized to any number of selections without replacement. Suppose that k cards are to be selected one at a time and removed from a deck of n cards $(k=1,2,\cdots,n)$. Then each possible outcome of this experiment will be a permutation of k cards from the deck, and the total number of these permutations will be $\mathrm{P}_n^k = n(n-1)\cdots(n-k+1)$. This number P_n^k is called the number of permutations of n elements taken k at a time.

When $k=n$, the number of possible outcomes of the experiment will be the number P_n^n of different permutations of all n cards. It is seen from the equation just derived that

$$\mathrm{P}_n^n = n(n-1)\cdots 1 = n!.$$

The symbol $n!$ is read n factorial. In general, the number of permutations of n different items is $n!$.

Here and elsewhere in the theory of probability, it is convenient to define $0!$ by the relation

$$0! = 1.$$

Suppose that there is a set of n distinct elements from which it is desired to choose a subset containing k elements ($1 \leqslant k \leqslant n$). We shall determine the number of different subsets that can be chosen. In this problem, the arrangement of the elements in a subset is irrelevant and each subset is treated as a unit. Such a subset is called a combination. No two combinations will consist of exactly the same elements. We shall denote by $\binom{n}{k}$ the number of combinations of n elements taken k at a time.

For example, if the set contains four elements a, b, c and d and if each subset consists of two of these elements, then the following six different combinations can be obtained: $\{a,b\}$, $\{a,c\}$, $\{a,d\}$, $\{b,c\}$, $\{b,d\}$ and $\{c, d\}$. Hence, $\binom{4}{2} = 6$. When combinations are considered, the subsets $\{a, b\}$ and $\{b, a\}$ are identical and only one of these subsets is counted.

Now the numerical value of $\binom{n}{k}$ for given integer n and k ($1 \leqslant k \leqslant n$) will be derived. It is known that the number of permutations of n elements taken k at a time is A_n^k. A list of these P_n^k permutations could be constructed as follows: First, a particular combination of k elements is selected. Each different permutation of these k will yield a permutation on the list. Since there are $k!$ permutations of these k elements, this particular combination will produce $k!$ permutations on the list. When a different combination of k elements is selected, $k!$ other permutations on the list will be obtained. Since each combination of k elements will yield $k!$ permutations on the list, the total number of permutations on the list must be $k!\binom{n}{k}$. Hence, it follows that $P_n^k = k!\binom{n}{k}$ from which

$$\binom{n}{k} = \frac{n!}{k!(n-k)!} = \frac{n(n-1)\cdots(n-k+1)}{k!}.$$

Unlike permutation, no matter what real number n is, the formula

$$\binom{n}{r} = \frac{n(n-1)\cdots(n-r+1)}{r!}$$

is well defined as long as r is a non-negative integer. Note that we do not require that n is a natural number or $n > r$. For instance,

$$\binom{-1}{r} = \frac{(-1)(-2)\cdots(-r)}{r!} = (-1)^r,$$

and by convention

$$\binom{n}{r} = 0,$$

where n is a natural number such that $n < r$.

The combination number $\binom{n}{r}$ is often called the binomial coefficient, since it appears in the following binomial expansion:

$$(a+b)^n = \sum_{i=1}^{n}\binom{n}{i}a^i b^{n-i}. \tag{1.30}$$

Many useful formulae about combinations can be derived from the above equation. For example, letting $a=b=1$ in (1.30), then it follows that

$$\binom{n}{0}+\binom{n}{1}+\cdots+\binom{n}{n}=2^n.$$

Also, substituting $a=-1$, $b=1$ into (1.30) yields

$$\binom{n}{0}-\binom{n}{1}+\cdots+(-1)^n\binom{n}{n}=0.$$

Two other formulas commonly used are

$$\binom{n}{k-1}+\binom{n}{k}=\binom{n+1}{k} \tag{1.31}$$

and

$$\binom{m+n}{k}=\sum_{i=0}^{k}\binom{m}{i}\binom{n}{k-i}. \tag{1.32}$$

To see this, notice an elementary identity $(1+x)^{m+n}=(1+x)^m(1+x)^n$, i.e.,

$$\sum_{k=0}^{m+n}\binom{m+n}{k}x^k=\sum_{i=0}^{m}\binom{m}{i}x^i\sum_{j=0}^{n}\binom{n}{j}x^j,$$

and compare the coefficients of x^k in both sides.

Moreover, letting $m=n$, $k=n$, we have by (1.32)

$$\sum_{i=0}^{k}\binom{n}{i}^2=\binom{2n}{n}.$$

Divide n different items into k groups such that the i-th group has r_i items, where $r_1+\cdots+r_k=n$. Then the number of possible ways is

$$\frac{n!}{r_1!\cdots r_k!},$$

which is just the coeffcient in front of $x_1^{r_1}\cdots x_k^{r_k}$ in the expansion of $(x_1+\cdots+x_k)^n$. $n!$ is usually approximated by $\sqrt{2\pi n}(n/\mathrm{e})^n$ from the well-known Stirling's formula:

$$n! = \sqrt{2\pi n}\left(\frac{n}{\mathrm{e}}\right)^n \mathrm{e}^{\theta_n/(12n)},\quad 0<\theta_n<1.$$

Exercises 1

1. Select randomly three digits from 1, 2, 3, 4 and 5 and use them to write a three-digit integer. Find the probability that the integer is even.

2. Three balls are drawn in succession without replacement from a box with 5 white

balls and 6 black ones. Find the probability that the first and third are black, and the second is white.

3. A box contains n white balls, n black balls and n red balls. Draw m balls at random. Find the probability that m_1 balls are white, m_2 are black and m_3 are red ($m_1 + m_2 + m_3 = m$).

4. A box contains N balls numbered $1, 2, \cdots, N$. Draw randomly n balls from the box in succession and with replacement. Find the probability that the numbers on the balls drawn are strictly increasing.

5. Place randomly a red rook and a black rook on chessboard. Find the probability that they can fight each other.

6. Consider an arbitrary group of 40 people. Find the probability that no two have the same birthday, that is, have been born on the same day of the same month but not necessarily in the same year.

7. Select randomly $2r$ ($2r \leqslant n$) shoes from n distinct pairs of shoes. Find the probability that

(1) there is no one pair of shoes;

(2) there is just one pair of shoes;

(3) there are just two pairs of shoes;

(4) there are just r pairs of shoes.

8. An elevator starts at the bottom floor with 7 passengers and stops at the 10th floor. Assume that each passenger leaves at each floor equally likely. Find the probability that no two passengers leave at the same floor.

9. Select at random 13 cards from a deck of 52 cards. Find the probability that it consists of 5 spades, 3 hearts, 3 diamonds, and 2 clubs.

10. Select at random 5 cards from a deck of 52 cards. Find the probabilities of the following events:

(1) royal flush (ten, jack, queen, king, ace in a single suit);

(2) four of a kind (four cards of equal face values);

(3) flush (five cards in a single suit);

(4) three of a kind and two of another kind.

11. Compare the probability that at least one six appears on rolling a dice four times and the probability that a double six appear in rolling two dice twenty four times. Which one is bigger?

12. A harbor can hold only a ship during a certain period. Some day two ships will arrive at the harbor independently and with equal probability at any time within 24 hours. If they will stay for 3 hours and 4 hours respectively, find the probability that one ship has to wait for some time.

13. Select randomly two points C and D in the line segment AB, and find the probability that AC, CD and DB can constitute a triangle.

14. Select randomly three points in the line segment [0, 1], and find the probability that three segments composed of 0 to the three points can constitute a triangle.

15. Toss a coin with diameter 1 on a piece of paper with squares. How small should the squares are such that the probability that the coin does not intersect with boundary lines of squares in the paper is less than 1%?

16. It is well known $P(\varnothing) = 0$. Now an event A is such that $P(A) = 0$; ask whether it is true that $A = \varnothing$? If yes, please give a reasonable explanation; otherwise, find a counterexample.

17. Let A, B, C and D be four events. Try to represent the following events:

(1) at least one of them occurs;

(2) just two of them occur;

(3) both A and B occur but neither does C nor D;

(4) none of them occurs;

(5) at most one occurs;

(6) at least two of them occur;

(7) at most two of them occur.

18. Let A, B and C be three events. Describe the probabilistic meaning of the following equations:

(1) $A \cup B \cup C = A$;

(2) $A \subset \overline{BC}$.

19. Select randomly one student from some class. Let A = {the student selected is male}, B = {the student selected does not like to sing} and C = {the student selected is an athlete}. Ask

(1) what do $AB\overline{C}$ and $A\overline{B}C$ mean respectively?

(2) under what condition $ABC = A$ holds?

(3) under what condition $\overline{C} \subset B$ holds?

(4) under what condition $A = B$ holds?

20. Four components A, B, C and D constitute a series-parallel circuit as in Figure 1-5. Let A, B, C and D be the events that the corresponding components work well, respectively.

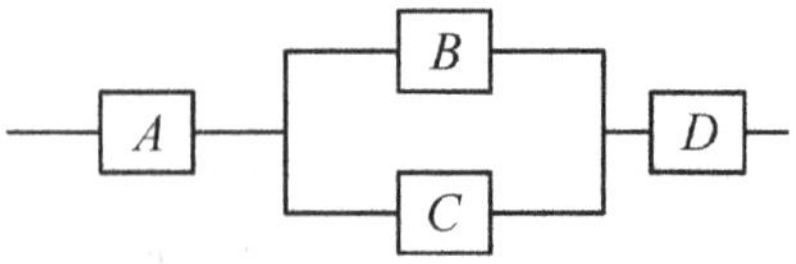

Figure 1-5

(1) Express the event that the circuit works in order in terms of A, B, C and D;

(2) Express the event that the circuit works out of order in terms of $\overline{A}$, $\overline{B}$, $\overline{C}$ and $\overline{D}$.

21. Show by the definition that

(1) $\overline{A \cup B} = \overline{A}\,\overline{B}$;

(2) $A(B \cup C) = AB \cup AC$.

22. A box contains n balls numbered $1, 2, \cdots, n$. Find the probability that

(1) two balls randomly selected from this box are labelled by 1 and 2 respectively;

(2) none of three balls randomly selected from this box is labelled by 1;

(3) three balls of five balls randomly selected from this box are labelled by 1, 2 and 3 respectively.

23. Prove the exclusion-inclusion formula of Section 1.3 by an induction argument.

24. Select randomly a term in the expansion of determinant of an n by n matrix. Find the probability that it contains at least an element of principal diagonal.

25. m students shall take the final exam. Before examination, each student randomly draws one paper from n different papers with replacement. Find the probability that at least one paper is not used after the exam is over.

26. Revisit Example 18 in Section 1.3, and find the probability that just k $(k \leqslant n)$ letters are put in the correct envelopes.

27. Calculate $P(A\overline{B})$ and $P(\overline{AB})$ given that $p = P(A)$, $q = P(B)$ and $r = P(A \cup B)$.

28. Suppose that A occurs for certain whenever A_1 and A_2 both occur. Show $P(A) \geqslant P(A_1) + P(A_2) - 1$.

29. Let A_1 and A_2 be two events, show that

(1) $P(A_1A_2) = 1 - P(\overline{A}_1) - P(\overline{A}_2) + P(\overline{A}_1\overline{A}_2)$;

(2) $1 - P(\overline{A}_1) - P(\overline{A}_2) \leqslant P(A_1A_2) \leqslant P(A_1 \cup A_2) \leqslant P(A_1) + P(A_2)$.

30. Show $P(AB) + P(AC) - P(BC) \leqslant P(A)$ for three arbitrary events A, B and C.

31. Find the smallest-field containing events A and B.

32. Show that the conditional probability defined in Definition 2 satisfies the axiomatic hypotheses with respect to probability.

33. A family has three children, one of which is a girl. Find the probability that there is at least one boy in this family.

34. There are m defective items in n products. Now select randomly two from them. Find the probability that

(1) the other is also defective given that at least one of two is defective;

(2) the other is defective given that at least one of two is nondefective.

35. Some factory uses three machines A, B, C to produce bolt. Their output are 25%, 35% and 40% and the ratios of defective bolt are 5%, 4% and 2% respectively.

(1) select randomly one from bolts, and find the probability that it is defective;

(2) select randomly one from bolts and find it is defective. Find the probability that it is from machine A.

36. Urn A has a white balls and b black balls; urn B has c white balls and d black balls. Now someone first selects randomly two balls from urn A and put them into urn B. Then he/she selects randomly two balls from urn B. Find the probability that they are both white.

37. A box contains a ($a \geqslant 3$) white balls and b black balls. Now A, B and C select a ball from it in turn without replacement. Find the probability that each of B and C gets a white ball using the formula for total probability.

38. Enemy plane will be shot down if either its first part is hit by one missile, or its second part by two missiles, or its third part by three missiles. The probability of shooting each part is proportional to its acreage and acreage of three parts are 0.1, 0.2 and 0.7 respectively. Given that two missiles hit an enemy plane, find the probability that it will be shot down.

39. 96% of a set of products are nondefective. Now use a simplified method of sampling. The probability that a true qualified product is affirmed qualified is 0.98 and the probability that a substandard product is diagnosed as qualified is 0.05. Find the probability that a product diagnosed as qualified is really qualified.

40. A box contains 9 balls, 3 of which are red. Now A and B draw one ball in turn and one will be by rule punished if he gets a red ball.

(1) What is the probability that A first draws and A will not be punished?

(2) What is the probability that B is not punished given that A first draws and A is not punished?

(3) What is the probability that B is not punished given that A first draws and A is punished?

(4) Is B drawing before A helpful to A?

(5) If A draws before B and B is not punished, then what is the probability that A is not punished?

41. 3 out of 8 guns are not adjusted and the rest are adjusted. The probability is 0.3 that a gunman hits the target using a not-adjusted gun, while the probability is 0.8 that he hits the target using an adjusted gun. One day he randomly selects one from the 8 guns and hits the target. Find the probability that the gun the gunman uses is an adjusted one.

42. Consider a fair regular octahedron. Its first, second, third and fourth faces are red; its first, second, third and fifth sides are white; its first, sixth, seventh and eighth sides are black. Now throw randomly this regular octahedron. Let A, B and C be the events that it faces red, white and black color respectively. Are A, B and C

independent?

43. Suppose that A, B and C are mutually independent. Show that $A \cup B$, $A \cap B$ and $A - B$ are independent of C.

44. Suppose that A, B and C are mutually independent. Show that $\overline{A}$, $\overline{B}$ and $\overline{C}$ are also mutually independent.

45. An urn contains two red balls, two white balls and one black ball. Three balls are drawn from the urn at random and with replacement. Find the probability that

(1) these three balls drawn are all red;

(2) none of the three balls is red;

(3) at least one of these three balls is red;

(4) these three balls are all distinct color.

46. Manufacturing a kind of parts goes through three procedures, and the ratios of defective products by each procedure are 2%, 3% and 5% respectively. Assume that these three procedures are independent of each other. Find the ratio of defective parts.

47. A gunman shoots at a target 3 times independently and the ratio of hitting the target is 0.4, 0.5 and 0.7 respectively in his first, second and third shoot. Find the probability that

(1) the gunman hits the target only one time;

(2) the gunman hits the target at least one time.

48. Tossing a coin n times independently, each falling a head with probability p. Find the probability that

(1) only one head appears in these n tosses;

(2) at least one head appears in these n tosses;

(3) at least two heads appear in these n tosses.

49. One interactive computer has 20 terminals, each used by different departments independently. The probability is 0.7 that each terminal is in use at a certain time. Find the probability that more than ten computers are in use simultaneously.

50. A component in an electric machine changes its status every ten thousandth of a second according to the following rules:

(1) if the component is now on, then it keeps on after ten thousandth of a second with probability $1-\alpha$ and turns off with probability α;

(2) if the component is now off, then it keeps off after ten thousandth of a second with probability $1-\beta$, and turns on with probability β. Suppose $0<\alpha<1$, $0<\beta<1$ and θ_n is the probability that a component will be off after n ten thousandth of a second. Give a recursive formula for θ_n.

51. In a Bernoulli model, A occurs with probability p. Find the probability that A occurs k times before $\overline{A}$ occurs m times.

52. A particle moves a step upward, downward, left or right equally likely each

time. Find the probability that it returns to the origin after $2n$ steps.

53. A, B and C play a game, each player winning with the same chance. By convention, the person who first wins three sections will be the champion. Now assume A has won the first and the third sections, and B has won the second section. What is the probability that C is the champion?

54. The probability that a person is O, A, B and AB of blood type is 0.46, 0.40, 0.11 and 0.03 respectively. Now sample randomly 5 people, and find the probability that

(1) two of them are of O type blood, while the other three are not;

(2) three of them are of O type blood, while the other two are of A type blood;

(3) none of them is of AB type blood.

55. Assume that a silkworm lays k spawns with probability $\lambda^k e^{-\lambda}/k!$, $(\lambda > 0)$, and assume that each spawn becomes an imago with probability p independently. Show that the probability is $(\lambda p)^r e^{-\lambda p}/r!$ that a silkworm will give birth to r baby silkworm.

56. A telephone operator receives k calls during a unit time with probability $P(k) = \lambda^k e^{-\lambda}/k!$ $(\lambda > 0)$, independently. Find the probability that the telephone operator receives s calls during two unit times.

57. (The ballot problem) A and B are two candidates for a certain position. Assume that A gets P votes and B gets Q votes, and $P > Q$. What is the probability that A gets more votes than B during counting?

Chapter 2

Random Variables and Distribution Functions

2.1 Discrete random variables

2.1.1 The concept of random variables

There are generally many outcomes in a random trial. How can we easily display these outcomes together with their corresponding probabilities, and study them through modern mathematical methods? The random variables discussed in the present chapter will provide us with such tools.

In many trials, outcomes can be expressed by the numerical variable which is defined as taking a sequence of values. For example,

(1) Let ξ be the nonnegative integers $0,1,2,\cdots$, and define it as the number of phone calls some operator receives during a particular interval of time. Then $\xi = 2$ stands for the event {there are two calls within this interval of time}, while $\xi = 0$ stands for the event {there are no call within this interval of time}.

(2) All possible values in measurement constitute a sample space $\{\omega: \omega \in (a,b)\}$ when we measure length. In turn we can directly use a variable η to express the outcome of measurement: $\eta \in [1.5, 2.5]$ stands for the event {the value of measurement is between 1.5 and 2.5}.

The values of the variables assumed above are determined by outcomes of trials, so we say they are functions of sample points, and call them random variables, denoted usually by $\xi,\eta,\zeta,\cdots$ or $X, Y, Z,\cdots$. In other words, a random variable ξ is just a function of ω: $\xi = \xi(\omega)$, $\omega \in \Omega$, $\xi(\omega) \in \mathbf{R}$.

We are not only concerned with what values random variables take, but also need to pay attention to probabilities that the variables take on a specific value. Since a random variable takes on values within real numbers, what one is usually interested in is the set of single points, an interval $[a, b)$, or the union or intersection of a number of such intervals. Therefore it is expected that $\{\omega:\xi(\omega) \in [a, b)\}$ is an event so that it is reasonable to talk about its probability. Moreover, it is required that the countable union, intersection and complement of all these intervals represent events. Then we can do computation for events and probabilities.

A Borel set B is by definition what we have by set operations like union, intersection, complement of left open and right closed intervals in the real line. For example, the set of single points, the bounded or unbounded open interval or closed interval or half open and half closed interval, and the union or intersection of finitely or countably many of intervals. A Borel field $\mathscr{B}$ consists of all Borel sets. The requirement above is equivalent to establishing a link between a Borel set B on the real line and an event $\{\omega:\xi(\omega) \in B\}$, and that between Borel field $\mathscr{B}$ and event field $\mathscr{F}$.

Taking the explanation above into account, we will give a mathematically rigorous description of a random variable.

Definition 1 *Suppose that* $\xi(\omega)$ *is a real function defined in a probability space* $\{\Omega, \mathscr{F}, P\}$ *and that for any Borel set B*

$$\xi^{-1}(B) := \{\omega : \xi(\omega) \in B\} \in \mathscr{F}. \tag{2.1}$$

Then we can say that ξ *is a random variable, and that* $\{P(\xi(\omega) \in B), B \in \mathscr{B}\}$ *is a probability distribution associated with* ξ.

The distribution function is a complete description of a random variable.

Like in function theory, we call the function ξ satisfying (2.1) to be measurable with respect to the σ-field $\mathscr{F}$. It is a rather complicated problem to write out the probability distribution of a random variable. In most cases, however, it is sufficient for one to calculate probability of (2.1), with a specific sequence of Borel sets in order to decide the whole probability distribution. We first introduce a simpler class of random variables.

2.1.2 Discrete random variables

Definition 2 *If a random variable* ξ *takes at most a set of countably many values (finite or infinite), then we call* ξ *a discrete random variable.*

For a discrete random variable ξ, let $\{x_i\}$ be the set of all possible values. A key problem is to provide the probability distribution $P(\xi = x_i)$ (simply written as $p(x_i)$ or p_i), $i = 1, 2, \cdots$.

$$\begin{bmatrix} x_1 & x_2 & \cdots & x_n & \cdots \\ p(x_1) & p(x_2) & \cdots & p(x_n) & \cdots \end{bmatrix} \tag{2.2}$$

is said to be distribution sequence of ξ. It contains two parts: (1) the values of ξ, and (2) the probabilities of ξ taking each possible value.

It is easy to show that such a distribution sequence has the following properties:

$$p(x_i) > 0, \ i = 1, 2, \cdots,$$

and

$$\sum_{i=1}^{\infty} p(x_i) = 1. \tag{2.3}$$

Having distribution sequence (2.2), we can calculate all probabilities of events related to ξ. In fact, in view of the countable additivity of probability, we have

$$P(\xi(\omega) \in B) = \sum_{x_i \in B} p(x_i) \tag{2.4}$$

for any Borel set B on the line.

Example 1 Suppose that the distribution sequence of random variable ξ is

$$\begin{pmatrix} -2 & -1 & 0 & 1 & 2 \\ \frac{a-1}{4} & \frac{a+1}{4} & 0.1 & 0.2 & 0.2 \end{pmatrix},$$

(1) find the constant a;

(2) find $P(-1 < \xi \leqslant 2)$.

Solution (1) Solving the equation

$$\frac{a-1}{4} + \frac{a+1}{4} + 0.1 + 0.2 + 0.2 = 1,$$

we have $a = 1$, so the distribution sequence is

$$\begin{pmatrix} -2 & -1 & 0 & 1 & 2 \\ 0 & 0.5 & 0.1 & 0.2 & 0.2 \end{pmatrix}.$$

(2) It follows from (2.4) that

$$\begin{aligned} P(-1 < \xi \leqslant 2) &= \sum_{-1 < x_i \leqslant 2} p(x_i) \\ &= 0.1 + 0.2 + 0.2 = 0.5. \end{aligned}$$

Example 2 Assume that the success probability is p in the Bernoulli probability model, and denote by ξ the number of times that an experiment is conducted until its r-th success. Calculate the distribution sequence of ξ.

Solution

$$\begin{aligned} P(\xi = k) &= P(\text{ there are } r-1 \text{ successes and } k-r \text{ failures} \\ &\quad \text{in the first } k-1 \text{ trials and success in the } k\text{-th trial}) \\ &= \binom{k-1}{r-1} p^{r-1} q^{k-r} p = \binom{k-1}{r-1} p^r q^{k-r}, \end{aligned}$$

where $k = r, r+1, r+2, \cdots$. This is called a Pascal distribution.

The following are some typical discrete random variables. They frequently appear in many ordinary situations, and play an extremely important role in theoretical research as well.

Ⅰ Degenerate distribution

Assume that a random variable ξ takes only one constant c, that is,

$$P(\xi = c) = 1. \tag{2.5}$$

This is referred to as a degenerate distribution, which is also called a single point distribution. In fact, $\{\xi = c\}$ is an event with probability 1; ξ is regarded as a constant. But sometimes we prefer to think of it as a (degenerate) random variable.

Ⅱ Two point distribution

If there are two possible values x_1, x_2 in an experiment, then the probability distribution is

$$\begin{pmatrix} x_1 & x_2 \\ p & q \end{pmatrix}, \quad p, q > 0, \quad p + q = 1. \tag{2.6}$$

This is called a two point distribution.

Each experiment has only two possible outcomes in the Bernoulli probability model—event A occurs or not. Although the outcome has nothing to do with a numerical value, we can quantize it. If we establish a correspondence between values of random variable and outcomes, then we get a random variable with a two point distribution. In particular, people usually denote this random variable by the indicator function of A, that is

$$\xi = \begin{cases} 1, & \text{if } A \text{ occurs}, \\ 0, & \text{otherwise}. \end{cases} \tag{2.7}$$

Its corresponding distribution sequence is

$$\begin{pmatrix} 0 & 1 \\ q & p \end{pmatrix}, \quad p, q > 0, \quad p + q = 1. \tag{2.8}$$

The name for this is the Bernoulli distribution. It is also called the 0 - 1 law. Thus any Bernoulli trial can be expressed by a Bernoulli distribution.

Ⅲ The binomial distribution

If a random variable ξ has the following distribution sequence:

$$P(\xi = k) = \binom{n}{k} p^k q^{n-k}, \quad p, q > 0, \quad p + q = 1, \tag{2.9}$$

where $k = 0, 1, 2, \cdots, n$, then we say that ξ obeys a binomial distribution, and simply write it as $\xi \sim B(n, p)$. Here n and p are its two parameters. $P(\xi = k)$ is just the probability $b(k; n, p)$ of k successes in the Bernoulli probability model discussed in Chapter 1. It appears in the binomial expansion of $(p + q)^n$ with 1 as the total sum.

The binomial distribution is one of the most important probability distributions in probability theory and has a wide variety of applications. To illustrate,

(1) To diagnose whether one gets some kind of disease is a Bernoulli trial. Whether people get such a disease or not is consided to be independent, and the probability that each person gets such a disease is approximately identical. Thus to check on the disease situations of n individuals somewhere, on a one-by-one basis, can be thought of as n repeated Bernoulli trials, and the number of diseased individuals obeys a binomial distribution.

(2) Consider an insurance company for some kind of disaster (for example, fire disaster). Suppose that a disaster befalls an individual is mutually independent and probabilities associated with such a disaster are equal. Assume that the probability this disaster befalls each individual is p, then the number of people who suffer this disaster among n people obeys the binomial distribution.

(3) Consider n machines of the same type. Assume that the probability that each breaks down is p during an interval of time, then the number of machines that break

down during this time period obeys the binomial distribution.

We shall state some important properties of the binomial distribution below.

(1) $$b(k;n,p) = b(n-k;n,1-p). \tag{2.10}$$

This follows directly from (2.9) and $\binom{n}{k} = \binom{n}{n-k}$. An alternative way of writing this is that the event {there appear k sucesses} is the same as the event {there appear $n-k$ failures}, while the probability that the latter occurs is $b(n-k;\ n,\ 1-p)$.

Computations of binomial distribution arise in many cases, so people usually use a table (see Appendix B) of binomial distribution values in practice. But the table is limited to the case of $p \leqslant 0.5$. One can use (2.10) to calculate in the case of $p > 0.5$.

(2) Monotonicity and the best possible number of successes.

Fix n, p. Since

$$\frac{b(k;n,p)}{b(k-1;n,p)} = \frac{(n-k+1)p}{kq} = 1 + \frac{(n+1)p-k}{kq},$$

then $b(k;\ n,\ p)/b(k-1;\ n,\ p) > 1$ when $k < (n+1)p$, so $b(k;\ n,\ p)$ increases; $b(k;\ n,\ p)/b(k-1;\ n,\ p) < 1$ when $k > (n+1)p$, so $b(k;\ n,\ p)$ decreases. $b(k;\ n,\ p) = b(k-1;\ n,\ p)$ attains its maximum when $(n+1)p$ is an integer and $k = (n+1)p$, and then we call $m = (n+1)p$ or $(n+1)p-1$ the best possible number of successes; when $(n+1)p$ is not an integer, the best possible number of successes is

$$m = [(n+1)p].$$

Table 2-1 gives a specific binomial distribution, where n is 20, p are 0.1, 0.3, 0.5 respectively, and they particularly display the property (2).

Table 2-1

k/p	0.1	0.3	0.5	k/p	0.1	0.3	0.5
0	0.1216	0.0008	—	11	—	0.0120	0.1602
1	0.2702	0.0068	—	12	—	0.0039	0.1201
2	0.2852	0.0278	0.0002	13	—	0.0010	0.0739
3	0.1901	0.0716	0.0011	14	—	0.0002	0.0370
4	0.0898	0.1304	0.0046	15	—	—	0.0148
5	0.0319	0.1789	0.0148	16	—	—	0.0046
6	0.0089	0.1916	0.0370	17	—	—	0.0011
7	0.0020	0.1643	0.0739	18	—	—	0.0011
8	0.0004	0.1144	0.1201	19	—	—	0.0002
9	0.0001	0.0654	0.1602	20	—	—	—
10	—	0.0308	0.1762				

As the computer technology developes rapidly, people would prefer use mathematical/statistical softwares to calculate the distribution values. See 12 at the Supplements and Remarks for more information.

(3) Recursive formula.

If $\xi \sim B(n, p)$, then

$$P(\xi = k+1) = \frac{p}{q}\frac{n-k}{k+1}P(\xi = k)$$

Starting with $P(\xi=0)=q^n$ and recursively employing the above equation, we can obtain each value of $P(\xi=k)$.

(4) Asymptotic behaviors as n goes to infinity.

Suppose that p depends on n, which we simply write as p_n, then we have the following theorem.

Theorem 1 *(Poisson) If there exists a positive constant λ such that $np_n \to \lambda$ as $n \to \infty$, then*

$$\lim_{n\to\infty} b(k;\ n,\ p) = \frac{\lambda^k}{k!}e^{-\lambda},\ k = 0,\ 1,\ 2, \cdots. \tag{2.11}$$

Proof Set $\lambda_n = np_n$, then $p_n = \lambda_n / n$. Thus we have

$$\begin{aligned} & b(k;n,p) \\ = & \binom{n}{k} p_n^k (1-p_n)^{n-k} \\ = & \frac{n(n-1)(n-2)\cdots(n-k+1)}{k!}\left(\frac{\lambda_n}{n}\right)^k\left(1-\frac{\lambda_n}{n}\right)^{n-k} \\ = & \frac{\lambda_n^k}{k!}\frac{n(n-1)(n-2)\cdots(n-k+1)}{n^k}\frac{\left(1-\frac{\lambda_n}{n}\right)^n}{\left(1-\frac{\lambda_n}{n}\right)^k} \\ \to & \frac{\lambda^k}{k!}e^{-\lambda} \quad (n \to \infty). \end{aligned}$$

This concludes the proof. Call the above theorem the Poisson approximation to the binomial distribution.

Usually, p is independent of n. But when n is very large (say $n \geqslant 50$), p is very small (say $p \leqslant 0.1$), and np is not very large, we take $\lambda = np$, and

$$b(k;n,p) \approx \frac{\lambda^k}{k!}e^{-\lambda} = \frac{(np)^k}{k!}e^{-\lambda}. \tag{2.12}$$

It is comparatively easy to compute the right hand side of (2.12), so the problem of computing $b(k;\ n,\ p)$ in this case is solved.

Example 3 Somebody shoots a target with the probability 0.001 of hitting it each time. Now he shoots 5 000 times. Calculate the probability that he hits the target twice or more times.

Solution Denote by ξ the number of times he hits, then $P(\xi = k) = b(k;\ 5\ 000,\ 0.001)$. Observe that $n = 5\ 000$, $p = 0.001$, and $np = 5$, so we use (2.12) to get

$$P(\xi = k) \approx \frac{5^k}{k!}e^{-5},$$

and

$$\sum_{k=2}^{5000} P(\xi = k) = 1 - P(\xi = 0) - P(\xi = 1)$$

$$\approx 1 - e^{-5} - 5e^{-5} \approx 0.9596.$$

The Poisson Theorem gives an approximation of the binomial distribution for large n and small p. The next theorem will give an approximation when n is large but p has a moderate value.

de Moivre-Laplace Theorem (de Moivre, 1733, Laplace, 1801) *Suppose* $\xi_n \sim B(n, p)$, *and that* $p = p_n$, $q = 1 - p$ *satisfies* $npq \to \infty$. *Let*

$$j = j(n), x = x(n) = \frac{j - np}{\sqrt{npq}}, p_n(x) = P(\xi_n = j)$$

Then for any finite interval $[a, b]$,

$$p_n(x) = P(\xi_n = j) \sim \frac{1}{\sqrt{2\pi npq}} e^{-x^2/2},$$

holds uniformly in $x \in [a, b]$ *whenever* j *is a function of* n *that* $x = x(n)$ *keeps its value in* $[a, b]$, *where* $a_n \sim b_n$ *means that* $a_n / b_n \to 1$.

de Moivre-Laplace Theorem tells us that when npq is large, but $(j - np)/\sqrt{npq}$ is not so large, we have the following approximation of the binomial distribution:

$$P(\xi_n = j) \approx \frac{1}{\sqrt{2\pi npq}} e^{-x^2/2}, x = \frac{j - np}{\sqrt{npq}}.$$

Furthermore, under the conditions of the de Moivre-Laplace Theorem, we have

$$P\left(a \leqslant \frac{(\xi_n - np)}{\sqrt{npq}} \leqslant b\right) \to \frac{1}{\sqrt{2\pi}} \int_a^b e^{-x^2/2} dx.$$

This is so called the de Moivre-Laplace central limit theorem. The general central limit theorem will be introduced in Chapter 4. The integral above is related another important distribution, normal distribution, which will be introduced later, and so this approximation of binomial distribution is called normal approximation. For the proof of the de Moivre-Laplace Theorem, the reader can refer to the Supplements and Remarks of this chapter.

Ⅳ The Poisson distribution

In the preceding subsection we have used the Poisson expression $e^{-\lambda}\lambda^k/k!$ merely as a convenient approximation to the binomial distribution in the case of large n and small p. We have here a special case of the remarkable fact that there exist a few distributions with great universality which occur in a surprisingly great variety of problems. The three principal distributions, with ramifications throughout probability theory, are the binomial distribution, the normal distribution (to be introduced in the following section) and the Poisson distribution, the merits of which we are going to discuss about.

Assume that a random variable ξ takes all nonnegative integers with probabilities

$$P(\xi=k)=\frac{\lambda^k}{k!}e^{-\lambda} \quad (\lambda>0), \quad k=0,1,2,\cdots, \tag{2.13}$$

then we say that ξ obeys a Poisson distribution, and simply write as $\xi \sim P(\lambda)$, where λ is the parameter of ξ. We shall show that λ is just the mean of ξ later on.

We note first that on adding the equation (2.13) for $k=0,1,\cdots$, we get $e^{-\lambda}$ times the Taylor series for e^{λ} on the right side. Hence for any fixed λ, the quantities on the right side of (2.13) add to one, and therefore it is possible to image an ideal experiment in which $e^{-\lambda}\lambda^k/k!$ is the probability of the event that there appear exactly k successes. The following examples will illustrate the wide range and the importance of various applications of Poisson distribution. The true nature of the Poisson distribution will become apparent only in connection with the theory of stochastic processes. See 2 at the Supplements and Remarks.

(1) In social daily life, the amounts of various service requirement, like the number of calls an operator receives during an interval of time, the number of passengers arriving at the bus stop, the number of customers coming to a supermarket or the number of goods sold by a supermarket, obey the Poisson law. Hence the Poisson distribution plays an important role in management science and operational research.

(2) In biology, with regard to the number of microorganism in some defined region, we can model the number of their offspring based on Poisson law.

(3) A radioactive substance emits α-particles, and the number of particles reaching a given portion of space during time t is the best-known example of random events obeying the Poisson law. Of course, the substance gradually decay, and in the long run the density of α-particles will decline. However, with radium it takes years before a decrease of matter can be detected; in case of relatively short periods the conditions can be considered to be constant and we have an ideal realization of the hypotheses which lead to the Poisson distribution.

Example 4 On a certain crossroad the flow of traffic may be assumed to be Possionian. Suppose that the probability that no automobile passes through within one minute is 0.4, find the probability that more than one automobile pass through within 1 minute.

Solution Denote by ξ the number of automobiles passing through the crossroad, and assume $\xi \sim P(\lambda)$. Note that $P(\xi=0)=e^{-\lambda}=0.4$, so we have $\lambda=\ln 5-\ln 2$. The desired probability is

$$\begin{aligned}P(\xi>1) &= \sum_{k=2}^{\infty} P(\xi=k)\\ &= 1-P(\xi=0)-P(\xi=1)\\ &= 1-e^{-\lambda}-\lambda e^{-\lambda}\\ &= \frac{3}{5}-\frac{2}{5}\ln\frac{5}{2}\approx 0.2335.\end{aligned}$$

The Poisson theorem tells us that, if each one of n independent events $A_1, \cdots, A_n$

will occur with a small probability p, then the number that these n events occur has approximately a Poisson distribution $P(np)$. This approximation is called Poisson approximation. In the Poisson approximation, the independence of the events $A_1, \cdots, A_n$ can be relaxed. In fact, it remains a good approximation even when the events are not independent, provided that their independence is weak. For instance, in Example 21 on the matching problem in Chapter 1, if denote A_i by the event that the i-th letter will be placed in its correct envelope, then it is easy to see that

$$P(A_i)=\frac{1}{n} \text{ and } P(A_i \mid A_j)=\frac{1}{n-1}\approx P(A_i),\ i\neq j.$$

Thus, we see that while the events $A_1, \cdots, A_n$ are not independent, their dependence, for large n, appears to be weak. They are likely "approximately" independent. Based on this fact, it seems reasonable to expect that the number of their occurrences will approximately have a Poisson distribution $P(\lambda)$ with $\lambda=n\times\frac{1}{n}=1$, and indeed this is veried in Example 27 of Section 1.4, Chapter 1.

In general, consider n events $A_1, \cdots, A_n$, with p_i equal to the probability that event A_i occurs. If all the p_i are small, and the events are either independent or at most "weakly dependent", then the number of these events that occur approximately has a Poisson distribution $P(\lambda)$ with $\lambda=\sum_{i=1}^{n} p_i$.

The above property provides a useful approximation techniques for various probabilities. For instance, let us consider the birthday problem presented in Example 8 of Section 1.2, Chapter 1. In this example we suppose that each of n people is equally likely to have any of the 365 days of the year as their birthday, and the problem is to determine the probability that a set of n people all have different birthdays. If denote A_{ij} by the event that the i-th people and the j-the people have the same birthday, then each one of $\binom{n}{2}$ events $\{A_{ij}; 1\leqslant i<j\leqslant n\}$ occurs with the same probability $P(A_{ij})=\frac{1}{365}$. Whereas the events $\{A_{ij}; 1\leqslant i<j\leqslant n\}$ are not independent, their independence appears likely to be weak. It is reasonable to approximate the distribution of the number that these events occur by a Poisson distribution $P(\lambda)$ with parameter $\lambda=\binom{n}{2}\frac{1}{365}$. Therefore, the probability that no two of n people have the same birthday is

$$P(\xi=0)\approx e^{-\lambda}=\exp\left\{-\frac{n(n-1)}{2\times 365}\right\},$$

which yields a result in agreement with the result of Example 8 of Section 1.2, Chapter 1.

V The geometric distribution

Assume that a random variable ξ takes on the set of positive integers with the probability

$$P(\xi = k) = pq^{k-1},\ p+q=1,\ p,q>0, \tag{2.14}$$

where $k = 1, 2, \cdots$, then we say that ξ obeys a geometric distribution.

Consider the Bernoulli probability model. If the probability of success is p each time, then the number of experiments required in order to attain the first success just obeys the geometric distribution. This is a special case of $r = 1$ in Example 2.

The geometric distribution has an interesting property—the memoryless property. If the first m experiments fail then the number η of experiments required after m experiments in order to attain success is identical to a geometric distribution (it looks as if the first m failures are forgotten!).

Indeed, let $\xi = \eta + m$, and then $\xi = m + k$ when $\eta = k$. It is obvious that ξ is a geometric distribution with parameter p, therefore

$$\begin{aligned} & P(\eta = k \mid \text{the first } m \text{ trials fail}) \\ = & \frac{P(\eta = k, \text{the first } m \text{ trials fail})}{P(\text{the first } m \text{ trials fail})} \\ = & \frac{P(\xi = m+k)}{q^m} = \frac{pq^{m+k-1}}{q^m} = pq^{k-1}. \end{aligned}$$

This is our desired result.

The above property is equivalent to

$$P(\xi > m+k \mid \xi > m) = P(\xi > k), m,k = 1,2,\cdots,$$

which can also be derived directly from the definition of the conditional distribution. On the contrary, if it is a random variable taking positive integer values and possessing the memoryless property (i. e., the above equation holds), then we can prove that ξ has a geometric distribution.

Ⅵ The hypergeometric distribution

Let n, N and M be positive integers with $n \leqslant N$ and $M \leqslant N$. The hypergeometric distribution is defined as follows

$$P(\xi = k) = \frac{\binom{M}{k}\binom{N-M}{n-k}}{\binom{N}{n}},$$

$$k = 0,1,2,\cdots,\min(n,M). \tag{2.15}$$

Consider a sampling inspection of product quality without replacement. If there are M defects in N products, then the number of defects found in n sampling products obeys a hypergeometric distribution.

Since $\sum\limits_{k=0}^{\min(n,M)} P(\xi = k) = 1$, we have shown a very useful combinatorial formula:

$$\sum_{i=0}^{n}\binom{M}{i}\binom{N-M}{n-i} = \binom{N}{n}.$$

There is a close relation between the binomial distribution and the hypergeometric distribution. If n, k are fixed in (2.15), and then as $N \to \infty, M / N \to p$ we have

$$\frac{\binom{M}{k}\binom{N-M}{n-k}}{\binom{N}{n}} \to \binom{n}{k} p^k q^{n-k} \qquad (N \to \infty).$$

Hence when N is sufficiently large, a hypergeometric distribution can be approximately calculated by using a binomial distribution as a proxy.

The hypergeometric distribution above can be extended to the following situation. Suppose that there are n_1 products of the first quality, n_2 products of the second quality, and $N - n_1 - n_2$ products of the third quality among N products. Now taking a sample of r products out of them, then the probability that there are k_1 first quality products, k_2 second quality products, and $r - k_1 - k_2$ third quality products is

$$\frac{\binom{n_1}{k_1}\binom{n_2}{k_2}\binom{N-n_1-n_2}{r-k_1-k_2}}{\binom{N}{r}},$$

where $k_1 \leqslant n_1$, $k_2 \leqslant n_2$, $r - k_1 - k_2 \leqslant N - n_1 - n_2$.

Example 5 A set of 52 cards consist of four colors, each having 13 cards. Then the probability that a subset of 13 cards consist of 5 spades, 4 hearts, 3 diamonds and 1 club is

$$\frac{\binom{13}{5}\binom{13}{4}\binom{13}{3}\binom{13}{1}}{\binom{52}{13}}.$$

2.2 Distribution functions and continuous random variables

2.2.1 Distribution functions

I Definitions

As known to us, the probability distribution of discrete random variable is described by a distribution sequence. But there is no distribution sequence for other types of random variables. If all possible values of a random variable consist of an interval, then we are not able to enumerate all these values and their probabilities. To this end, we need to introduce a new method of expressing a probability distribution, which is ideally applicable to all random variables.

In Section 2.1, we defined a probability distribution to be all the probabilities: $\{P(\xi \in B)\}$, where B is any Borel set in $\mathbf{R}$. Now define $B = (-\infty, x]$ as a Borel set. It is certainly reasonable to talk about the probability of event $\{\xi \leqslant x\} = \{\omega : \xi(\omega) \leqslant x\}$. If we define such probabilities for all real numbers x, then it is easy to calculate the probability of event $\{a < \xi \leqslant b\}$ for any pair of real numbers a, b $(a < b)$:

$$P(a < \xi \leqslant b) = P(\xi \leqslant b) - P(\xi \leqslant a). \tag{2.16}$$

Furthermore, since any Borel set B is reproduced by the union or intersection or complement of a finite (or countable) number of left open and right closed intervals, then one can calculate $P(\xi \in B)$ by (2.16). So, $P(\xi \leqslant x)$, for any real number x, will stand for the probability distribution of ξ.

Definition 3 *Let ξ be a random variable on a probability space $(\Omega, \mathscr{F}, P)$. We define its distribution function as*

$$F(x) = P(\xi \leqslant x), \quad -\infty < x < \infty. \tag{2.17}$$

For any given random variable ξ, its distribution function is uniquely determined and is a real function, so we can use such a powerful tool as real analysis to study these random variables.

The probability $P(\xi(\omega) \in B)$ can be expressed in term of distribution function for any Borel set B. In fact, it follows from (2.16)

$$P(a < \xi \leqslant b) = F(b) - F(a), \tag{2.18}$$

from which we obtain probabilities of other events using operation rules of events and probabilities. For instance,

$$\begin{aligned}
P(\xi < a) &= \lim_{b \to a-0} P(\xi \leqslant b) = F(a-0);\\
P(\xi = a) &= P(\xi \leqslant a) - P(\xi < a)\\
&= F(a) - F(a-0);\\
P(\xi > a) &= 1 - P(\xi \leqslant a) = 1 - F(a);\\
P(a < \xi < b) &= P(\xi < b) - P(\xi \leqslant a)\\
&= F(b-0) - F(a).
\end{aligned}$$

Example 6 Suppose that a random variable ξ is distributed as Bernoulli distribution:

$$\begin{pmatrix} 0 & 1 \\ q & p \end{pmatrix}, \qquad p, q > 0, \ p + q = 1. \tag{2.19}$$

Determine its distribution function $F(x)$, and calculate $P(-1 < \xi < 0.5)$.

Solution When $x < 0$, $P(\xi \leqslant x) = 0$ (null event);

when $0 \leqslant x < 1$, $P(\xi \leqslant x) = P(\xi = 0) = q$;

when $x \geqslant 1$,

$$\begin{aligned}
P(\xi \leqslant x) &= P(\xi = 0) + P(\xi = 1)\\
&= q + p = 1.
\end{aligned}$$

Hence we obtain the following distribution function:

$$F(x) = \begin{cases} 0, & x < 0, \\ q, & 0 \leqslant x < 1, \\ 1, & x \geqslant 1. \end{cases}$$

Next, let us compute $P(-1 < \xi < 0.5)$.

$$P(-1 < \xi < 0.5) = F(0.5 - 0) - F(-1) = q.$$

In Example 6, we give a distribution function of the Bernoulli distribution, which is a step function with leaps at $x = 0$ and $x = 1$.

In general, suppose that ξ has the distribution sequence

$$\begin{pmatrix} x_1 & x_2 & \cdots & x_k & \cdots \\ p(x_1) & p(x_2) & \cdots & p(x_k) & \cdots \end{pmatrix}, \tag{2.20}$$

where $x_1 < x_2 < \cdots < x_k < \cdots$, then its distribution function is

$$F(x) = \begin{cases} 0, & x < x_1, \\ p(x_1), & x_1 \leqslant x < x_2, \\ \vdots & \vdots \\ \sum\limits_{i \leqslant k} p(x_i), & x_k \leqslant x < x_{k+1}, \\ \vdots & \vdots \end{cases}$$

This is a step function, which is constant in $[x_k, x_{k+1})$, but leaps at x_k with height $p(x_k)$, $k = 1, 2, \cdots$ (see Figure 2-1).

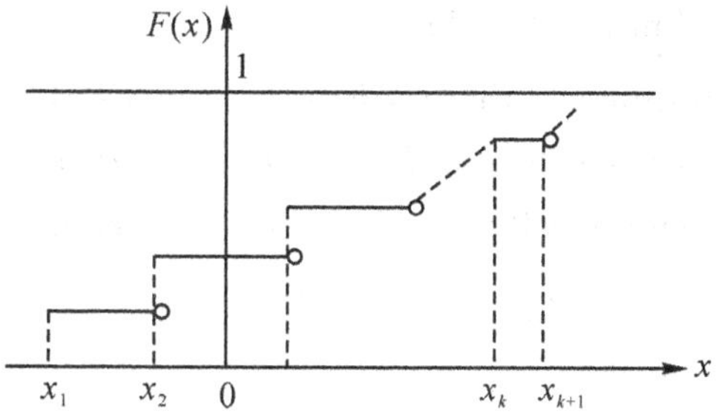

Figure 2-1

Example 7 Choose randomly a point P in a triangle $\triangle ABC$, and let ξ be the distance from P to line segment BC. Calculate the distribution function of ξ.

Solution Let h be the height of line segment BC (see Figure 2-2). It is obvious that $P(\xi \leqslant x) = 0$ when $x < 0$. When $0 \leqslant x < h$, make a line segment DE paralleling line segment BC with distance x to BC inside $\triangle ABC$. Then the event, $\{\xi \leqslant x\}$, occurs if and only if P falls within $DBCE$. It follows from the definition of geometric probability that

$$P(\xi \leqslant x) = \frac{\text{area of } DBCE}{\text{area of } \triangle ABC} = 1 - \left(1 - \frac{x}{h}\right)^2.$$

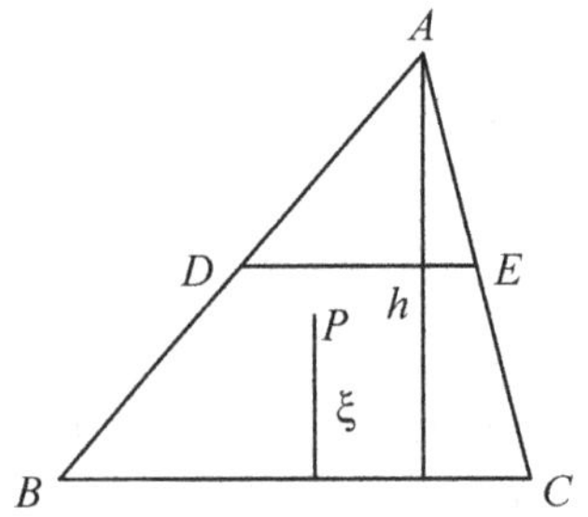

Figure 2-2

When $x \geqslant h$, $\{\xi \leqslant x\}$ must occur, so $P(\xi \leqslant x) = 1$. In summary, the distribution function is

$$F(x) = \begin{cases} 0, & x < 0, \\ 1 - \left(1 - \dfrac{x}{h}\right)^2, & 0 \leqslant x < h, \\ 1, & x \geqslant h. \end{cases}$$

Ⅱ Properties

Since a distribution function is defined by the probability of event $\{\xi \leqslant x\}$, it obviously follows that $0 \leqslant F(x) \leqslant 1$. In addition, the distribution function F possesses the following three fundamental properties:

(1) $F(x)$ is monotonic non-decreasing in x, that is, if $a < b$ then $F(a) \leqslant F(b)$;

(2) $\lim\limits_{x \to -\infty} F(x) = 0$, $\lim\limits_{x \to \infty} F(x) = 1$;

(3) $F(x)$ is right continuous, that is, $F(x+0) = F(x)$.

Proof (1) It follows from the fact $F(b) - F(a) = P(a < \xi \leqslant b) \geqslant 0$.

(2) Since $F(x)$ is monotonically non-decreasing and bounded, then

$$F(-\infty) = \lim_{n \to \infty} F(-n)$$

exists and is finite. Noting that

$$\{\xi \leqslant -(n+1)\} \subset \{\xi \leqslant -n\}, \quad \bigcap_{n=1}^{\infty} \{\xi \leqslant -n\} = \varnothing.$$

In turn, applying the probability continuity theorem, we have

$$\begin{aligned} \lim_{n \to \infty} F(-n) &= \lim_{n \to \infty} P(\xi \leqslant -n) \\ &= P\left(\bigcap_{n=1}^{\infty} \{\xi \leqslant -n\}\right) \\ &= P(\varnothing) = 0; \end{aligned}$$

similarly, $\{\xi \leqslant n\} \subset \{\xi \leqslant (n+1)\}$ and $\bigcup\limits_{n=1}^{\infty} \{\xi \leqslant n\} = \Omega$, so we have

$$\begin{aligned} \lim_{n \to \infty} F(n) &= \lim_{n \to \infty} P\{\xi \leqslant n\} \\ &= P\left(\bigcup_{n=1}^{\infty} \{\xi \leqslant n\}\right) \\ &= P(\Omega) = 1. \end{aligned}$$

(3) By the monotonicity of $F(x)$ again, it suffices to show $\lim_{n\to\infty} F(x+1/n) = F(x)$. Note that

$$\{\xi \leqslant x+1/n\} \subset \{\xi \leqslant x+1/(n-1)\}$$

and

$$\bigcap_{n=1}^{\infty} \{\xi \leqslant x+1/n\} = \{\xi \leqslant x\},$$

then

$$\begin{aligned}\lim_{n\to\infty} F(x+1/n) &= \lim_{n\to\infty} P(\xi \leqslant x+1/n)\\ &= P(\xi \leqslant x) = F(x).\end{aligned}$$

Thus we have proven that a distribution function has the three useful properties listed above. On the contrary, if a real function possesses these three properties it must be a distribution function of a random variable.

Example 8 Suppose a random variable has the distribution function as follows:

$$F(x) = \begin{cases} 0, & x \leqslant -1, \\ a + b\arcsin x, & -1 < x \leqslant 1, \\ 1, & x > 1. \end{cases}$$

Find constants a, b.

Solution $F(x)$ is assumed to satisfy the three properties given above. We firstly note that $F(-\infty) = 0$, $F(+\infty) = 1$. Secondly, we know that if $b > 0$ then $F(x)$ is non-decreasing within each interval, so $F(x)$ is non-decreasing whenever $0 \leqslant a + b\arcsin x \leqslant 1$.

Thus it remains to determining the right continuity of $F(x)$. We need only to look at $x = -1$ and $x = 1$. Since $F(-1+0) = F(-1)$ and $F(1+0) = F(1)$, we have

$$a - \frac{b\pi}{2} = 0, \quad a + \frac{b\pi}{2} = 1,$$

and thus, $a = 1/2$ and $b = 1/\pi$.

2.2.2 Continuous random variables and density functions

Definition 4 *Suppose that a random variable ξ has distribution function $F(x)$ and there exists a non-negative integrable function $p(x)$ such that $F(x)$ can be written as*

$$F(x) = \int_{-\infty}^{x} p(y)\,dy, \quad -\infty < x < \infty, \tag{2.21}$$

then ξ is called a continuous random variable, and $p(x)$ is called the probability density function of ξ, or more simply its density function.

$F(x)$, in (2.21) is said to be absolutely continuous.

By definition, the distribution function $F(x)$ of a continuous random variable possesses the following useful properties.

(1) As given by the theory of real functions, if $F(x)$ is absolutely continuous, then it must be continuous everywhere, and is differentiable at continuity points of

$p(x)$. Also

$$F'(x) = p(x). \tag{2.22}$$

(2) It is easily seen from (2.21) that for a continuous random variable ξ we can directly calculate the probability that ξ lies in an interval $(a, b]$ once $p(x)$ is given,

$$\begin{aligned} P(a < \xi \leqslant b) &= F(b) - F(a) \\ &= \int_{-\infty}^{b} p(y)\mathrm{d}y - \int_{-\infty}^{a} p(y)\mathrm{d}y \\ &= \int_{a}^{b} p(y)\mathrm{d}y. \end{aligned} \tag{2.23}$$

From this we can calculate the probability that ξ lies in every Borel set B in $\mathbf{R}$ in terms of $p(x)$, i. e.,

$$P(\xi \in B) = \int_{B} p(x)\mathrm{d}x.$$

(3) In particular, for an arbitrary constant c, we have

$$\begin{aligned} P(\xi = c) &= F(c) - F(c-0) \\ &= \lim_{h \to 0^+} \int_{c-h}^{c} p(y)\mathrm{d}y = 0. \end{aligned} \tag{2.24}$$

Therefore it does not make sense to consider the probability of ξ at a single point for a continuous random variable. This is one of reasons that we cannot describe a continuous random variable by using its distribution sequence. On the other hand, the event $\{\xi = c\}$ probably happens. This implies that an event A of zero probability is not necessarily empty. Similarly $P(A) = 1$ does not mean $A = \Omega$.

The density function possesses the following properties:

(1) $p(x) \geqslant 0$, $\quad$ (2.25)

(2) $\int_{-\infty}^{+\infty} p(x)\mathrm{d}x = 1.$ $\quad$ (2.26)

The latter follows from the fact that $F(+\infty) = 1$. On the contrary, if a function defined in $(-\infty, \infty)$ satisfies (2.25) and (2.26) then it can be considered to be a density function of some random variable.

Example 9 Is it possible for $F(x)$ of Example 8 to be the distribution function of a continuous random variable?

Solution Observe that $F(x)$ is differentiable everywhere except at $x = -1$ and $+1$. Denote by $p(x)$ its derivative. When $-1 < x < 1$, $p(x) = F'(x) = \dfrac{1}{\pi\sqrt{1-x^2}}$; $p(x) = 0$ otherwise. So $p(x)$ satisfies (2.25) and (2.26). This implies $F(x)$ is a distribution function of the continuous random variable with the density function $p(x)$.

We remark that there are other types of random variables, besides discrete and continuous random variables. For instance,

$$F(x)=\begin{cases}0, & x<0,\\ (1+x)/2, & 0\leqslant x<1,\\ 1, & x\geqslant 1\end{cases}$$

is a distribution function. But it is neither discrete nor continuous. In fact, it is a mixture of $F_1(x)$ and $F_2(x)$:

$$F(x)=\frac{1}{2}(F_1(x)+F_2(x)),$$

where $F_1(x)$ is a degenerate distribution at $x=0$ and $F_2(x)$ is uniform on $[0, 1]$.

In addition, there exists a distribution function which is continuous but not absolutely continuous. Fortunately, distribution functions most commonly used are either discrete or continuous. In the sequel, we use distribution function in the study of general random variables; we mainly use distribution sequence for discrete random variables, whereas we use density function primarily for continuous random variables. Unless specifically stated, we shall not mention other types of variables.

2.2.3 Typical continuous random variables

Ⅰ The uniform distribution

Let $a<b$. A random variable ξ is said to obey a uniform distribution on the interval $[a, b]$ if its density function is

$$p(x)=\begin{cases}1/(b-a), & a\leqslant x\leqslant b,\\ 0, & \text{otherwise.}\end{cases}$$

Simply write $\xi\sim U(a,b)$. When $x<a$, clearly $P(\xi\leqslant x)=0$; when $a\leqslant x<b$,

$$\begin{aligned}P(\xi\leqslant x)&=\int_{-\infty}^{x}p(y)\mathrm{d}y\\&=\int_{a}^{x}\frac{1}{b-a}\mathrm{d}y\\&=\frac{x-a}{b-a};\end{aligned}$$

when $x\geqslant b$,

$$\begin{aligned}P(\xi\leqslant x)&=\int_{-\infty}^{x}p(y)\mathrm{d}y\\&=\int_{a}^{b}\frac{1}{b-a}\mathrm{d}y=1.\end{aligned}$$

Hence the distribution function is

$$F(x)=\begin{cases}0, & x<a,\\ (x-a)/(b-a), & a\leqslant x<b,\\ 1, & x\geqslant b.\end{cases}$$

This is a uniform distribution function on the interval $[a, b]$, which corresponds to a geometric probability with sample space $[a, b]$, and ξ is just the position of points thrown at random in the interval $[a, b]$.

Consider a real number x and its approximate value $\hat{x}$ by rounding off to the n-th decimal place, and then the error $\varepsilon = \hat{x} - x$ between the true value x and its approximate value $\hat{x}$ is generally represented as being uniformly distributed over the interval $(-0.5 \cdot 10^{-n}, 0.5 \cdot 10^{-n}]$, from which we can do an error analysis. This is applicable for data obtained by a wide range of operations, and is very important for solving problems related to computer science since the length of bytes in the computer is always finite.

Ⅱ The normal distribution

If a random variable ξ has density function

$$p(x) = \frac{1}{\sqrt{2\pi}\sigma} e^{-\frac{(x-a)^2}{2\sigma^2}}, \qquad -\infty < x < \infty, \tag{2.27}$$

then we call ξ a normal random variable, and simply write $\xi \sim N(a, \sigma^2)$, where $-\infty < a < \infty$, $\sigma > 0$, are two parameters. The corresponding distribution function is

$$F(x) = \frac{1}{\sqrt{2\pi}\sigma} \int_{-\infty}^{x} e^{-\frac{(t-a)^2}{2\sigma^2}} dt, \qquad -\infty < x < \infty. \tag{2.28}$$

Let us verify that $p(x)$ defined in (2.27) is really a density function. Obviously, $p(x) > 0$. Also,

$$\left(\frac{1}{\sqrt{2\pi}\sigma} \int_{-\infty}^{+\infty} e^{-\frac{(t-a)^2}{2\sigma^2}} dt\right)^2 = \left(\frac{1}{\sqrt{2\pi}} \int_{-\infty}^{+\infty} e^{-\frac{t^2}{2}} dt\right)^2$$
$$= \frac{1}{2\pi} \int_{-\infty}^{+\infty} \int_{-\infty}^{+\infty} e^{-\frac{t^2+s^2}{2}} dt ds.$$

The above double integral can be expressed in terms of polar coordinates as follows

$$\frac{1}{2\pi} \int_0^{2\pi} d\theta \int_0^{+\infty} e^{-\frac{r^2}{2}} r dr = 1,$$

that is, $\int_{-\infty}^{+\infty} p(x) dx = 1$.

The normal distribution is one of the most important distributions in probability theory. It, together with the binomial and the Poisson distribution, plays an important role in both theory and applications. On the one hand, the normal distribution is widely used in practice. Generally speaking, if an argument is affected by a lot of random factors, none of which plays a significant role, then this argument may be thought of as a normal random variable. For instance, the process of measurement is affected by various factors like equipment, even an observer's vision and psychology, and outside perturbations, the measurement result is approximately a normal random variable. Here the parameter a is a true value, and the error of measurement is also a normal random variable. In fact, the normal distribution was first introduced by Gauss in the descriptive process of measurement error in the 19th century, with the result that the normal distribution is referred to as error distribution or Gaussian distribution. In addition, the height and weight of an adult man, the falling position of a shell and the size of some kind of products are approximately normally distributed. On the other hand, a lot of distributions can be approximated by normal distributions under suitable conditions, while other distributions can be derived by the use of normal distributions. Therefore, the normal distribution is very important in the study of probability theory. Let us first examine the graph of its density function.

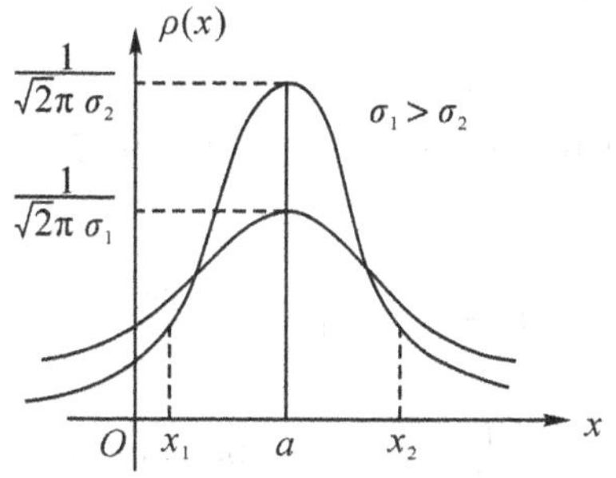

Figure 2-3

Note that if x_1 and x_2 are symmetric about the line $x = a$, that is, $a - x_1 = x_2 - a$, then $p(x_1) = p(x_2)$, so $p(x)$ is symmetric about the line $x = a$.

When $x > a$, $p(x)$ decreases; when $x < a$, $p(x)$ increases. $p(x) \to 0$ as $x \to \pm\infty$, and $p(x)$ attains its maximum at $x = a$ $\left(p(x)\text{ has the maximum value } \frac{1}{\sqrt{2\pi}\sigma}\right)$. So, the larger σ is, and the lower the highest point is. Because the area between the density curve and the x - axis is equal to 1, that is, $\int_{-\infty}^{+\infty} p(x)\mathrm{d}x = 1$, the graph of $p(x)$ becomes flatter as σ increases, which means that ξ takes values far from a with larger probability. Conversely, the graph of $p(x)$ is steeper as σ decreases, meaning ξ concentrates around a with larger probability.

The normal distribution $N(a, \sigma^2)$ with $a = 0$ and $\sigma = 1$ is called a standard normal distribution. Note that the density curve is symmetric about the y -axis at this time. Let $\varphi(x)$ and $\Phi(x)$ be its density function and distribution function respectively. Then,

$$\varphi(x) = \frac{1}{\sqrt{2\pi}} e^{-\frac{x^2}{2}}, \ \Phi(x) = \int_{-\infty}^{x} \varphi(t)\mathrm{d}t,$$

$$-\infty < x < \infty. \tag{2.29}$$

It is not easy to compute a probability using the formula (2.29), so a table is normally used for the calculation of such probabilities. It is generally enough to have such a table (see Appendix B) for values associated with the standard normal distribution function $\Phi(x)$. We shall explain how to use this table below.

(1) Assume $\xi \sim N(0,1)$.

For $x \geqslant 0$, the table gives the value of $\Phi(x)$ at some specific point x. One may use linear interpolation to compute other values of $\Phi(x)$ in the case of $x \geqslant 0$.

When $x < 0$, let $y = -x$. Since $\varphi(x)$ is symmetric about the line $x = 0$, then

$$\int_{-\infty}^{x} \varphi(t)\,dt = \int_{y}^{+\infty} \varphi(t)\,dt = 1 - \int_{-\infty}^{y} \varphi(t)\,dt,$$

in other words,

$$\Phi(x) = 1 - \Phi(y). \tag{2.30}$$

This, together with the value of $\Phi(y)$ in the table, gives $\Phi(x)$.

(2) Assume $\xi \sim N(a, \sigma^2)$.

Set $\eta = \dfrac{\xi - a}{\sigma}$ (a standardized random variable of ξ). Then η is normal $N(0, 1)$. Indeed, by definition we have

$$P(\eta \leqslant x) = P(\xi \leqslant \sigma x + a) = \int_{-\infty}^{\sigma x + a} \frac{1}{\sqrt{2\pi}\,\sigma} e^{-\frac{(t-a)^2}{2\sigma^2}}\,dt$$

$$= \frac{1}{\sqrt{2\pi}} \int_{-\infty}^{x} e^{-u^2/2}\,du = \Phi(x).$$

Example 10 Let $\xi \sim N(0,1)$.

(1) Find $P(-1 < \xi < 3)$.

(2) Suppose that $P(\xi < \lambda) = 0.9755$, find λ.

Solution (1)

$$P(-1 < \xi < 3) = \Phi(3) - \Phi(-1) = \Phi(3) + \Phi(1) - 1$$
$$= 0.9987 + 0.8413 - 1 = 0.8400.$$

(2) Note that $\Phi(\lambda) = 0.9755$, which lies between $\Phi(1.96) = 0.9750$ and $\Phi(1.98) = 0.9762$. Since $\Phi(x)$ is non-decreasing, then λ is between 1.96 and 1.98. By using a linear interpolation,

$$\lambda \approx 1.96 + \frac{\Phi(\lambda) - \Phi(1.96)}{\Phi(1.98) - \Phi(1.96)} \cdot (1.98 - 1.96)$$

$$\approx 1.968.$$

Example 11 Suppose that $\xi \sim N(2, 9)$, calculate $P(5 < \xi < 20)$.

Solution Let $\eta = (\xi - 2)/3$, and then $\eta \sim N(0, 1)$. Hence

$$P(5 < \xi < 20) = P\left(\frac{5-2}{3} < \frac{\xi - 2}{3} < \frac{20-2}{3}\right)$$
$$= P(1 < \eta < 6) = \Phi(6) - \Phi(1)$$
$$\approx 1 - 0.8413 = 0.1587.$$

Example 12 Suppose that $\xi \sim N(a, \sigma^2)$, find $P(|\xi - a| < \sigma)$, $P(|\xi - a| < 2\sigma)$ and $P(|\xi - a| < 3\sigma)$.

Solution Let $\eta = (\xi - a)/\sigma$, and then $\eta \sim N(0,1)$. Hence

$$P(|\xi - a| < \sigma) = P(|\eta| < 1) = 2\Phi(1) - 1 \approx 0.6827.$$

Similarly,

$$P(|\xi - a| < 2\sigma) = P(|\eta| < 2) \approx 0.9545,$$
$$P(|\xi - a| < 3\sigma) = P(|\eta| < 3) \approx 0.9973.$$

This shows that 99.73% of the values of the normal random variable described lie in the range $(a-3\sigma, a+3\sigma)$, and therefore the probability of ξ lying outside this interval is very small, and we can look on it as zero. This is referred to as the 3σ principle by those using normal random variable characteristics in everyday applications.

Example 13 There are two routes by bus from a city's southern district to a train station located in the city's northern area. The first route is shorter, but the ride encounters heavy traffic, so the time τ required is $N(50, 100)$; the second ride is a bit longer, but unexpected traffic jams seldom occur and the time τ required is $N(16, 60)$.

(1) If one has 70 minutes, then which routes should be chosen?

(2) What is it about if one has 65 minutes?

Solution A basic idea is that one should choose a route that has a higher probability of arriving at the train station within the time period specified.

(1) The probability one can arrive at the train station within 70 minutes taking the first route is

$$P(\tau \leqslant 70) = \Phi\left(\frac{70-50}{10}\right) = \Phi(2) = 0.9772;$$

similarly, the probability one can arrive at the train station within 70 minutes using the second route is

$$P(\tau \leqslant 70) = \Phi\left(\frac{70-60}{4}\right) = \Phi(2.5) = 0.9938.$$

Hence one should choose the second route.

(2) The probability one can arrive at the train station within 65 minutes taking the first route is

$$P(\tau \leqslant 65) = \Phi(1.5) = 0.9332;$$

similarly, the probability that one can arrive at the train station within 65 minutes using the second route is

$$P(\tau \leqslant 65) = \Phi(1.25) = 0.8944.$$

Hence it is wiser to choose the first route.

Ⅲ The exponential distribution

The exponential distribution has the density function

$$p(x) = \begin{cases} \lambda e^{-\lambda x}, & x \geqslant 0; \\ 0, & x < 0. \end{cases} \quad (\lambda > 0) \tag{2.31}$$

It is easy to prove that $p(x)$ defined above satisfies the two conditions required of a density function. Now let us calculate its distribution function.

When $x<0$, $P(\xi \leqslant x) = \int_{-\infty}^{x} 0 \mathrm{d}t = 0$; when $x \geqslant 0$, $P(\xi \leqslant x) = \int_0^x \lambda e^{\lambda t} \mathrm{d}t = 1 - e^{-\lambda x}$.

So the distribution function desired is

$$F(x) = \begin{cases} 1 - e^{-\lambda x}, & x \geqslant 0; \\ 0, & x < 0. \end{cases}$$

The exponential distribution possesses a memoryless property similar to that of the geometric distribution. In fact, suppose that a random variable ξ is an exponential distribution with parameter λ. Then for any $s,\ t>0$,

$$P(\xi>s+t\mid\xi>s)=\frac{P(\xi>s+t,\xi>s)}{P(\xi>s)}$$

$$=\frac{e^{-\lambda(s+t)}}{e^{-\lambda s}}=e^{-\lambda t}=P(\xi>t).$$

We remark that the exponential distribution is a unique continuous distribution with such a memoryless property.

Ⅳ The Gamma distribution

The Gamma distribution has the following density function

$$p(x)=\begin{cases}\dfrac{\lambda^r}{\Gamma(r)}x^{r-1}e^{-\lambda x}, & x\geqslant 0,\\ 0, & x<0.\end{cases}\quad(\lambda>0,r>0),\tag{2.32}$$

where $\Gamma(r)$ is the first type of Euler integral. Simply write $\Gamma(\lambda,\ r)$ for a Gamma distribution with parameters λ, r. When r is an integer, we call it an Erlang distribution. Note that the distribution with $r=1$ is just an exponential distribution.

Ⅴ The Weibull distribution

If a random variable has density function

$$p(x)=\begin{cases}\dfrac{\alpha}{\sigma}\left(\dfrac{x-\mu}{\sigma}\right)^{\alpha-1}\exp\left\{-\left(\dfrac{x-\mu}{\sigma}\right)^{\alpha}\right\}, & x>\mu,\\ 0, & x\leqslant\mu,\end{cases}$$

then ξ is said to be a Weibull random variable with parametersu, σ and α, written by $\xi\sim Weib(\mu,\sigma,\alpha)$. The distribution function of the Weibull distribution is

$$F(x)=\begin{cases}1-\exp\left\{-\left(\dfrac{x-\mu}{\sigma}\right)^{\alpha}\right\}, & x>\mu,\\ 0, & x\leqslant\mu.\end{cases}$$

When $\alpha=1$ and $\mu=0$, the Weibull distribution is the exponential distribution. The Weibull distribution is widely used in engineering practice due to its versatility. In particular, it is widely used, in the field of life phenomena, as the distribution of the lifetime of some object. It is named after Swedish mathematician Waloddi Weibull, who proposed it for the interpretation of fatigue data.

Ⅵ The Pareto distribution

A random variable ξ is said to have a Pareto distribution with parameters $x_0>0$ and $\alpha>0$, if its density function is given by

$$p(x)=\begin{cases}\alpha x_0^{\alpha}x^{-(\alpha+1)}, & x>x_0,\\ 0, & x\leqslant x_0.\end{cases}$$

The Pareto distribution is used in description of social, scientic, geophysical, actuarial phenomena, and many other types of observable phenomena. It is characterized by a scale parameter x_0 and a shape parameter α, which is known as the tail index. When this distribution is used to model the distribution of wealth, then the parameter α is called the Pareto index. The Pareto distribution is named after the Italian economist and sociologist Vilfredo Pareto. Originally applied to describing the distribution of wealth in a society, fitting the trend that a large portion of wealth is held by a small fraction of the population, the Pareto distribution has colloquially become known and referred to as the Pareto principle, or "80-20 rule", and is sometimes called the "Matthew principle". This rule states that, for example, 80% of the wealth of a society is held by 20% of its population. However, the Pareto distribution only produces this result for a particular power value, $\alpha(\alpha=\log_4 5)$.

The Pareto distribution has a discrete version as follows.

$$P(\xi = k) = \frac{C}{k^{\alpha}}, k = 1,2,\cdots,$$

where

$$C = \left(\sum_{k=1}^{\infty} \frac{1}{k^{\alpha}}\right)^{-1}.$$

This distribution is named zeta distribution to the fact that the function

$$\xi(s) = \sum_{k=1}^{\infty} \frac{1}{k^{s}}$$

is known in mathematical disciplines as the Riemann zeta function (after the German mathematician G. F. B. Riemann). The zeta distribution was used by Pareto to describe the distribution of family incomes in a given country. However, it was G. K. Zipf who applied these distribution in a wide variety of different areas and, in doing so, popularized their use. So, the zeta distribution is sometimes called the Zipf distribution.

Ⅶ The β distribution

A random variable ξ is said to have a β distribution with parameters a and b, written as $\xi \sim \beta(a,b)$, if it has a density function as

$$p(x) = \begin{cases} \dfrac{1}{B(a,b)} x^{a-1}(1-x)^{b-1}, & 0 \leqslant x \leqslant 1; \\ 0, & \text{otherwise}, \end{cases}$$

where $a,b>0$, $B(a,b)=\int_0^1 x^{a-1}(1-x)^{b-1}\,dx$ is β integral, and $B(a,b)=\dfrac{\Gamma(a+b)}{\Gamma(a)\Gamma(b)}$.

When $a=b=1$, the β distribution is a uniform distribution on the interval $[0,1]$. The distribution can be used to model a random phenomenon whose set of possible values is some finite interval $[c, d]$—which by letting c denote the origin and taking $d-c$ as a unit measurement can be transformed into the interval $[0,1]$. The β distribution has close relationship with the binomial distribution and the Γ distribution.

When $a=b$, the β density is symmetric about $\frac{1}{2}$, giving more and more weight to regions about $\frac{1}{2}$ as the common value a increases. When $b>a$, the density is skewed to the left (in the sense that the smaller values become more likely), and it is skewed to the right when $a>b$.

Ⅷ The Cauchy distribution

The Cauchy distribution has the following density function

$$p(x)=\frac{1}{\pi}\frac{1}{1+(x-\theta)^2}, -\infty<x<\infty,$$

where θ is a parameter with $-\infty<\theta<\infty$.

2.3 Random vectors

One often wants to study simultaneously several arguments in one random experiment. For instance, one needs to determine the position of a variable along with the position coordinates; one needs to consider the amount of a commodity supply, as well as the income of consumers and the prices of market in the study of model of market supply.

Definition 5 *If random variables $\xi_1(\omega), \xi_2(\omega), \cdots, \xi_n(\omega)$ are defined on a common probability space $(\Omega, \mathscr{F}, P)$, then we call*

$$\xi(\omega)=(\xi_1(\omega), \xi_2(\omega), \cdots, \xi_n(\omega))' \tag{2.33}$$

an n-dimensional random vector.

Note that each coordinate of a random vector is a random variable, so we can study each coordinate separately. There also exists a close connection between coordinates, however. This is of great significance to many problems. In the sequel, we focus on a 2-dimensional case, but most of the results obtained in such a case can be extended to any n-dimensional situation.

2.3.1 Discrete random vectors

If a random vector takes only a finitely many or countably many pairs of values, then we call it a discrete random vector. For such a random vector one needs only to enumerate all the possible pairs of values and their corresponding probabilities in order to express its probability distribution.

Example 14 There are two white balls and three black balls in a box. We draw two balls out of the box consecutively, one at a time. Suppose that ξ represents the number of white balls in the first draw, and η the number of white balls in the second draw. Calculate the joint probability distribution either (1) with replacement or (2) without replacement.

Solution (1) Observe that the values ξ and η take possibly are 0 and 1. We can enumerate their matching and corresponding probabilities as follows.

$\{\xi=0,\eta=0\}$ stands for the event that the results of both the first and second draw are black balls. Since drawing each time is independent and the probability of drawing a black ball is 3/5. Hence

$$P(\xi=0,\eta=0)=P(\xi=0)P(\eta=0)=\frac{3}{5}\cdot\frac{3}{5}.$$

Similarly, we have

$$P(\xi=0,\eta=1)=\frac{3}{5}\cdot\frac{2}{5},$$

$$P(\xi=1,\eta=0)=\frac{2}{5}\cdot\frac{3}{5},$$

and

$$P(\xi=1,\eta=1)=\frac{2}{5}\cdot\frac{2}{5}.$$

(2) In this case the values that ξ and η may take possibly are still 0 and 1. But since we draw without replacement, ξ and η are not independent. Using the multiplication formula we have

$$P(\xi=0,\eta=0)=P(\xi=0)P(\eta=0\mid\xi=0)=\frac{3}{5}\cdot\frac{2}{4}.$$

Similarly, we have

$$P(\xi=0,\eta=1)=\frac{3}{5}\cdot\frac{2}{4},$$

$$P(\xi=1,\eta=0)=\frac{2}{5}\cdot\frac{3}{4},$$

and

$$P(\xi=1,\eta=1)=\frac{2}{5}\cdot\frac{1}{4}.$$

To summarize as Table 2-2(1) and Table 2-2(2):

Table 2-2(1)

ξ	η	
	0	1
0	$\frac{3}{5}\cdot\frac{3}{5}$	$\frac{3}{5}\cdot\frac{2}{5}$
1	$\frac{2}{5}\cdot\frac{3}{5}$	$\frac{2}{5}\cdot\frac{2}{5}$

Table 2-2(2)

ξ	η	
	0	1
0	$\frac{3}{5}\cdot\frac{2}{4}$	$\frac{3}{5}\cdot\frac{2}{4}$
1	$\frac{2}{5}\cdot\frac{3}{4}$	$\frac{2}{5}\cdot\frac{1}{4}$

Generally speaking, the joint distribution array of a 2-dimensional discrete random vector is

$$P(\xi=x_i,\eta=y_j)=p_{ij},\quad i,j=1,\ 2,\cdots,\tag{2.34}$$

and written as $p(x_i,y_j),i,j=1,\ 2,\cdots$, or as Table 2-3.

Table 2-3

ξ	η				
	y_1	y_2	$\cdots$	y_j	$\cdots$
x_1	p_{11}	p_{12}	$\cdots$	p_{1j}	$\cdots$
x_2	p_{21}	p_{22}	$\cdots$	p_{2j}	$\cdots$
$\vdots$	$\vdots$	$\vdots$		$\vdots$	
x_i	p_{i1}	p_{i2}	$\cdots$	p_{ij}	$\cdots$
$\vdots$	$\vdots$	$\vdots$		$\vdots$	

The joint distribution array of an n-dimensional discrete random vector is

$$P(\xi_1 = x_1(i_1), \xi_2 = x_2(i_2), \cdots, \xi_n = x_n(i_n)) = p_{i_1 i_2 \cdots i_n}, \quad i_1, i_2, \cdots, i_n = 1, 2, \cdots \tag{2.35}$$

These joint distributions possess many properties similar to those of the distribution of 1-dimensional discrete random variable. For example, in Formula (2.34), p_{ij} must satisfy

$$p_{ij} \geqslant 0, \quad i, j = 1, 2, \cdots; \quad \sum_i \sum_j p_{ij} = 1. \tag{2.36}$$

Having known the joint distribution, the probabilities of some events can be calculated in terms of the joint distribution. To illustrate, consider a 2-dimensional discrete random vector (ξ, η). For any 2-dimensional Borel set B_2,

$$P((\xi, \eta) \in B_2) = \sum_{(x_i, y_j) \in B_2} p_{ij}. \tag{2.37}$$

In addition, ξ and η have their own distributions as two 1-dimensional discrete random variables. Next we write out these 1-dimensional distributions. Note that ξ can only take $x_1, x_2, \cdots, x_i, \cdots$, and the event $\{\xi = x_i\}$ is a sum of the mutually disjoint events $\{(\xi = x_i, \eta = y_j), j = 1, 2, \cdots\}$. Hence we have

$$\begin{aligned} P(\xi = x_i) &= \sum_{j=1}^{\infty} P(\xi = x_i, \eta = y_j) \\ &= \sum_{j=1}^{\infty} p_{ij} =: p_{i\cdot}, \qquad i = 1, 2, \cdots, \end{aligned}$$

where $p_{i\cdot}$ is the sum of p_{ij} taken over as the second index j. Similarly, we have

$$\begin{aligned} P(\eta = y_j) &= \sum_{i=1}^{\infty} P(\xi = x_i, \eta = y_j) \\ &= \sum_{i=1}^{\infty} p_{ij} =: p_{\cdot j}, \qquad j = 1, 2, \cdots, \end{aligned}$$

where $p_{\cdot j}$ is the sum of p_{ij} taken over as the first index i. The above two summations express the distribution sequences of ξ and η respectively, which are just the sums of the rows and columns in the diagram. We usually write these totals along the right hand side and the bottom of the diagram respectively, and call them the marginal distributions.

Example 15 Calculate the marginal distributions in Example 14.

Solution Recall (1) of Example 14.

$$P(\xi = 0) = \frac{3}{5} \cdot \frac{3}{5} + \frac{3}{5} \cdot \frac{2}{5} = \frac{3}{5}.$$

Other probabilities can be similarly obtained. See Table 2-4(1) and Table 2-4(2).

Table 2-4(1)

ξ	η = 0	η = 1	$p_{i\cdot}$
0	$\frac{3}{5} \cdot \frac{3}{5}$	$\frac{3}{5} \cdot \frac{2}{5}$	$\frac{3}{5}$
1	$\frac{2}{5} \cdot \frac{3}{5}$	$\frac{2}{5} \cdot \frac{2}{5}$	$\frac{2}{5}$
$p_{\cdot j}$	$\frac{3}{5}$	$\frac{2}{5}$	

Table 2-4(2)

ξ	η = 0	η = 1	$p_{i\cdot}$
0	$\frac{3}{5} \cdot \frac{2}{4}$	$\frac{3}{5} \cdot \frac{2}{4}$	$\frac{3}{5}$
1	$\frac{2}{5} \cdot \frac{3}{4}$	$\frac{2}{5} \cdot \frac{1}{4}$	$\frac{2}{5}$
$p_{\cdot j}$	$\frac{3}{5}$	$\frac{2}{5}$	

We remark that the joint distributions in Table 2-4(1) and Table 2-4(2) of Example 14 are different, but their marginal distributions are identical. This shows that the joint distributions are not determined uniquely by their marginal distributions, and one needs to take the relation between coordinates into account.

2.3.2 Joint distribution functions

As in the 1-dimensional case, for a general n-dimensional random vector $\boldsymbol{\xi} = (\xi_1, \xi_2, \cdots, \xi_n)'$, one cannot express its probability distribution using a distribution sequence, but instead one needs to make use of its distribution function. Since $\{\xi_i(\omega) \leqslant x_i\} \in \mathscr{F}$ for any n real numbers $x_1, x_2, \cdots, x_n$, for an n-dimensional interval $c_n = \prod_{i=1}^{n} (-\infty, x_i]$ in $\mathbf{R}^n$ we have

$$\{\xi(\omega) \in c_n\} = \bigcap_{i=1}^{n} \{\xi_i(\omega) \leqslant x_i\} \in \mathscr{F}. \tag{2.38}$$

One can further prove that for any Borel set B_n in $\mathbf{R}^n$ the probability of the event $\{\xi \in B_n\}$ can be expressed in terms of (2.38). Because of this, it is therefore plausible to express the probability distribution of ξ by (2.38).

Definition 6 *Let $(\xi_1, \cdots, \xi_n)'$ be a random vector. Its joint distribution function is defined to be*

$$F(x_1, \cdots, x_n) = P(\xi_1 \leqslant x_1, \cdots, \xi_n \leqslant x_n) \tag{2.39}$$

for any $(x_1, \cdots, x_n) \in \mathbf{R}^n$.

For the 2-dimensional random vector (ξ, η), distribution function $F(x, y) = P(\xi \leqslant x, \eta \leqslant y)$ (where $(x, y) \in \mathbf{R}^2$) represents the probability that the point (ξ, η) lies in the shadowed area as shown in Figure 2-4.

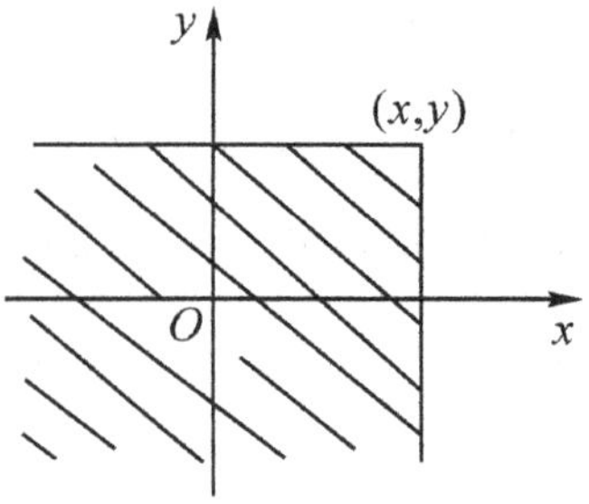

Figure 2-4

Having it, one can calculate directly the probability that (ξ, η) lies in rectangle region I: $a_1 < x \leqslant b_1$, $a_2 < y \leqslant b_2$, as follows:

$$P((\xi,\eta) \in I) = F(b_1, b_2) - F(a_1, b_2) - F(b_1, a_2) + F(a_1, a_2). \tag{2.40}$$

The bivariate distribution function possesses some properties similar to those of univariate distribution function:

(1) monotonically non-decreasing in each argument;

(2) right continuous in each argument;

(3) for any (x, y),

$$F(x, -\infty) = 0,\ F(-\infty, y) = 0,$$
$$F(+\infty, +\infty) = 1. \tag{2.41}$$

In addition, since the probability in equation (2.40) is necessarily non-negative, then

(4) for any real numbers $a_1 < b_1$, $a_2 < b_2$ it follows

$$F(b_1, b_2) - F(a_1, b_2) - F(b_1, a_2) + F(a_1, a_2) \geqslant 0. \tag{2.42}$$

Next, we turn to the relation between the joint distribution function and its marginal distribution functions. Note that the distribution function of ξ is

$$\begin{aligned} F_\xi(x) &= P(\xi \leqslant x, -\infty < \eta < +\infty) \\ &= F(x, +\infty),\ x \in \mathbf{R}. \end{aligned} \tag{2.43}$$

Similarly, we have

$$F_\eta(y) = F(+\infty, y), \quad y \in \mathbf{R}. \tag{2.44}$$

Hence given the joint distribution function we can determine its associated marginal distribution functions. It is not hard to extend the above statements to the case of n-dimensional distribution functions.

2.3.3 Continuous random vectors

Definition 7 *If there exists a non-negative integrable function $p(x_1, \cdots, x_n)$, such that the distribution function $F(x_1, \cdots, x_n)$ can be written as*

$$F(x_1, \cdots, x_n) = \int_{-\infty}^{x_1} \cdots \int_{-\infty}^{x_n} p(y_1, \cdots, y_n) \mathrm{d}y_1 \cdots \mathrm{d}y_n, \tag{2.45}$$

then we call F a distribution of continuous type, and call p a joint probability density *function*.

It is obvious that the density function p satisfies the following conditions:

(1) $p(x_1, \cdots, x_n) \geqslant 0$;

(2) $\int_{-\infty}^{+\infty} \cdots \int_{-\infty}^{+\infty} p(y_1, \cdots, y_n) \mathrm{d}y_1 \cdots \mathrm{d}y_n = 1.$ (2.46)

For any continuous random vector, its distribution function is continuous in each argument, and $F(x_1, \cdots, x_n)$ is differentiable with respect to each argument at any continuity point $(x_1, \cdots, x_n)$ of $p(x_1, \cdots, x_n)$:

$$\frac{\partial^n F}{\partial x_1 \cdots \partial x_n} = p(x_1, \cdots, x_n). \tag{2.47}$$

Since $p(x_1, \cdots, x_n)$ is integrable, it is almost also surely continuous. Hence there is a one to one correspondence between distribution function of continuous random vector and its density function except at a set of measure zero. We can express the joint probability distribution using the joint density function. In fact, for any Borel set $B_n \in \mathscr{B}^n$, we have

$$P(\xi(\omega) \in B_n) = \int \cdots \int_{(x_1, \cdots, x_n) \in B_n} p(x_1, \cdots, x_n) \mathrm{d}x_1 \cdots \mathrm{d}x_n. \tag{2.48}$$

(The proof is omitted.)

Suppose that $(\xi_1, \cdots, \xi_n)$ is a continuous random vector, what can we say about its marginal distribution functions? Let us consider the 2-dimensional case again. Suppose that (ξ, η) has density function $p(x, y)$ and distribution function $F(x, y)$, then the marginal distribution function of ξ is as follows:

$$\begin{aligned} F_\xi(x) &= F(x, +\infty) = \int_{-\infty}^{x} \int_{-\infty}^{+\infty} p(u, v) \mathrm{d}u \mathrm{d}v \\ &= \int_{-\infty}^{x} \left(\int_{-\infty}^{+\infty} p(u, v) \mathrm{d}v \right) \mathrm{d}u. \end{aligned} \tag{2.49}$$

We set

$$p_\xi(u) = \int_{-\infty}^{+\infty} p(u, v) \mathrm{d}v, \tag{2.50}$$

then

$$F_\xi(x) = \int_{-\infty}^{x} p_\xi(u) \mathrm{d}u. \tag{2.51}$$

By definition, it follows from (2.51) that ξ is a continuous random variable with (2.50) as its density function. Similarly, η is also a continuous random variable with the density function

$$p_\eta(v) = \int_{-\infty}^{+\infty} p(u, v) \mathrm{d}u. \tag{2.52}$$

$p_\xi(x)$ and $p_\eta(y)$ are, by definition, the marginal densities of (ξ, η) or $p(x, y)$.

Example 16 Suppose that a random vector (ξ, η) has the density function as follows

$$p(x, y) = \begin{cases} A\mathrm{e}^{-2(x+y)}, & x > 0, y > 0, \\ 0, & \text{otherwise}. \end{cases}$$

(1) determine the constant A;

(2) find the distribution function;

(3) find the marginal densities;

(4) find $P(\xi < 1, \eta < 2)$;

(5) find $P(\xi + \eta < 1)$.

Solution (1) It follows by property (2.46) of a joint probability density that

$$1 = \int_0^{+\infty}\int_0^{+\infty} A e^{-2(x+y)}\,dx dy = \frac{A}{4},$$

which implies $A = 4$.

(2) It follows by definition that

$$F(x,y) = \int_{-\infty}^{x}\int_{-\infty}^{y} p(u,v)\,du dv.$$

We calculate $F(x, y)$ in detail below. When $x \leqslant 0$ or $y \leqslant 0$, $p(x,y) = 0$, so $F(x, y) = 0$; when $x > 0$ and $y > 0$, we have

$$\begin{aligned} F(x,y) &= \int_{-\infty}^{x}\int_{-\infty}^{0} 0\,du dv + \int_{-\infty}^{0}\int_{0}^{y} 0\,du dv + \int_0^x\int_0^0 4e^{-2(u+v)}\,du dv \\ &= (1-e^{-2x})(1-e^{-2y}). \end{aligned}$$

(3) Since $F(x, y)$ is now known, we first calculate its marginal distributions by (2.43) and (2.44), and then calculate marginal density by (2.22). In particular, the marginal distribution function of ξ is

$$F_\xi(x) = F(x,\infty) = \begin{cases} 1-e^{-2x}, & x > 0; \\ 0, & x \leqslant 0. \end{cases}$$

So

$$p_\xi(x) = F_\xi{}'(x) = \begin{cases} 2e^{-2x}, & x > 0; \\ 0, & x \leqslant 0. \end{cases}$$

Similarly,

$$p_\eta(y) = F_\xi{}'(y) = \begin{cases} 2e^{-2y}; & y > 0; \\ 0, & y \leqslant 0. \end{cases}$$

We can also directly calculate the marginal densities from $p(x, y)$ by using (2.50) and (2.51).

(4) $P(\xi < 1, \eta < 2) = F(1,2) = (1-e^{-2})(1-e^{-4})$.

(5) It follows from (2.48) that

$$\begin{aligned} P(\xi+\eta<1) &= \iint\limits_{x+y<1} p(x,y)\,dx dy \\ &= \iint\limits_{x+y<1, x>0, y>0} 4e^{-2(x+y)}\,dx dy \\ &= \int_0^1\left(\int_0^{1-x} 4e^{-2(x+y)}\,dy\right)dx \\ &= 1-3e^{-2}. \end{aligned}$$

Next we introduce two typical continuous random vectors.

Ⅰ The n-dimensional uniform distribution

The n-dimensional uniform distribution has the following density function

$$p(x_1,\cdots,x_n)=\begin{cases}A, & (x_1,\cdots,x_n)\in G;\\ 0, & \text{otherwise},\end{cases} \tag{2.53}$$

where G is a Borel set in $\mathbf{R}^n$. It immediately follows that $A=1/S_G$, where S_G is the measure of G (as G is a 2- or 3-dimensional region, S_G is its area or volume).

The coordinates of point randomly chosen in a region in the plane obey 2-dimensional uniform distribution.

Example 17 Suppose that (ξ,η) obeys the uniform distribution in the unit disk $x^2+y^2\leqslant 1$. Find its marginal densities.

Solution Observe that the joint density is

$$p(x,y)=\begin{cases}\dfrac{1}{\pi}, & x^2+y^2\leqslant 1;\\ 0, & \text{otherwise.}\end{cases}$$

It obviously follows that $p(x,y)=0$ as $|x|>1$. Now consider the case $|x|\leqslant 1$.

$$\begin{aligned}p_\xi(x)&=\int_{-\infty}^{+\infty}p(x,y)\,\mathrm{d}y\\ &=\int_{-\sqrt{1-x^2}}^{\sqrt{1-x^2}}\frac{1}{\pi}\,\mathrm{d}y\\ &=\frac{2}{\pi}\sqrt{1-x^2}.\end{aligned}$$

That is

$$p_\xi(x)=\begin{cases}\dfrac{2}{\pi}\sqrt{1-x^2}, & |x|\leqslant 1;\\ 0, & |x|>1.\end{cases}$$

Similarly,

$$p_\eta(y)=\begin{cases}\dfrac{2}{\pi}\sqrt{1-y^2}, & |y|\leqslant 1;\\ 0, & |y|>1.\end{cases}$$

Hence one easily concludes that marginal distributions are not uniform though (ξ,η) is jointly uniform.

Ⅱ The n-dimensional normal distribution

Suppose that $\boldsymbol{B}=(b_{ij})$ is an $n\times n$ positive definite symmetric matrix. Let $|\boldsymbol{B}|$ be its determinant, and $\boldsymbol{B}^{-1}$ its inverse. Let $\boldsymbol{x}=(x_1,\cdots,x_n)'$, $\boldsymbol{a}=(a_1,\cdots,a_n)'$. Call

$$p(\boldsymbol{x})=\frac{1}{(2\pi)^{n/2}|\boldsymbol{B}|^{1/2}}\exp\left\{-\frac{1}{2}(\boldsymbol{x}-\boldsymbol{a})'\boldsymbol{B}^{-1}(\boldsymbol{x}-\boldsymbol{a})\right\} \tag{2.54}$$

an n-dimensional normal density function.

If a random vector $\boldsymbol{\xi}$ has this density function, then $\boldsymbol{\xi}$ is said to be multi-normal distributed, written as $\boldsymbol{\xi}\sim N(\boldsymbol{a},\boldsymbol{B})$.

For verifying $\int_{-\infty}^{\infty} \cdots \int_{-\infty}^{\infty} p(\boldsymbol{x})\,\mathrm{d}\boldsymbol{x} = 1$, where $\mathrm{d}\boldsymbol{x} = \mathrm{d}x_1 \cdots \mathrm{d}x_n$, we first consider the special case of $\boldsymbol{a}=\boldsymbol{0}$ and $\boldsymbol{B}=\boldsymbol{I}$, where $\boldsymbol{I}$ is an $n\times n$ identity matrix. In this case,

$$p(\boldsymbol{x}) = \frac{1}{(2\pi)^{n/2}}\exp\left\{-\frac{1}{2}\boldsymbol{x}'\boldsymbol{x}\right\} = \prod_{i=1}^{n}\frac{1}{\sqrt{2\pi}}\mathrm{e}^{-\frac{x_i^2}{2}}.$$

So,

$$\int_{-\infty}^{\infty}\cdots\int_{-\infty}^{\infty} p(\boldsymbol{x})\,\mathrm{d}\boldsymbol{x} = \prod_{i=1}^{n}\int_{-\infty}^{\infty}\frac{1}{\sqrt{2\pi}}\mathrm{e}^{-\frac{x_i^2}{2}}\,\mathrm{d}x_i = 1.$$

In the general case, one can find an $n\times n$ positive definite symmetric matrix $\boldsymbol{L}$ such that $\boldsymbol{B}=\boldsymbol{L}\boldsymbol{L}$. Then $\boldsymbol{B}^{-1}=\boldsymbol{L}^{-1}\boldsymbol{L}^{-1}$, $|\boldsymbol{L}| = |\boldsymbol{B}|^{1/2}$. Let $\boldsymbol{y}=\boldsymbol{L}^{-1}(\boldsymbol{x}-\boldsymbol{a})$. Then $\boldsymbol{y}'=(\boldsymbol{x}-\boldsymbol{a})'\boldsymbol{L}^{-1}$. And so,

$$\begin{aligned}\int_{-\infty}^{\infty}\cdots\int_{-\infty}^{\infty} p(\boldsymbol{x})\,\mathrm{d}\boldsymbol{x} &= \int_{-\infty}^{\infty}\cdots\int_{-\infty}^{\infty}\frac{1}{(2\pi)^{n/2}\,|\boldsymbol{B}|^{1/2}}\exp\left\{-\frac{1}{2}\boldsymbol{y}'\boldsymbol{y}\right\}|\boldsymbol{L}|\,\mathrm{d}\boldsymbol{y}\\ &= \int_{-\infty}^{\infty}\cdots\int_{-\infty}^{\infty}\frac{1}{(2\pi)^{n/2}}\exp\left\{-\frac{1}{2}\boldsymbol{y}'\boldsymbol{y}\right\}\mathrm{d}\boldsymbol{y} = 1.\end{aligned}$$

For $n = 1$, set $\boldsymbol{B} = \sigma^2$ and $\boldsymbol{a} = a$. Then (2.54) becomes

$$p(x) = \frac{1}{\sqrt{2\pi}\sigma}\exp\left\{-\frac{(x-a)^2}{2\sigma^2}\right\},$$

which is just the 1-dimensional normal density function introduced in Section 2.2.

For $n = 2$, set

$$\boldsymbol{B} = \begin{pmatrix}\sigma_1^2 & r\sigma_1\sigma_2\\ r\sigma_1\sigma_2 & \sigma_2^2\end{pmatrix},$$

where $\sigma_1, \sigma_2 > 0$, $|r| < 1$. We have

$$\boldsymbol{B}^{-1} = \frac{1}{|\boldsymbol{B}|}\begin{pmatrix}\sigma_2^2 & -r\sigma_1\sigma_2\\ -r\sigma_1\sigma_2 & \sigma_1^2\end{pmatrix}.$$

Also, set $\boldsymbol{x} = (x, y)$, $\boldsymbol{a} = (a, b)$. Then (2.54) becomes

$$\begin{aligned}p(x,y) = &\frac{1}{2\pi\sigma_1\sigma_2\sqrt{1-r^2}}\exp\left\{-\frac{1}{2(1-r^2)}\right.\\ &\left.\cdot\left[\frac{(x-a)^2}{\sigma_1^2} - \frac{2r(x-a)(y-b)}{\sigma_1\sigma_2} + \frac{(y-b)^2}{\sigma_2^2}\right]\right\},\end{aligned} \tag{2.55}$$

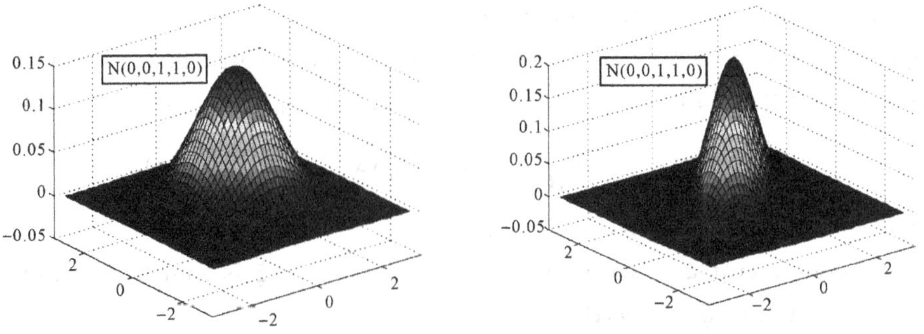

Figure 2-5

and simply write $(\xi,\eta) \sim N(a,b,\sigma_1^2,\sigma_2^2,r)$. The Figure 2-5 gives the graphs of the densities of bivariate normal distributions $N(0,0,1,1,0)$ and $N(0,0,1,1,0.7)$ respectively.

Some simple computation gives

$$p(x,y) = \frac{1}{\sqrt{2\pi}\sigma_1}\exp\left\{-\frac{(x-a)^2}{2\sigma_1^2}\right\} \cdot \frac{1}{\sqrt{2\pi}\sigma_2\sqrt{1-r^2}}\exp\left\{-\frac{\left[y-b-\frac{r\sigma_2}{\sigma_1}(x-a)\right]^2}{2\sigma_2^2(1-r^2)}\right\}, \tag{2.56}$$

of which the first part is the density function of $N(a,\sigma^2)$, and the second part is a normal density for each fixed x so that the integral is just equal to 1. Hence the marginal density of ξ is

$$p_\xi(x) = \int_{-\infty}^{+\infty} p(x,y)\mathrm{d}y = \frac{1}{\sqrt{2\pi}\sigma_1}\exp\left\{-\frac{(x-a)^2}{2\sigma_1^2}\right\}.$$

This shows that $\xi \sim N(a,\sigma_1^2)$. Similarly, $\eta \sim N(b,\sigma_2^2)$.

The above statement implies that the marginal distributions of a 2-dimensional normal distribution are still normal distributions, and do not depend upon r. But the converse is not necessarily correct. That is, (ξ,η) is not necessarily jointly normal even though each marginal distribution is normal.

Example 18 Suppose that (ξ,η) has the joint density function

$$p(x,y) = \frac{1}{2\pi}\mathrm{e}^{-\frac{x^2+y^2}{2}}(1+\sin x\sin y),$$

where $-\infty < x, y < \infty$. Find its marginal distributions.

Solution

$$\begin{aligned} p_\xi(x) &= \int_{-\infty}^{+\infty} p(x,y)\mathrm{d}y \\ &= \frac{1}{\sqrt{2\pi}}\mathrm{e}^{-\frac{x^2}{2}}\int_{-\infty}^{+\infty}\frac{1}{\sqrt{2\pi}}\mathrm{e}^{-\frac{y^2}{2}}\mathrm{d}y \\ &\quad + \frac{1}{2\pi}\mathrm{e}^{-\frac{x^2}{2}}\sin x\int_{-\infty}^{+\infty}\mathrm{e}^{-\frac{y^2}{2}}\sin y\mathrm{d}y \\ &= \frac{1}{\sqrt{2\pi}}\mathrm{e}^{-\frac{x^2}{2}}, \quad -\infty < x < \infty. \end{aligned}$$

Similarly,

$$p_\eta(y) = \frac{1}{\sqrt{2\pi}}\mathrm{e}^{-\frac{y^2}{2}}, \quad -\infty < y < \infty.$$

Hence both ξ and η obey the normal distribution, but their joint distribution does not.

2.4 Independence of random variables

Now we will extend the independence of random events to random variables. If (ξ, η) is a discrete random vector with the joint distribution sequence given by (2.34) in Section 2.3, then it is natural to define the independence of ξ and η to be the independence of all events $\{\xi = x_i\}$ and $\{\eta = y_j\}$.

Definition 8 *Suppose that the joint distribution sequence of a discrete random vector* (ξ, η) *satisfies*

$$P(\xi = x_i, \eta = y_j) = P(\xi = x_i)P(\eta = y_j), \quad i, j = 1, 2, \cdots, \tag{2.57}$$

then we call ξ *and* η *mutually independent. If* (2.57) *is invalid for at least one pair* (i, j), *then we call* ξ *and* η *dependent.*

In accordance with the notation in Section 2.3, (2.57) can be written as

$$p_{ij} = p_{i\cdot} \cdot p_{\cdot j}, \quad i, j = 1, 2, \cdots.$$

Consider Examples 14 and 15 in Section 2.3. In case (1), ξ and η are independent of each other, while they are dependent in case (2). What features does the corresponding joint distribution function have when ξ and η are independent? For any x and y,

$$\begin{aligned} P(\xi \leqslant x, \eta \leqslant y) &= \sum_{x_i \leqslant x} \sum_{y_j \leqslant y} P(\xi = x_i, \eta = y_j) \\ &= \sum_{x_i \leqslant x} P(\xi = x_i) \sum_{y_j \leqslant y} P(\eta = y_j) \\ &= P(\xi \leqslant x) P(\eta \leqslant y). \end{aligned}$$

That is,

$$F(x, y) = F_\xi(x) \cdot F_\eta(y). \tag{2.58}$$

On the contrary, if (2.58) holds for all x, y, then (2.57) is valid. So, (2.57) and (2.58) are equivalent for discrete random variables. Motivated by this fact, it is reasonable to define the independence for general random variables by (2.58).

Definition 9 *Suppose that* $F(x, y)$, $F_\xi(x)$ *and* $F_\eta(y)$ *are the joint distribution function and marginal distribution functions of* (ξ, η) *respectively. If* (2.58) *is valid for all* x, y, *then we say* ξ *and* η *are independent.*

In contrast to (2.57), we have the following theorem for continuous random variables.

Theorem 2 *Suppose that* $p(x, y)$, $p_\xi(x)$ *and* $p_\eta(y)$ *are the joint density function and marginal density functions of* (ξ, η) *respectively. Then* ξ *and* η *are independent if and only if*

$$p(x, y) = p_\xi(x) \cdot p_\eta(y). \tag{2.59}$$

Proof For any x, y, it follows

$$
\begin{aligned}
F(x,\ y) &= F_\xi(x)\cdot F_\eta(y)\\
&\Leftrightarrow \int_{-\infty}^{x}\int_{-\infty}^{y} p(u,v)\,du\,dv = \int_{-\infty}^{x} p_\xi(u)\,du \int_{-\infty}^{y} p_\eta(v)\,dv \\
&\Leftrightarrow \int_{-\infty}^{x}\int_{-\infty}^{y} p(u,v)\,du\,dv = \int_{-\infty}^{x}\int_{-\infty}^{y} p_\xi(u)p_\eta(v)\,du\,dv \\
&\Leftrightarrow p(x,y) = p_\xi(x)p_\eta(y).
\end{aligned} \tag{2.60}
$$

This is the desired conclusion.

Note that ξ and η are not independent in Examples 17 and 18 from the previous section. If ξ and η are independent, then it is seen from (2.57), (2.58) and (2.59) that the joint distribution is uniquely determined by their marginal distributions.

Example 19 Suppose $(\xi,\eta)\sim N(a,b,\sigma_1^2,\sigma_2^2,r)$. Find out the necessary and sufficient condition for ξ, η to be independent.

Solution Note that $\xi\sim N(a,\sigma_1^2)$ and $\eta\sim N(b,\sigma_2^2)$. By definition,

$$
\begin{aligned}
\xi,\ \eta \text{ are independent} &\Leftrightarrow p(x,y) = p_\xi(x)\cdot p_\eta(y)\\
&= \frac{1}{2\pi\sigma_1\sigma_2}\exp\left\{-\frac{1}{2}\left[\frac{(x-a)^2}{\sigma_1^2}+\frac{(y-b)^2}{\sigma_2^2}\right]\right\}\\
&\Leftrightarrow\ r=0.
\end{aligned}
$$

We can readily extend the above definitions and theorems to n random variables. For instance, suppose that $F(x_1,\cdots,x_n)$, $F_1(x_1),\cdots,F_n(x_n)$ are joint distribution function and marginal distribution functions of $\xi_1,\cdots,\xi_n$, then we call them mutually independent if

$$
F(x_1,\cdots,x_n) = F_1(x_1)\cdots F_n(x_n). \tag{2.61}
$$

Corollary If $\xi_1,\cdots,\xi_n$ are mutually independent, then so are any r random variables $(2\leqslant r<n)$.

Proof It suffices to prove the statement for $\xi_1,\cdots,\xi_{n-1}$. The other is similar.

$$
\begin{aligned}
F(x_1,\cdots,x_{n-1}) &= P(\xi_1\leqslant x_1,\cdots,\xi_{n-1}\leqslant x_{n-1})\\
&= P(\xi_1\leqslant x_1,\cdots,\xi_{n-1}\leqslant x_{n-1},\xi_n<+\infty)\\
&= P(\xi_1\leqslant x_1)\cdots P(\xi_{n-1}\leqslant x_{n-1})P(\xi_n<+\infty)\\
&= F_1(x_1)\cdots F_{n-1}(x_{n-1}).
\end{aligned}
$$

But the converse conclusion is not necessarily true by the definition of independence of n events.

The independence of random variables is a very important concept. People have deeply investigated its various equivalence relations and properties. Besides the above, we have the following properties.

(1) A necessary and sufficient condition for $\xi_1,\cdots,\xi_n$ to be independent is

$$
P(\xi_1\in B_1,\cdots,\xi_n\in B_n) = P(\xi_1\in B_1)\cdots P(\xi_n\in B_n)
$$

for all 1-dimensional Borel sets $B_1,\cdots,B_n$.

(2) An n-dimensional random vector $\boldsymbol{\xi}$ and an m-dimensional random vector $\boldsymbol{\eta}$ are independent if

$$P(\boldsymbol{\xi} \in A, \boldsymbol{\eta} \in B) = P(\boldsymbol{\xi} \in A)P(\boldsymbol{\eta} \in B),$$

where A, B are arbitrary n-dimensional and m-dimensional Borel sets respectively.

(3) If two random vectors are independent, then so are their sub-vectors.

Example 20 Suppose that ξ is a constant a, show ξ and η are independent for any random variable η.

Proof Since η is a general random variable, we use (2.58) below. Observe that the distribution function of ξ is

$$F_\xi(x) = \begin{cases} 0, & x < a; \\ 1, & x \geqslant a. \end{cases}$$

When $x < a$, $\{\xi \leqslant x\}$ is empty, and $\{\xi \leqslant x, \eta \leqslant y\}$ is also empty for any y. Thus

$$\begin{aligned} F(x,y) &= P(\xi \leqslant x, \eta \leqslant y) = 0 \\ &= P(\xi \leqslant x)P(\eta \leqslant y) = F_\xi(x)F_\eta(y). \end{aligned}$$

When $x \geqslant a$, $\{\xi \leqslant x\} = \{\xi < +\infty\}$, so

$$\begin{aligned} F(x, y) &= P(\xi \leqslant x, \eta \leqslant y) = P(\xi < +\infty, \eta \leqslant y) \\ &= P(\eta \leqslant y) = F_\xi(x)F_\eta(y). \end{aligned}$$

Thus (2.58) holds for any x and y. That is, ξ and η are independent.

2.5 Conditional distribution

In Chapter 1, we have defined the conditional probability. With a similar argument, we can consider the conditional distribution of a random variable given the value of another random variable. We consider the definition starting with discrete random variable.

2.5.1 Discrete case

Suppose that (ξ, η) has a joint distribution sequence $P(\xi = x_i; \eta = y_j) = p_{ij}, i, j = 1, 2, \cdots$. Conditional on the event $\xi = x_i$ with $P(\xi = x_i) > 0$, the conditional probability of $\eta = y_j$ is obtained as

$$P(\eta = y_j \mid \xi = x_i) = \frac{P(\xi = x_i, \eta = y_j)}{P(\xi = x_i)} = \frac{p_{ij}}{p_{i\cdot}}, \quad j = 1, 2, \cdots, \tag{2.62}$$

which is a distribution sequence of the random variable η.

Denition 10 *(2.62) is called the conditional distribution sequence of given* $\xi = x_i$, *written as* $p_{\eta|\xi}(y_j \mid x_i)$. *And*

$$P(\eta \leqslant y \mid \xi = x_i) = \sum_{j: y_j \leqslant y} p_{\eta|\xi}(y_j \mid x_i)$$

is called the conditional distribution function of given $\xi = x_i$.

From the definition of the conditional distribution and the definition of the independence ξ and η, it follows that ξ and η are independent if and only if

$$P(\eta = y_j \mid \xi = x_i) = P(\eta = y_j)$$

for any $i, j \geqslant 1$.

Example 21 Let p be the probability of each trial being a success in repeated independent Bernoulli trials, and S_n be the number of trials when the n-th success appears.

(1) Find the conditional distribution of S_{n+1} given $S_n = t$.

(2) Find the conditional distribution of S_n given $S_{n+1} = w$.

Solution For $t \leqslant w$, the event $\{S_n = t, S_{n+1} = w\}$ means that, in w trials, the t-th trial and the w-trial result in successes, $n-1$ trials among the first trial to the $(t-1)$-th trial result in successes, and the remainders result in failures. So,

$$P(S_n = t, S_{n+1} = w) = p \cdot p \cdot \binom{n-1}{t-1} p^{n-1} q^{w-(n+1)} = \binom{n-1}{t-1} p^{n+1} q^{w-(n+1)}.$$

Note

$$P(S_n = t) = \binom{n-1}{t-1} p^n q^{t-n}.$$

Hence, given $S_n = t$, the conditional distribution sequence of S_{n+1} is obtained as

$$P(S_{n+1} = w \mid S_n = t) = \frac{P(S_n = t,\ S_{n+1} = w)}{P(S_n = t)} = pq^{w-t-1}.$$

This means, given the condition $S_n = t$, $S_{n+1} - S_n$ has a geometric distribution.

While, given $S_{n+1} = w$, the conditional distribution of S_n is obtained as

$$\begin{aligned} P(S_n = t \mid S_{n+1} = w) &= \frac{P(S_n = t,\ S_{n+1} = w)}{P(S_{n+1} = w)} \\ &= \frac{\binom{n-1}{t-1} p^{n+1} q^{w-(n+1)}}{\binom{n}{w-1} p^{n+1} q^{w-(n+1)}} \\ &= \frac{\binom{n-1}{t-1}}{\binom{n}{w-1}},\ t = n, \cdots, w-1. \end{aligned}$$

This conditional distribution does not depend on the parameter p.

2.5.2 Continuous case

Suppose that the random vector (ξ, η) has a joint density function $p(x, y)$ and a joint distribution function $F(x, y)$. In the continuous case, $P(\xi = x) = 0$ for any x, and so the conditional probability $P(\eta \leqslant y \mid \xi = x)$ will have no mean. So, we need to define the condition distribution by the density function. Given the condition that $\xi = x$, the conditional distribution of η can be understood as

$$\begin{aligned}P(\eta\leqslant y\mid\xi=x)&=\lim_{\Delta x\to 0}P(\eta\leqslant y\mid x<\xi\leqslant x+\Delta x)\\&=\lim_{\Delta x\to 0}\frac{P(x<\xi\leqslant x+\Delta x,\eta\leqslant y)}{P(x<\xi\leqslant x+\Delta x)}\\&=\lim_{\Delta x\to 0}\frac{F(x+\Delta x,y)-F(x,y)}{F_\xi(x+\Delta x)-F_\xi(x)}.\end{aligned}$$

By dividing the numerator and denominator by x respectively, and taking limits, the above equation is

$$\frac{\frac{\partial F(x,y)}{\partial x}}{F'(x)}=\frac{\int_{-\infty}^{y}p(x,v)\mathrm{d}v}{p_\xi(x)}=\int_{-\infty}^{y}\frac{p(x,v)}{p_\xi(x)}\mathrm{d}v.$$

The expressions above show that the conditional distribution function is also of the continuous type. So, we introduce the following definition.

Definition 11 *Let $p(x,y)$ be the joint density function of a random vector (ξ,η) and $p_\xi(x)=\int_{-\infty}^{\infty}p(x,y)\mathrm{d}y$ is the marginal density function of ξ. Suppose that $p_\xi(x)$ is positive at point x. Then the following function of y as*

$$P(\eta\leqslant y\mid\xi=x)=\int_{-\infty}^{y}\frac{p(x,v)}{p_\xi(x)}\mathrm{d}v,\ y\in\mathbf{R}$$

is called the conditional distribution function of η given $\xi=x$, and write as $F_{\eta|\xi}(y|x)$. The function of y as

$$p_{\eta|\xi}(y\mid x)=\frac{p(x,y)}{p_\xi(x)},\ y\in\mathbf{R}\tag{2.63}$$

is called the conditional density function of η given $\xi=x$.

When $p_\xi(x)=\int_{-\infty}^{\infty}p(x,y)\mathrm{d}y=0$, $p(x,y)=0$ for all y, and the right hand of (2.63) is the type of $\frac{0}{0}$. In such case, $p_{\eta|\xi}(y|x)$ is usually defined to be 0.

Similarly, when $p_\eta(y)>0$, the conditional density function of ξ given $\eta=y$ is defined as

$$p_{\eta|\xi}(y\mid x)=\frac{p(x,y)}{p_\eta(y)},\ x\in\mathbf{R}.$$

By the formula of the conditional density function,

$$p(x,y)=p_{\xi|\eta}(x\mid y)p_\eta(y).$$

Hence

$$p_{\eta|\xi}(y\mid x)=\frac{p_{\xi|\eta}(x\mid y)p_\eta(y)}{\int_{-\infty}^{\infty}p_{\xi|\eta}(x\mid v)p_\eta(v)\mathrm{d}v},\tag{2.64}$$

which is the continuous version of the Bayes's formula.

Example 22 Suppose $(\xi,\eta)\sim N(a,b,\sigma_1^2,\sigma_2^2,r)$. Find the conditional density function $p_{\eta|\xi}(y|x)$.

Solution The joint density function of (ξ,η) is as follows.

$$p(x,y)=\frac{1}{2\pi\sigma_1\sigma_2\sqrt{1-r^2}}\exp\left\{-\frac{1}{2(1-r^2)}\left[\frac{(x-a)^2}{\sigma_1^2}-2r\frac{(x-a)(y-b)}{\sigma_1\sigma_2}+\frac{(y-b)^2}{\sigma_2^2}\right]\right\}.$$

In the following lines for deriving the conditional density function of η given $\xi=x$, we will consider the factors which do not depend on the variable y as constants C_is. The last constant can be derived as the normalization constant such that $\int_{-\infty}^{\infty} p_{\eta|\xi}(y|x)\mathrm{d}y=1$.

$$\begin{aligned}p_{\eta|\xi}(y\mid x)&=\frac{p(y\mid x)}{\int_{-\infty}^{\infty}p(x,v)\mathrm{d}v}=C_1p(x,y)\\&=C_2\exp\left\{-\frac{1}{2(1-r^2)}\left[\frac{(y-b)^2}{\sigma_2^2}-2r\frac{(x-a)(y-b)}{\sigma_1\sigma_2}\right]\right\}\\&=C_3\exp\left\{-\frac{1}{2(1-r^2)}\left(\frac{y-b}{\sigma_2}-r\frac{x-a}{\sigma_1}\right)^2\right\}\\&=C_3\exp\left\{-\frac{1}{2\sigma_2^2(1-r^2)}\left[y-b-r\frac{\sigma_2}{\sigma_1}(x-a)\right]^2\right\}.\end{aligned}$$

The above calculation can be simply written as

$$p_{\eta|\xi}(y\mid x)\overset{\infty}{y}p(x,y)\overset{\infty}{y}\cdots\overset{\infty}{y}\exp\left\{-\frac{1}{2\sigma_2^2(1-r^2)}\left[y-b-r\frac{\sigma_2}{\sigma_1}(x-a)\right]^2\right\}$$

where $f(x,y)\overset{\infty}{y}g(x,y)$ means that the proportion of $f(x,y)$ and $g(x,y)$ does not depend on the variable y. By recalling the normal density, we find that $p_{\eta|\xi}(y|x)$ is a normal density and so

$$p_{\eta|\xi}(y\mid x)=\frac{1}{\sqrt{2\pi}\sigma_2\sqrt{1-r^2}}\exp\left\{-\frac{1}{2\sigma_2^2(1-r^2)}\left[y-b-r\frac{\sigma_2}{\sigma_1}(x-a)\right]^2\right\}.$$

Hence, given the condition $\xi=x$, the conditional distribution of the bivariate normal distribution is a normal distribution $N\left(b+r\frac{\sigma_2}{\sigma_1}(x-a),(1-r^2)\sigma_2^2\right)$ with the first parameter $m=b+r\frac{\sigma_2}{\sigma_1}(x-a)$ being a linear function of x and the second parameter independent of x (see Figure 2-6). This is an important fact used in various statistical problems.

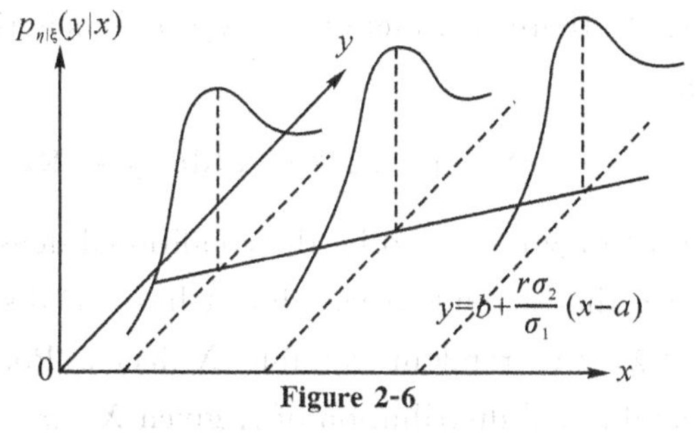

Figure 2-6

The above argument is also valid for multi-dimensional random vectors. Led $\boldsymbol{X}$ be a n-dimensional random vector and $\boldsymbol{Y}$ a m-dimensional random vector. For n-dimensional real vectors $\boldsymbol{x}=(x_1,\cdots,x_n)'$ and $\boldsymbol{y}=(y_1,\cdots,y_n)'$, we denote $\mathrm{d}\boldsymbol{x}$ to be $\mathrm{d}x_1\cdots\mathrm{d}x_n$ and $\boldsymbol{x}\leqslant\boldsymbol{y}$ to be

$$x_1\leqslant y_1, x_2\leqslant y_2,\cdots,x_n\leqslant y_n.$$

Definition 12 *Suppose that* $\boldsymbol{X}=(X_1,\cdots,X_n)'$ *and* $\boldsymbol{Y}=(Y_1,\cdots,Y_m)'$ *are two random vectors with joint density function* $p(\boldsymbol{x},\boldsymbol{y})$, *and so the density function of* $\boldsymbol{Y}$ *is the marginal density function as*

$$p_{\boldsymbol{Y}}(\boldsymbol{y})=\int_{R^n}p(\boldsymbol{x},\boldsymbol{y})\mathrm{d}\boldsymbol{x}.$$

If $p_{\boldsymbol{Y}}(\boldsymbol{y})$ *is nonzero at point* $\boldsymbol{y}$, *then the following function of* $\boldsymbol{x}$ as

$$P(\boldsymbol{X}\leqslant\boldsymbol{x}\mid\boldsymbol{Y}=\boldsymbol{y})=\int_{\boldsymbol{u}\leqslant\boldsymbol{x}}\frac{p(\boldsymbol{u},\boldsymbol{y})}{p_{\boldsymbol{Y}}(\boldsymbol{y})}\mathrm{d}\boldsymbol{u},\boldsymbol{x}\in\mathbf{R}^n$$

is called the conditional distribution function of $\boldsymbol{X}$ *given* $\boldsymbol{Y}=\boldsymbol{y}$, *and written as* $F_{\boldsymbol{X}|\boldsymbol{Y}}(\boldsymbol{x}|\boldsymbol{y})$. *The function of* $\boldsymbol{x}$ *as*

$$p_{\boldsymbol{X}|\boldsymbol{Y}}(\boldsymbol{x}|\boldsymbol{y})=\frac{p(\boldsymbol{x},\boldsymbol{y})}{p_{\boldsymbol{Y}}(\boldsymbol{y})},\boldsymbol{x}\in\mathbf{R}^n$$

is called the conditional density function of $\boldsymbol{X}$ *given* $\boldsymbol{Y}=\boldsymbol{y}$.

2.5.3 The general case

In general, suppose that (ξ,η) is a random vector. For given x, if the limit

$$\lim_{\varepsilon\to0^+}\frac{P(\eta\leqslant y,x-\varepsilon<\xi\leqslant x+\varepsilon)}{P(x-\varepsilon<\xi\leqslant x+\varepsilon)}$$

exists and is finite for any $y\in\mathbf{R}$, then we call the limit

$$F_{\eta|\xi}(y\mid x)=\lim_{\varepsilon\to0^+}\frac{P(\eta\leqslant y,x-\varepsilon<\xi\leqslant x+\varepsilon)}{P(x-\varepsilon<\xi\leqslant x+\varepsilon)},\ y\in\mathbf{R} \tag{2.65}$$

as the conditional distribution function of η given $\xi=x$. If there is y_j, $j=1,2,\cdots$, such that $F_{\eta|\xi}(y|x)$ has the following representation

$$F_{\eta|\xi}(y\mid x)=\sum_{j:y_j\leqslant y}p_{\eta|\xi}(y\mid x),\ y\in\mathbf{R},$$

then the sequence $p_{\eta|\xi}(y|x)$, $j=1,2,\cdots$, is said to be the conditional distribution sequence of η given $\xi=x$. If there is function $p_{\eta|\xi}(y|x)$ such that $F_{\eta|\xi}(y|x)$ has the following representation

$$F_{\eta|\xi}(y\mid x)=\int_{-\infty}^{y}p_{\eta|\xi}(v\mid x)\mathrm{d}v,\ y\in\mathbf{R},$$

then the function $p_{\eta|\xi}(y|x)$ of y is said to be the conditional density of η given $\xi=x$.

Example 23 Suppose that a random variable Λ has a Γ distribution $\Gamma(b,a)$, and given the condition $\Lambda=\lambda$, the random variable X has a Poisson distribution with parameter λ. Find the conditional distribution of Λ given $X=x$.

Solution In this example, Λ is a continuous random variable and X is a discrete random variable. For $x=0,1,\cdots$, we have

$$P(X=x \mid \Lambda=\lambda)=\frac{\lambda^{x}}{x!}\mathrm{e}^{-\lambda}.$$

By the definition,

$$P(X=x \mid \Lambda=\lambda)=\lim_{\Delta\lambda\to 0}\frac{P(X=x, \Lambda\in(\lambda,\lambda+\Delta\lambda])}{P(\Lambda\in(\lambda,\lambda+\Delta\lambda])}.$$

That is,

$$\begin{aligned}P(X=x, \Lambda\in(\lambda,\lambda+\Delta\lambda]) &= P(X=x \mid \Lambda=\lambda)P(\Lambda\in(\lambda,\lambda+\Delta\lambda])+o(\Delta\lambda)\\ &= P(X=x \mid \Lambda=\lambda)p_{\Lambda}(\lambda)\Delta\lambda+o(\Delta\lambda).\end{aligned}$$

So

$$P(X=x, \Lambda\leqslant y)=\int_{-\infty}^{y}P(X=x \mid \Lambda=\lambda)p_{\Lambda}(\lambda)\mathrm{d}\lambda.$$

Hence

$$P(\Lambda\leqslant y \mid X=x)=\frac{P(X=x, \Lambda\leqslant y)}{P(X=x)}=\int_{-\infty}^{y}\frac{P(X=x \mid \Lambda=\lambda)p_{\Lambda}(\lambda)}{P(X=x)}\mathrm{d}\lambda$$

Therefore, given the condition $X=x$, the density function of is obtained as

$$p_{\Lambda|X}(\lambda \mid x)=\frac{P(X=x \mid \Lambda=\lambda)p_{\Lambda}(\lambda)}{P(X=x)}$$

$$\infty_{\lambda}\lambda^{x}\mathrm{e}^{-\lambda}\lambda^{(b-1)}\mathrm{e}^{-\lambda a}\infty_{\lambda}\lambda^{x+b-1}\mathrm{e}^{-(a+1)\lambda}, \ \lambda>0.$$

By normalizing the last function such that its integral is 1, we obtain

$$p_{\Lambda|X}(\lambda \mid x)=\frac{(a+1)^{x+b}}{\Gamma(x+b)}\lambda^{x+b-1}\mathrm{e}^{-(a+1)\lambda}, \ \lambda>0.$$

So, given the condition $X=x$, Λ has a Γ distribution $\Gamma(x+b, a+1)$.

2.5.4 The conditional probability given a random variable

Let X be a random variable with D being the range of its values, and A be an event. Then

$$g(x)=\lim_{\varepsilon\to 0^{+}}\frac{P(A, x-\varepsilon<X\leqslant x+\varepsilon)}{P(x-\varepsilon<X\leqslant x+\varepsilon)}, \ x\in D \tag{2.66}$$

is a real function defined on D denoted by $P(A \mid X=x)$. Then we can define a random variable $g(X)$ as the function of X, and call $g(X)$ the conditional probability of event A given the random variable X. This conditional probability is denoted by $P(A \mid X)$. It should be mentioned that, as a function of X, the conditional probability $P(A \mid X)$ is a random variable.

Due to the definition,

$$P(A \mid X)=g(X) \text{ if and only if } P(A \mid X=x)=g(x), \ x\in D.$$

If X is a discrete random variable with its distribution sequence $p_X(x)$, then from the total probability formula we have

$$\begin{aligned}P(A) &= \sum_{i}P(A \mid X=x_i)P(X=x_i)\\ &= \sum_{i}g(x_i)p_X(x_i).\end{aligned}$$

If X is a continuous random variable with its density function $p_X(x)$, then

$$P(A \mid X = x) \approx \frac{P(A, X \in (\lambda, \lambda + \Delta\lambda])}{P(X \in (\lambda, \lambda + \Delta\lambda])}.$$

That is

$$P(A, X \in (\lambda, \lambda + \Delta\lambda]) = P(A \mid X = x) p_X(x) \Delta\lambda + o(\Delta\lambda).$$

Hence

$$P(A) = P(A, -\infty < X < \infty) = \int_{-\infty}^{\infty} P(A \mid X = x) p_X(x) \mathrm{d}x$$

$$= \int_{-\infty}^{\infty} g(x) p_X(x) \mathrm{d}x, \qquad (2.67)$$

which is the continuous version of the total probability formula.

Example 24 Let $U_1, U_2, \cdots$ be a sequence of independent random variable which are uniformly distributed on interval $(0,1]$. Denote

$$\xi = \min\{n \geqslant 1 : U_1 + \cdots + U_n > 1\}.$$

Find the distribution of ξ.

Solution To find the probability $P(\xi > n)$, we denote

$$\xi(x) = \min\{n \geqslant 1 : U_1 + \cdots + U_n > x\}, \quad 0 < x \leqslant 1.$$

Then

$$P(\xi(x) > 1) = P(U_1 \leqslant x) = x,$$

$$\begin{aligned}
P(\xi(x) > n+1) &= P(U_1 + \cdots + U_{n+1} \leqslant x) \\
&= \int_{-\infty}^{\infty} P(U_1 + \cdots + U_{n+1} \leqslant x \mid U_1 = y) p_{U_1}(y) \mathrm{d}y \\
&= \int_0^1 P(U_1 + \cdots + U_{n+1} \leqslant x \mid U_1 = y) \mathrm{d}y = \int_0^1 P(U_2 + \cdots + U_{n+1} \leqslant x - y) \mathrm{d}y \\
&= \int_0^x P(U_1 + \cdots + U_n \leqslant x - y) \mathrm{d}y = \int_0^x P(U_1 + \cdots + U_n \leqslant u) \mathrm{d}u \\
&= \int_0^x P(\xi(u) > n) \mathrm{d}u.
\end{aligned}$$

By induction,

$$P(\xi(x) > n) = \frac{x^n}{n!}.$$

So

$$P(\xi > n) = \frac{1}{n!} \text{ and } P(\xi = n) = \frac{1}{(n-1)!} - \frac{1}{n!}.$$

2.6 Functions of random variables

People often need to study functions of random variables. For example, the kinetic energy of molecule $T = mv^2/2$ is a function of speed — a random variable v; the distribution χ_n^2 often used in mathematical statistics, its corresponding random variable is $\chi_n^2 = \xi_1^2 + \cdots + \xi_n^2$, where the ξ_i is independent and identically distributed as $N(0,1)$. Thus χ_n^2 is a function of $\xi_1, \cdots, \xi_n$.

In general, if ξ is a random variable, $y = g(x)$ a real function, then $\eta = g(\xi)$ is a function of ξ. It is natural to raise the following two problems: (1) Is η a random variable? (2) If so, is there any connection between the distribution functions of ξ and η? The same problem arises in functions of several random variables.

It is relatively easy to solve the first problem. Indeed, if $\eta = g(\xi)$ is a random variable then it must satisfy (2.1) in Section 2.1. We need only to make some restrictions on $g(x)$.

Definition 13 *Suppose that $g(x)$ is a 1-dimensional real function, $\mathscr{B}$ is the Borel σ-field in* **R**. *If for any $B \in \mathscr{B}$,*

$$\{x : g(x) \in B\} = g^{-1}(B) \in \mathscr{B}, \tag{2.68}$$

that is, the pre-image of an arbitrary Borel set under g is also a Borel set, then we call $g(x)$ a Borel function.

It is well known in real function theory that all piecewise continuous functions and piecewise monotone functions are Borel functions. So most of functions commonly used are Borel functions.

Now we answer the first question. If ξ is a random variable defined on a probability space $(\Omega, \mathscr{F}, P)$, $f(x)$ is a Borel function. Let $\eta = f(\xi)$, then for an arbitrary $B \in \mathscr{B}$, by (2.68) above and (2.1) of Section 2.1, and we obtain

$$\begin{aligned}\{\omega : \eta(\omega) \in B\} &= \{\omega : f(\xi(\omega)) \in B\} \\ &= \{\omega : \xi(\omega) \in f^{-1}(B)\} \in \mathscr{F},\end{aligned}$$

so η is a random variable.

We can similarly define n variate Borel functions. If $f(x_1, \cdots, x_n)$ is a Borel function, then $\eta = f(\xi_1, \cdots, \xi_n)$ is a random variable. Functions we shall study in the sequel are like this.

Let us turn to the second problem.

2.6.1 Functions of discrete random variables

This is a simple case, so we only illustrate it with some examples.

Example 25 Suppose that ξ has distribution sequence

$$\begin{pmatrix} -1 & 0 & 1 & 2 \\ \frac{1}{4} & \frac{1}{2} & \frac{1}{8} & \frac{1}{8} \end{pmatrix}.$$

Let $\eta = 2\xi - 1, \zeta = \xi^2$, and find the distribution sequences of η and ζ.

Solution Since η takes only a finite number of values, $-3, -1, 1, 3$, it is enough to compute their corresponding probabilities. Clearly

$$P(\eta = -3) = P(\xi = -1) = \frac{1}{4}.$$

Similarly, we can obtain the probabilities of the other events. In summary, we have the distribution of η as follows:

$$\begin{pmatrix} -3 & -1 & 1 & 3 \\ \frac{1}{4} & \frac{1}{2} & \frac{1}{8} & \frac{1}{8} \end{pmatrix}.$$

Possible values of ζ are 0, 1, 4. But noting that $\{\zeta = 1\} = \{\xi = 1\} \cup \{\xi = -1\}$, then we have

$$P(\zeta = 1) = P(\xi = 1) + P(\xi = -1) = \frac{1}{8} + \frac{1}{4} = \frac{3}{8}.$$

Other computations are similar, and we have the distribution sequence of ζ:

$$\begin{pmatrix} 0 & 1 & 4 \\ \frac{1}{2} & \frac{3}{8} & \frac{1}{8} \end{pmatrix}.$$

In general, assume that ξ is such that

$$P(\xi = x_i) = p(x_i), \quad i = 1,2,\cdots,$$

then the distribution of $\eta = f(\xi)$ is

$$P(\eta = y_j) = \sum_{f(x_i)=y_j} p(x_i), \quad j = 1,2,\cdots.$$

Example 26 Assume that $\xi \sim B(n_1, p)$, $\eta \sim B(n_2, p)$, and that ξ and η are independent. Find the distribution of $\zeta = \xi + \eta$.

Solution Since ξ and η take the values $0,1,2,\cdots, n_1$ and $0,1,2,\cdots,n_2$ respectively, then the values ζ possibly takes are $0,1,2,\cdots, n_1 + n_2$, and by (2.61) of Section 2.4, it follows that

$$\begin{aligned} P(\zeta = r) &= \sum_{k=0}^{r} P(\xi = k, \eta = r-k) \\ &= \sum_{k=0}^{r} P(\xi = k) P(\eta = r-k) \\ &= \sum_{k=0}^{r} \binom{n_1}{k} p^k q^{n_1-k} \binom{n_2}{r-k} p^{r-k} q^{n_2-r+k} \\ &= p^r q^{n_1+n_2-r} \sum_{k=0}^{r} \binom{n_1}{k} \binom{n_2}{r-k} \\ &= \binom{n_1+n_2}{r} p^r q^{n_1+n_2-r}, \end{aligned}$$

where we used some properties for combinatorial quantities. The computation shows $\xi + \eta \sim B(n_1 + n_2, p)$. This statement manifests an important feature of the binomial distribution: the sum of two independent binomial random variables with common second parameter is also a binomial distribution whose first parameter is just the sum of the first parameters of two summands. This is often called the regenerativity (additivity) property of binomial distributions. This claim is almost obvious if one uses probabilistic intuition: ξ and η are the numbers of successes in independent n_1 and n_2 Bernoulli trials, respectively, when combining the two trials together, $\zeta = \xi + \eta$ is just the number of successes in $n_1 + n_2$ Bernoulli trials.

The formula obtained in the computation above

$$P(\zeta = r) = \sum_{k=0}^{r} P(\xi = k) P(\eta = r - k) \tag{2.69}$$

is called the discrete convolution formula.

2.6.2 Functions of continuous random variables

Suppose that ξ has the density function $p(x)$. What we want is to calculate the distribution function $G(y)$ of $\eta = f(\xi)$. In fact, $G(y) = P(\eta \leqslant y) = P(f(\xi) \leqslant y)$. Note that $D = \{x : f(x) \leqslant y\}$ is a 1-dimensional Borel set, so

$$G(y) = P(\xi \in D) = \int_{x \in D} p(x)\,\mathrm{d}x. \tag{2.70}$$

It is normally hard to decide whether η is a continuous random variable or not; and even if so, it is complex to calculate its density function under general circumstances. However, we can directly derive the density function $g(y)$ of η in some special and typical cases.

Theorem 3 *Suppose that $f(x)$ is strictly monotone, and its inverse $f^{-1}(y)$ has the continuous derivative. Then $\eta = f(\xi)$ is a continuous random variable with density function*

$$g(y) = \begin{cases} p(f^{-1}(y)) \mid (f^{-1}(y))' \mid, & y \in \text{the range of } f(x), \\ 0, & \text{otherwise.} \end{cases} \tag{2.71}$$

Proof Without loss of generality, assume that $f(x)$ is strictly increasing, and

$$A < f(x) < B \quad \text{for } -\infty < x < \infty.$$

Obviously, if $y \leqslant A$ then $G(y) = 0$, so $g(y) = 0$. As $A < y < B$, $\{\eta \leqslant y\} = \{f(\xi) \leqslant y\} = \{\xi \leqslant f^{-1}(y)\}$, thus

$$G(y) = P(\eta \leqslant y) = \int_{-\infty}^{f^{-1}(y)} p(x)\,\mathrm{d}x.$$

Letting $x = f^{-1}(v)$, we have

$$G(y) = \int_{A}^{y} p(f^{-1}(v))(f^{-1}(v))'\mathrm{d}v = \int_{-\infty}^{y} g(v)\,\mathrm{d}v,$$

where $g(v)$ is as in (2.71). As $y \geqslant B$, $G(y) = 1$, so $g(y) = 0$. This completes the proof of Theorem 3.

Theorem 4 *If $y = f(x)$ is piecewise strictly monotone in disjoint intervals $I_1, I_2, \cdots$, which is a partition of Ω, and its inverse $h_i(y)$ in the i-th interval I_i is continuously differentiable, then $\eta = f(\xi)$ is a continuous random variable, whose density function is*

$$g(y) = \begin{cases} \sum_i p(h_i(y)) \mid h'_i(y) \mid, & y \in \text{the definition domain of each } h_i, \\ 0, & \text{otherwise.} \end{cases} \tag{2.72}$$

Proof Observe that $\{f(\xi) \leqslant y\} = \{\xi \in \sum_i E_i(y)\}$, where $E_i(y)$ is the set of x in I_i such that $f(x) \leqslant y$. In a similar way to proof of (2.71), we obtain

$$P(\eta \leqslant y) = P(\xi \in \sum_i E_i(y)) = \sum_i \int_{E_i(y)} p(x)\mathrm{d}x$$
$$= \sum_i \int_{-\infty}^{y} p(h_i(u)) \mid h'_i(u) \mid \mathrm{d}u$$
$$= \int_{-\infty}^{y} \sum_i p(h_i(u)) \mid h'_i(u) \mid \mathrm{d}u,$$

from which (2.72) immediately follows.

Example 27 Assume that $\xi \sim N(0,1)$, calculate the density function of $\eta = \xi^2$.

Solution Note that $y = x^2$ is piecewise monotone. Its inverse is $x = h_1(y) = -\sqrt{y}$ as $x \in I_1 = (-\infty, 0)$; $x = h_2(y) = \sqrt{y}$ as $x \in I_2 = (0, \infty)$. Hence for $y > 0$

$$g_\eta(y) = \frac{1}{\sqrt{2\pi}} \exp\left\{-\frac{(-\sqrt{y})^2}{2}\right\} \cdot \left|\frac{-1}{2\sqrt{y}}\right|$$
$$+ \frac{1}{\sqrt{2\pi}} \exp\left\{-\frac{(-\sqrt{y})^2}{2}\right\} \cdot \frac{1}{2\sqrt{y}}$$
$$= \frac{1}{\sqrt{2\pi y}} \mathrm{e}^{-\frac{y}{2}};$$

while for $y \leqslant 0$, $g_\eta(y) = 0$.

Call the above η a χ_1^2 distribution, which will be investigated in detail later on.

Example 28 Assume that ξ has a continuous distribution function, $F(x)$, calculate the distribution of $\theta = F(\xi)$.

Solution Note that $0 \leqslant F(x) \leqslant 1$. When $x < 0$, $P(\theta \leqslant x) = 0$; when $x \geqslant 1$, $P(\theta \leqslant x) = 1$.

Let us turn to the case $0 \leqslant x < 1$. Consider the inverse function of $F(x)$. Since $y = F(x)$ is not necessarily strictly increasing, that is, several different x corresponding to a common y, for the sake of clarity, define for $0 \leqslant y \leqslant 1$ (see Figure 2-7)

$$F^{-1}(y) = \sup\{x : F(x) < y\}$$

as the inverse of $F(x)$. F^{-1} is called the generalized inverse function of $F(x)$. Due to the definition F^{-1} and the properties of a distribution, it can be veried that the generalized inverse function F^{-1} has the following properties.

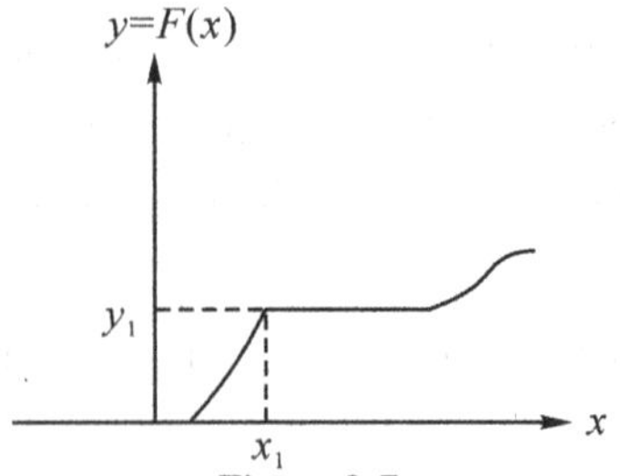

Figure 2-7

(1) $F^{-1}(y)$ $(0<y<1)$ is a non-decreasing function of y;

(2) $F(F^{-1}(y))\geqslant y$. If $F(x)$ is continuous at point $x=F^{-1}(y)$, then $F(F^{-1}(y))=y$;

(3) $F^{-1}(y)\leqslant x$ if and only if $y\leqslant F(x)$.

Now, as $F(x)$ is a continuous distribution function, we have

$$\begin{aligned} P(\theta\leqslant y) &= P(F(\xi)\leqslant y) = P(\xi\leqslant F^{-1}(y)) \\ &= F(F^{-1}(y)) = y, \end{aligned}$$

that is, θ is uniform on the interval $[0,1]$.

Example 29 (The converse to Example 28). Assume the random variable θ is uniform on the interval $[0,1]$ and a function $F(x)$ possesses the three properties required by a distribution function. Calculate the distribution of $\xi = F^{-1}(\theta)$.

Solution By definition, the distribution function of ξ is

$$P(\xi\leqslant x) = P(F^{-1}(\theta)\leqslant x) = P(\theta\leqslant F(x)).$$

Since $0\leqslant F(x)\leqslant 1$, and $\theta \sim U[0,1]$, then the above is equal to $F(x)$ for any x.

This example implies that whatever $F(x)$ is there must exist a random variable with $F(x)$ as its distribution function as long as $F(x)$ possesses the same three properties required of a distribution function.

Example 30 (The truncated random variable) Suppose that ξ is a random variable with distribution function $F_\xi(x)$, a and b are two arbitrary real numbers with $a<b$. We call

$$\eta = \xi\, I(a\leqslant\xi\leqslant b)$$

the truncated random variable of ξ. A very interesting case is $a=-c$, $b=c$ for some positive number c. In this case, η has the following distribution function:

$$F_\eta(x) = \begin{cases} 0, & x<-c; \\ F_\xi(x) - F_\xi(-c-0), & -c\leqslant x<0; \\ F_\xi(x)+1-F_\xi(c), & 0\leqslant x<c; \\ 1, & x\geqslant c. \end{cases}$$

The truncation method is very useful in theory of probability and its applications.

2.6.3 Functions of continuous random vectors

Suppose that $(\xi_1,\cdots,\xi_n)$ is a continuous random vector with the density function $p(x_1,\cdots,x_n)$ and $f(x_1,\cdots, x_n)$ is a Borel function on $\mathbf{R}^n$. Let $\eta = f(\xi_1,\cdots,\xi_n)$, and then the distribution function of η is determined by the following

$$\begin{aligned} F_\eta(y) &= P(f(\xi_1,\cdots,\xi_n)\leqslant y) \\ &= \int\cdots\int_{f(x_1,\cdots,x_n)\leqslant y} p(x_1,\cdots,x_n)\mathrm{d}x_1\cdots\mathrm{d}x_n. \end{aligned} \tag{2.73}$$

Let us look at some special cases below.

I $\eta = \xi_1 + \xi_2$

$$F_\eta(y) = \iint\limits_{x_1+x_2\leqslant y} p(x_1, x_2)\mathrm{d}x_1\mathrm{d}x_2$$
$$= \int_{-\infty}^{+\infty}\mathrm{d}x_1\int_{-\infty}^{y-x_1} p(x_1, x_2)\,\mathrm{d}x_2.$$

Making a change of variable $x_2 = z - x_1$ and exchanging the order of integration, we obtain

$$F_\eta(y) = \int_{-\infty}^{+\infty}\mathrm{d}x_1\int_{-\infty}^{y} p(x_1, z - x_1)\mathrm{d}z$$
$$= \int_{-\infty}^{y}\left(\int_{-\infty}^{+\infty} p(x_1, z - x_1)\mathrm{d}x_1\right)\mathrm{d}z.$$

This shows that η is a continuous random variable, and has the density function

$$p_\eta(z) = \int_{-\infty}^{+\infty} p(x, z - x)\mathrm{d}x. \tag{2.74}$$

In particular, when ξ_1 and ξ_2 are independent of each other and have density functions $p_1(x)$ and $p_2(x)$ respectively, the density function of $\xi_1 + \xi_2$ is

$$p_\eta(z) = \int_{-\infty}^{+\infty} p_1(x)p_2(z - x)\mathrm{d}x. \tag{2.75}$$

Similarly, we have

$$p_\eta(z) = \int_{-\infty}^{+\infty} p_1(z - x)p_2(x)\mathrm{d}x. \tag{2.76}$$

(2.75) and (2.76) are called convolution formulae, which are very similar to that in the discrete case.

Example 31 Suppose that ξ and η are independent identically distributed random variables with a common distribution $N(0,1)$. Calculate the density function of $\zeta = \xi + \eta$.

Solution Use the convolution formula (2.75). For an arbitrary $z \in \mathbf{R}$,

$$p_\zeta(z) = \int_{-\infty}^{+\infty} \frac{1}{\sqrt{2\pi}}\mathrm{e}^{-\frac{x^2}{2}}\frac{1}{\sqrt{2\pi}}\mathrm{e}^{-\frac{(z-x)^2}{2}}\mathrm{d}x$$
$$= \frac{1}{\sqrt{2\pi}\sqrt{2}}\mathrm{e}^{-\frac{z^2}{4}}\int_{-\infty}^{+\infty}\frac{\sqrt{2}}{\sqrt{2\pi}}\mathrm{e}^{-(\sqrt{2}x-\frac{z}{\sqrt{2}})^2/2}\,\mathrm{d}x.$$

Note that the integrand in the last equation is just the density function of $N(z/2, 1/2)$. Thus we have

$$p_\zeta(z) = \frac{1}{\sqrt{2\pi}\sqrt{2}}\mathrm{e}^{-\frac{z^2}{4}},$$

which implies $\zeta = \xi + \eta \sim N(0, 2)$.

In the sequel, we shall use a simpler method to prove: if ξ, η are independent, and $\xi \sim N(a_1, \sigma_1^2)$ and $\eta \sim N(a_2, \sigma_2^2)$, then $\xi + \eta \sim N(a_1 + a_2, \sigma_1^2 + \sigma_2^2)$. The above example is its special case. Like Example 26, a normal distribution possesses a regenerative property in two parameters.

Example 32 Suppose that ξ and η are independent random variables with the following density functions:

$$p_\xi(x) = \begin{cases} a\mathrm{e}^{-ax}, & x > 0; \\ 0, & x \leqslant 0 \end{cases}$$

with $a>0$ and

$$p_\eta(x) = \begin{cases} b\mathrm{e}^{-bx}, & x > 0; \\ 0, & x \leqslant 0 \end{cases}$$

with $b>0$. Calculate the density function of $\zeta = \xi + \eta$.

Solution Observe that $p_\xi(x)p_\eta(z-x) \neq 0$ if and only if $x > 0$ and $z - x > 0$, that is $z > x > 0$. Hence, it is easy to see from (2.75) that $p_\zeta(z) = 0$ when $z \leqslant 0$;

$$p_\zeta(z) = \int_0^z a\mathrm{e}^{-ax} b\mathrm{e}^{-b(z-x)}\mathrm{d}x = ab\mathrm{e}^{-bz}\int_0^z \mathrm{e}^{-(a-b)x}\mathrm{d}x$$

when $z > 0$.

We take the following two cases into account:

(1) If $a = b$, then $p_\zeta(z) = a^2 z\mathrm{e}^{-bz}$;

(2) If $a \neq b$, then

$$p_\zeta(z) = \frac{ab}{a-b}(\mathrm{e}^{-bz} - \mathrm{e}^{-az}).$$

Ⅱ $\eta = \xi_1/\xi_2$

$$F_\eta(y) = P\left(\frac{\xi_1}{\xi_2} \leqslant y\right) = \iint_{x_1/x_2 \leqslant y} p(x_1, x_2)\mathrm{d}x_1\mathrm{d}x_2$$

$$= \int_0^{+\infty}\mathrm{d}x_2\int_{-\infty}^{yx_2} p(x_1, x_2)\mathrm{d}x_1 + \int_{-\infty}^0 \mathrm{d}x_2\int_{yx_2}^{+\infty} p(x_1, x_2)\mathrm{d}x_1.$$

Letting $x_1 = zx_2$ and exchanging the order of integration, we obtain

$$F_\eta(y) = \int_{-\infty}^y \left[\int_0^{+\infty} p(zx_2, x_2)x_2\,\mathrm{d}x_2 - \int_{-\infty}^0 p(zx_2, x_2)x_2\mathrm{d}x_2\right]\mathrm{d}z = \int_{-\infty}^y p_\eta(z)\mathrm{d}z. \tag{2.77}$$

This shows that $\eta = \xi_1/\xi_2$ is a continuous random variable, and has the density function

$$p_\eta(z) = \int_{-\infty}^{+\infty} p(zx, x)\,|x|\,\mathrm{d}x. \tag{2.78}$$

Example 33 Suppose that ξ and η are independent identically distributed random variables with a common distribution $U(0,a)$. Calculate the density function of ξ/η.

Solution Observe that

$$p_\xi(x) = p_\eta(x) = \begin{cases} \dfrac{1}{a}, & 0 \leqslant x \leqslant a; \\ 0, & \text{otherwise}. \end{cases}$$

Since ξ, η are independent, only when $0 \leqslant xz \leqslant a$ and $0 \leqslant x \leqslant a$

$$p(zx, x) = p_\xi(zx)p_\eta(x) = \frac{1}{a^2} \neq 0.$$

Here the required region corresponds to the shaded part in Figure 2-8.

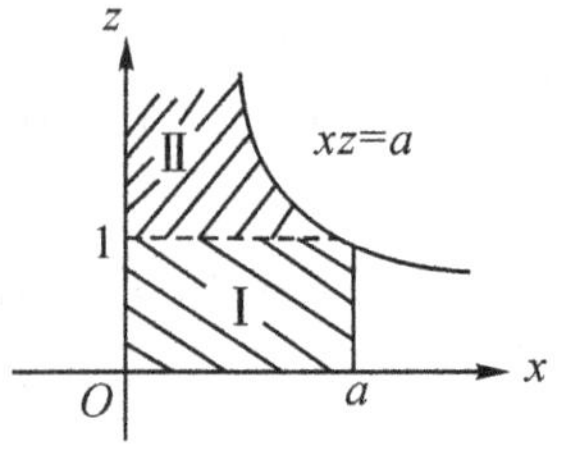

Figure 2-8

When $z<0$, it follows that for any x

$$p(zx,x)=0,$$

which implies from (2.78) that $p_{\xi/\eta}(z)=0$; when $0\leqslant z<1$, it follows obviously $0\leqslant xz\leqslant a$, so we have

$$p_{\xi/\eta}(z)=\int_0^a \frac{1}{a^2}x\mathrm{d}x=\frac{1}{2}.$$

When $z\geqslant 1$, the integral becomes

$$p_{\xi/\eta}(z)=\int_0^{a/z} \frac{1}{a^2}x\mathrm{d}x=\frac{1}{2z^2}.$$

Ⅲ Distributions of order statistics

Suppose that $\xi_1,\cdots,\xi_n$ are independent identically distributed random variables with the common distribution function $F(x)$. We rearrange $\xi_1(\omega),\cdots,\xi_n(\omega)$ in increasing order to obtain $\xi_1^*,\cdots,\xi_n^*$, which are usually called order statistics. They obviously satisfy $\xi_1^*\leqslant\cdots\leqslant\xi_n^*$. By definition, it follows $\xi_1^*=\min\{\xi_1,\cdots,\xi_n\}$, $\xi_n^*=\max\{\xi_1,\cdots,\xi_n\}$.

Now let us compute the distributions of ξ_1^*, ξ_n^* and (ξ_1^*,ξ_n^*). These will be very useful in applied statistics.

(1) The distribution of ξ_n^*.

$$\begin{aligned}P(\xi_n^*\leqslant x)&=P(\xi_1\leqslant x,\xi_2\leqslant x,\cdots,\xi_n\leqslant x)\\&=P(\xi_1\leqslant x)P(\xi_2\leqslant x)\cdots P(\xi_n\leqslant x)\\&=[F(x)]^n.\end{aligned}\tag{2.79}$$

(2) The distribution of ξ_1^*.

For this, we consider the complement event $\{\xi_1^*>x\}$ of $\{\xi_1^*\leqslant x\}$.

$$\begin{aligned}P(\xi_1^*>x)&=P(\xi_1>x,\xi_2>x,\cdots,\xi_n>x)\\&=P(\xi_1>x)P(\xi_2>x)\cdots P(\xi_n>x)\\&=[1-F(x)]^n.\end{aligned}\tag{2.80}$$

Hence we have

$$P(\xi_1^*\leqslant x)=1-[1-F(x)]^n.\tag{2.81}$$

(3) The joint distribution of (ξ_1^*,ξ_n^*).

$$\begin{aligned}F(x,y)&=P(\xi_1^*\leqslant x,\xi_n^*\leqslant y)\\&=P(\xi_n^*\leqslant y)-P(\xi_1^*>x,\xi_n^*\leqslant y)\\&=[F(y)]^n-P(\bigcap_{i=1}^n(x<\xi_i\leqslant y)).\end{aligned}$$

So, when $x < y$

$$F(x,y) = [F(x)]^n - [F(y) - F(x)]^n,$$

and when $x \geqslant y$

$$F(x,y) = [F(y)]^n. \tag{2.82}$$

In particular, if $(\xi_1, \cdots, \xi_n)$ is a continuous random vector with the joint density function $p(x_1, x_2, \cdots, x_n)$, then the above random variables (vectors) are still continuous, and we can obtain their density functions by taking the derivatives of corresponding distribution functions.

2.6.4 Transforms of random vectors

Suppose that $(\xi_1, \cdots, \xi_n)$ has the density function $p(x_1, \cdots, x_n)$ and $y_1 = f_1(x_1, \cdots, x_n), \cdots, y_m = f_m(x_1, \cdots, x_n)$ are m measurable functions. Let $\eta_1 = f_1(\xi_1, \cdots, \xi_n), \cdots, \eta_m = f_m(\xi_1, \cdots, \xi_n)$. Then $(\eta_1, \cdots, \eta_m)$ is a random vector. We want to compute the marginal distributions and the joint distribution as well. In a similar way to (2.73), we specify the joint distribution function as follows:

$$\begin{aligned} G(y_1, \cdots, y_m) &= P(\eta_1 \leqslant y_1, \cdots, \eta_m \leqslant y_m) \\ &= \int \cdots \int_D p(x_1, \cdots, x_n)\, \mathrm{d}x_1 \cdots \mathrm{d}x_n, \end{aligned} \tag{2.83}$$

where D is an n-dimensional domain: $\{(x_1, \cdots, x_n) : f_1(x_1, \cdots, x_n) \leqslant y_1, \cdots, f_m(x_1, \cdots, x_n) \leqslant y_m\}$.

Theorem 5 *If $m = n$, $\{f_j, j=1, \cdots, n\}$ has unique inverse functions $x_i = x_i(y_1, \cdots, y_n)$, $i = 1, \cdots, n$, and*

$$J = \frac{\partial(x_1, \cdots, x_n)}{\partial(y_1, \cdots, y_n)} \neq 0,$$

then $(\eta_1, \cdots, \eta_n)$ is continuous random vectors with density functions $q(y_1, \cdots, y_n)$ as follows:

$$q(y_1, \cdots, y_n) = p(x_1(y_1, \cdots, y_n), \cdots, x_n(y_1, \cdots, y_n))\, |J|, \tag{2.84}$$

where $(y_1, \cdots, y_n) \in$ the range domain of $(f_1, \cdots, f_n)$; otherwise, $q(y_1, \cdots, y_n) = 0$.

Proof Making a change of variables in (2.83)

$$u_1 = f_1(x_1, \cdots, x_n), \cdots, u_n = f_n(x_1, \cdots, x_n),$$

we obtain

$$G(y_1, \cdots, y_n) = \int_{-\infty}^{y_1} \cdots \int_{-\infty}^{y_n} q(u_1, \cdots, u_n)\, \mathrm{d}u_1 \cdots \mathrm{d}u_n.$$

Hence $q(y_1, \cdots, y_n)$ is the joint density of $(\eta_1, \cdots, \eta_n)$.

Example 34 Suppose that ξ and η are independent with exponential distribution of parameter 1. Calculate the joint density of $\alpha = \xi + \eta$ and $\beta = \xi / \eta$, and calculate the densities of α, β respectively.

Solution Observe first that the joint density of (ξ, η) is as follows: for $x > 0$ and $y > 0$

$$p(x,y) = e^{-(x+y)};$$

otherwise, $p(x,y)=0$. Also, it is easy to see that the functions $u = x + y$ and $v = x/y$ have inverse functions $x = uv/(1+v)$ and $y = u/(1+v)$. When $x,y > 0$, we have $u,v > 0$ and

$$J^{-1} = \frac{\partial(u,v)}{\partial(x,y)} = \begin{vmatrix} 1 & 1 \\ 1/y & -x/y^2 \end{vmatrix} = -\frac{x+y}{y^2} = -\frac{(1+v)^2}{u}.$$

Hence we have

$$|J| = \frac{u}{(1+v)^2}.$$

It follows from (2.84) that the joint density of (α,β) is

$$q(u,v) = \begin{cases} \dfrac{ue^{-u}}{(1+v)^2}, & u > 0, v > 0; \\ 0, & \text{otherwise.} \end{cases}$$

Next we calculate the marginal densities of (α,β). It is easy to know

$$p_\alpha(u) = \begin{cases} ue^{-u}, & u > 0; \\ 0, & u \leqslant 0, \end{cases}$$

and

$$p_\beta(v) = \begin{cases} \dfrac{1}{(1+v)^2}, & v > 0; \\ 0, & v \leqslant 0. \end{cases}$$

Furthermore we know that α and β are independent.

In the above example we could also calculate the distribution functions of $\xi + \eta$ and ξ/η by the approach introduced in Subsection 2.5.3, but the computation above is obviously easier.

This example tells us that (1) in order to decide whether functions of random vectors are independent, we can use the transform of random vectors to calculate their joint distribution, and then verify various necessary and sufficient conditions for independence; (2) in order to calculate the distribution of a function of random vectors we can introduce extra appropriate functions and calculate their joint distribution. It turns out that the desired distribution can be obtained as a marginal distribution.

Example 35 Suppose that X is a random variable, let $Y = 2X$. Then we can express the joint distribution function $F_{X,Y}(x,y)$ of (X,Y) by using the distribution function $F_X(x)$ of X. Indeed, if $y \geqslant 2x$, then

$$P(X \leqslant x, Y \leqslant y) = P(X \leqslant x) = F_X(x).$$

If $y < 2x$, then

$$P(X \leqslant x;\ Y \leqslant y) = P(Y \leqslant y) = P\left(X \leqslant \frac{y}{2}\right) = F_X\left(\frac{y}{2}\right).$$

So, we have

$$F_{X,Y}(x,y)=\begin{cases}F_X(x), & y\geqslant 2x;\\ F_X\left(\frac{y}{2}\right), & y<2x.\end{cases}$$

Example 36 Suppose that ξ and η are independent, and ξ follows a normal distribution $N(0,1)$, and η is uniform over $(0,\pi)$. Calculate the density function of $\alpha=\xi+a\cos\eta$, where a is a numerical constant.

Solution Observe that the joint density of (ξ,η) can be written as

$$p(x,y)=\begin{cases}\dfrac{1}{\pi\sqrt{2\pi}\sigma}\mathrm{e}^{-x^2/2\sigma^2}, & -\infty<x<\infty, 0<y<\pi;\\ 0, & \text{otherwise.}\end{cases}$$

Let $\beta=\eta$, the equations corresponding to $\alpha=\xi+a\cos\eta$, $\beta=\eta$ are

$$\begin{cases}u=x+a\cos y,\\ v=y,\end{cases}$$

which has a unique solution

$$\begin{cases}x=u-a\cos v,\\ y=v,\end{cases}$$

and $J=1$. Thus we obtain $q(u,v)=p(u-a\cos v, v)$. As a consequence, α has the density function

$$\begin{aligned}p_\alpha(u)&=\int_{-\infty}^{+\infty}p(u-a\cos v, v)\,\mathrm{d}v\\ &=\frac{1}{\pi\sqrt{2\pi}\sigma}\int_0^\pi \mathrm{e}^{-\frac{(u-a\cos v)^2}{2\sigma^2}}\,\mathrm{d}v.\end{aligned}$$

Example 37 Let $\boldsymbol{\xi}=(\xi_1,\cdots,\xi_n)'\sim N(\mu,\Sigma)$ and $\boldsymbol{\eta}=(\eta_1,\cdots,\eta_n)'=\boldsymbol{C}\boldsymbol{\xi}+\boldsymbol{a}$, where $\boldsymbol{C}$ is an $n\times n$ reversible matrix. Find the distribution of $\boldsymbol{\eta}$.

Solution The density function of $\boldsymbol{\xi}$ is as

$$p_\xi(x)=\frac{1}{(2\pi)^{\frac{n}{2}}|\Sigma|^{\frac{1}{2}}}\exp\left\{-\frac{1}{2}(x-u)'\Sigma^{-1}(x-u)\right\}.$$

Let $\boldsymbol{y}=\boldsymbol{C}\boldsymbol{x}+\boldsymbol{a}$, then $\boldsymbol{x}=\boldsymbol{C}^{-1}(\boldsymbol{y}-\boldsymbol{a})$. Hence, the density function of $\boldsymbol{\eta}$ is obtained as

$$\begin{aligned}p_\eta(y)&=p_\xi(C^{-1}(y-a))\,|C^{-1}|\\ &=\frac{1}{(2\pi)^{\frac{n}{2}}|\Sigma|^{\frac{1}{2}}|C|}\exp\left\{-\frac{1}{2}(C^{-1}(y-a)-\mu)'\Sigma^{-1}(C^{-1}(y-a)-\mu)\right\}\\ &=\frac{1}{(2\pi)^{\frac{n}{2}}|(C\Sigma C'|^{\frac{1}{2}}}\exp\left\{-\frac{1}{2}(y-a-Cu)'(C^{-1})'\Sigma^{-1}(y-a-C\mu)\right\}\\ &=\frac{1}{(2\pi)^{\frac{n}{2}}|(C\Sigma C'|^{\frac{1}{2}}}\exp\left\{-\frac{1}{2}(y-C\mu-a)'(C\Sigma C')^{-1}(y-C\mu-a)\right\}.\end{aligned}$$

Therefore, $\eta=C\xi+a\sim N(C\mu+a, C\Sigma C')$.

This example shows that, a reversible transform of a multi-normal random vector is also a multi-normal random vector. In particular, if $\xi \sim N(\mu, \Sigma)$, then $\eta = (\Sigma^{1/2})^{-1}(\xi - \mu) \sim N(0, I)$, i. e., $\eta_1, \cdots, \eta_n$ are independent one dimensional standard normal random variables, where $\Sigma^{1/2} = L$ is a positive definite symmetric matrix such that $\Sigma = LL$.

Example 38 Suppose that ξ and η are independent, identically distributed with a common normal distribution $N(0,1)$. Let $\xi = \rho\cos\varphi, \eta = \rho\sin\varphi$. Prove that $\rho = \rho(\xi,\eta)$ and $\varphi = \varphi(\xi,\eta)$ are independent.

Proof First, we use (2.84) to calculate the joint density of (ρ,φ). Let

$$\begin{cases} r = \rho(x,y), \\ \theta = \varphi(x,y), \end{cases} \qquad \begin{cases} x = r\cos\theta, \\ y = r\sin\theta. \end{cases}$$

Note that there is a one to one correspondence between $\{-\infty < x < \infty, -\infty < y < \infty, (x,y) \neq (0,0)\}$ and $\{r > 0, 0 \leqslant \theta < 2\pi\}$ and $J = r$. Also, by the hypothesis the joint density of (ξ,η) is

$$p(x,y) = \frac{1}{2\pi}\exp\left(-\frac{x^2+y^2}{2}\right).$$

Consequently, the joint density of (ρ,φ) is

$$q(r,\theta) = \begin{cases} \frac{1}{2\pi} r e^{-\frac{r^2}{2}}, & r > 0, 0 \leqslant \theta < 2\pi; \\ 0, & \text{otherwise}. \end{cases}$$

Next we show ρ and φ are independent. It is not hard to see that $q(r,\theta)$ can be written as a product of $R(r)$ and $\Theta(\theta)$, where

$$R(r) = \begin{cases} r e^{-\frac{r^2}{2}}, & r > 0; \\ 0, & \text{otherwise}, \end{cases}$$

$$\Theta(\theta) = \begin{cases} \frac{1}{2\pi}, & 0 \leqslant \theta < 2\pi; \\ 0, & \text{otherwise}. \end{cases}$$

Since $R(r)$, $\Theta(\theta)$ are density functions of ρ, φ respectively, ρ, φ are independent. We remark that ρ is often referred to as a Rayleigh distribution, and φ is a uniform distribution over $(0, 2\pi)$.

Conversely, if ξ_1 and ξ_2 are independent and uniformly distributed over $(0,1)$, let

$$\eta_1 = (-2\ln\xi_1)^{1/2}\cos(2\pi\xi_2), \quad \eta_2 = (-2\ln\xi_1)^{1/2}\sin(2\pi\xi_2).$$

Then η_1 and η_2 are independent and each follows a normal distribution $N(0,1)$. This is one of basic approaches to generating normal random numbers.

From the exercises in this chapter one sees that functions of ξ and η are likely dependent although ξ and η themselves are independent. However, in some special cases, the independence property is preserved.

Example 39 Suppose that X and Y are independent random variables. Assume that the random variable Z depends only on X, and W on Y, that is, $Z = g(X), W = h(Y)$ for g, h. If $g(x) = z$, $h(y) = w$ have a unique solution and the derivatives of g and h exist, then Z and W are independent.

Proof Define for any real numbers z and w

$$A = \{x : g(x) \leqslant z\}, \quad B = \{y : h(y) \leqslant w\}.$$

Since g and h are Borel measurable, then so are A and B.

Also it follows from the independence of X and Y that

$$P(X \in A, Y \in B) = P(X \in A)P(Y \in B).$$

Thus we have

$$\begin{aligned} P(Z \leqslant z, W \leqslant w) &= P(g(x) \leqslant z, h(Y) \leqslant w) \\ &= P(X \in A, Y \in B) = P(X \in A)P(Y \in B) \\ &= P(Z \leqslant x)P(W \leqslant w). \end{aligned}$$

Hence Z and W are independent of each other.

More generally, we have the following theorem.

Theorem 6 *Let* $1 \leqslant n_1 < n_2 < \cdots < n_k \leqslant n$. *Assume that* f_1 *is a Borel function of* n_1 *arguments*, $\cdots$, f_k *a Borel function of* $n_k - n_{k-1}$ *arguments. If* $X_1, \cdots, X_n$ *are independent, then so are* $f_1(X_1, \cdots, X_{n_1}), f_2(X_{n_1+1}, \cdots, X_{n_2}), \cdots, f_k(X_{n_{k-1}+1}, \cdots, X_{n_k})$. In particular, when $f_1, \cdots, f_k$ are functions of a single argument, $f_1(X_1), \cdots, f_k(X_k)$ are independent.

We remark that the converse statement is not valid. In fact, there exists such an example that ξ^2 and η^2 are independent but ξ and η are not.

2.6.5 Important distributions in statistics

In the present subsection we shall introduce three important distributions widely used in mathematical statistics, χ^2, t and F distributions. They have close relationship with normal distributions so that their densities can be derived as functions of normal random variables.

Let us start with a wider class of distributions than χ^2 distributions— the Gamma distribution, whose density is given by Formula (2.32) in Section 2.2. The following is an important property.

Lemma 1 *(Additivity of Gamma distribution) The Gamma distribution* $\Gamma(\lambda, r)$ *possesses the additivity property for its second parameter* r. *That is, if* ξ_1 *and* ξ_2 *are independent, and* $\xi_1 \sim \Gamma(\lambda, r_1), \xi_2 \sim \Gamma(\lambda, r_2)$, *then* $\xi_1 + \xi_2 \sim \Gamma(\lambda, r_1 + r_2)$.

Proof We shall compute the density of $\eta = \xi_1 + \xi_2$ using the convolution formula (2.76) as follows. When $z \leqslant 0$, $p_\eta(z) = 0$; when $z > 0$,

$$p_\eta(z) = \int_0^z \frac{\lambda^{r_1}}{\Gamma(r_1)} x^{r_1 - 1} \mathrm{e}^{-\lambda x} \frac{\lambda^{r_2}}{\Gamma(r_2)} (z - x)^{r_2 - 1} \mathrm{e}^{-\lambda(z-x)} \, \mathrm{d}x.$$

Substituting in $x = zt$ and making use of the second type of Euler integral

$$\begin{aligned} B(r_1, r_2) &=: \int_0^1 t^{r_1 - 1}(1 - t)^{r_2 - 1} \mathrm{d}t \\ &= \frac{\Gamma(r_1)\Gamma(r_2)}{\Gamma(r_1 + r_2)}, \end{aligned}$$

we have

$$p_\eta(z) = \frac{\lambda^{r_1+r_2}}{\Gamma(r_1+r_2)} z^{r_1+r_2-1} e^{-\lambda z}. \tag{2.85}$$

Therefore, $\eta \sim \Gamma(\lambda, r_1+r_2)$.

When $r=1$, $\Gamma(\lambda,1)$ is the exponential distribution. Another special case is $\lambda = 1/2$ and $r=n/2$ (n a positive integer).

Ⅰ The χ^2 distribution

Call $\Gamma(1/2, n/2)$ a χ_n^2 distribution, where n is called its degree of freedom. The density function is

$$p(x) = \begin{cases} \dfrac{(1/2)^{n/2}}{\Gamma(n/2)} x^{n/2-1} e^{-x/2}, & x>0; \\ 0, & x \leqslant 0. \end{cases} \tag{2.86}$$

Theorem 7 (1) *The χ_n^2 distribution possesses the additivity property. That is, if $\xi_1 \sim \chi_{n_1}^2$, $\xi_2 \sim \chi_{n_2}^2$, and ξ_1 and ξ_2 are independent, then $\xi_1+\xi_2 \sim \chi_{n_1+n_2}^2$.*

(2) *Suppose that $\xi_1, \cdots, \xi_n$ are independent standard normal random variables, then*

$$\eta = \xi_1^2 + \cdots + \xi_n^2 \sim \chi_n^2. \tag{2.87}$$

Proof (1) follows from the additivity property of Gamma distributions.

Prove (2). Let $\eta_i = \xi_i^2, i=1,\cdots,n$. Since the $\{\xi_i\}$ are independent, so are the $\{\eta_i\}$. Also, we have calculated the density of η_i in Example 24. Since $\Gamma(1/2) = \sqrt{\pi}$, $\eta_i \sim \chi_1^2$. By (1), and by an induction argument, we have

$$\sum_{i=1}^n \xi_i^2 = \sum_{i=1}^n \eta_i \sim \chi_n^2.$$

The above theorem displays the essence of χ^2 distribution, and the freedom degree n of χ^2 distribution is just the number of independent normal variables in the sum $\sum_{i=1}^n \xi_i^2$.

Ⅱ The t-distribution

Theorem 8 *If ξ and η are independent, and $\xi \sim N(0,1)$, $\eta \sim \chi_n^2$, then the random variable $T = \dfrac{\xi}{\sqrt{\eta/n}}$ has the density*

$$p(x) = \frac{\Gamma[(n+1)/2]}{\sqrt{n\pi}\,\Gamma(n/2)} (1+x^2/n)^{-(n+1)/2}, \quad -\infty < x < \infty. \tag{2.88}$$

We call the distribution of the random variable T above a $t(n)$ distribution with n as its degree of freedom.

To prove Theorem 8, we can use Theorem 3 to calculate the density of $\theta = \sqrt{\eta/n}$, and then use equation (2.78) for quotient to calculate the density of $T = \xi/\theta$. The details refer to Supplements and Remarks 10 at the end of this chapter.

Ⅲ The F-distribution

Theorem 9 *Suppose that ξ and η are independent, and $\xi \sim \chi_m^2$, $\eta \sim \chi_n^2$, then the* *random variable* $F = \dfrac{\xi/m}{\eta/n}$ *has the density*

$$p(x) = \begin{cases} \dfrac{\Gamma[(m+n)/2]}{\Gamma(m/2)\Gamma(n/2)} m^{m/2} n^{n/2} \dfrac{x^{m/2-1}}{(mx+n)^{(m+n)/2}}, & x > 0; \\ 0, & x < 0. \end{cases} \tag{2.89}$$

We call the distribution of the random variable F above an $F(m,n)$ distribution with m and n as its first and second degrees of freedom respectively.

In order to prove Theorem 9, we can first use Theorem 3 to calculate the densities of $\xi_1 = \xi/m$ and $\eta_1 = \eta/n$, and then use equation (2.78) for quotient to calculate the density of $F=\xi_1/\eta_1$. See the Supplements and Remarks 10 at the end of this chapter.

The F-distribution possesses the following properties:

(1) If $F \sim F(m,n)$, then $1/F \sim F(n,m)$. It immediately follows from the definition of F.

(2) If $T \sim t(n)$, then $T^2 \sim F(1,n)$.

Proof Write by Theorem 8, $T=\xi/\sqrt{\eta/n}$, where ξ and η are independent and $\xi \sim N(0,1)$, $\eta \sim \chi_n^2$. Note that $T^2 = \xi^2/(\eta/n)$. Also, $\xi^2 \sim \chi_1^2$ and ξ^2, η are independent. Hence $T^2 \sim F(1,n)$.

Supplements and Remarks

1. Before the middle of the 19-th century, the great interest of probability theory was still concentrated on the computation of the probabilites of random events. Russian mathematicians, Chebyshev (1821—1894), Markov (1856—1922), Lyapunov (1857—1918), and others first clearly introduced the concept of the random variable and widely used and studied it.

2. The proof of de Moivre-Laplace Theorem

We first show that

$$p_n(x) = P(\xi_n = j) \sim \frac{1}{\sqrt{2\pi npq}} e^{-x^2/2} \text{ uniformly in } x \in [a,b].$$

Let $k=n-j$. Then

$$j = np + x\sqrt{npq} \to \infty, k = nq - x\sqrt{npq} \to \infty.$$

Note

$$p_n(x) = \frac{n!}{j!k!} p^j q^k.$$

By the Stirling formula

$$m! = \sqrt{2\pi m} m^m e^{-m} e^{\theta_m}, \ 0 < \theta_m < \frac{1}{12m},$$

we obtain

$$p_n(x) = \frac{\sqrt{2\pi n} n^n e^{-n}}{\sqrt{2\pi j} j^j e^{-j} \sqrt{2\pi k}\, k^k e^{-k}} p^j q^k e^{\theta_n - \theta_j - \theta_k}$$

$$= \frac{1}{\sqrt{2\pi}} \sqrt{\frac{n}{jk}} \left(\frac{np}{j}\right)^j \left(\frac{nq}{k}\right)^k e^{\theta},$$

where

$$|\theta| < \frac{1}{12}\left(\frac{1}{n} + \frac{1}{j} + \frac{1}{k}\right).$$

Note that

$$\frac{jk}{n} = n\left(p + x\frac{pq}{n}\right)\left(q - x\frac{pq}{n}\right)$$

$$= npq\left(1 + x(q-p)\sqrt{\frac{1}{npq}} - x^2\frac{pq}{n}\right) \sim npq,$$

holds uniformly in $x \in [a,b]$, and $\frac{j}{np} = 1 + x\sqrt{\frac{q}{np}}$, $\frac{k}{nq} = 1 - x\sqrt{\frac{p}{nq}}$. We have

$$\log\left(\frac{np}{j}\right)^j \left(\frac{nq}{k}\right)^k = -j\log\frac{j}{np} - k\log\frac{k}{nq}$$

$$= -(np + x\sqrt{npq})\left[x\sqrt{\frac{q}{np}} - \frac{1}{2}\frac{qx^2}{np} + O\left(\left(\frac{q}{np}\right)^{3/2}\right)\right]$$

$$-(np - x\sqrt{npq})\left[-x\sqrt{\frac{q}{np}} - \frac{1}{2}\frac{px^2}{np} + O\left(\left(\frac{p}{np}\right)^{3/2}\right)\right]$$

$$= \frac{x^2}{2} + O\left(\frac{1}{\sqrt{npq}}\right) \text{ uniformly in } x \in [a,b].$$

It follows that

$$p_n(x) \sim \frac{1}{\sqrt{2\pi npq}}\left(\frac{np}{j}\right)^j \left(\frac{nq}{k}\right)^k \sim \frac{1}{\sqrt{2\pi npq}} e^{-x^2/2} \text{ uniformly in } x \in [a,b].$$

Next, we show that

$$P\left(a \leqslant \frac{\xi_n - np}{\sqrt{npq}} \leqslant b\right) \to \frac{1}{\sqrt{2\pi}}\int_a^b e^{-x^2/2}\,dx.$$

Denote $x_{nj} = \frac{1}{\sqrt{npq}}$ and $N_n = \{j : x_{nj} \in [a,b]\}$. Then $\# N_n \sim (b-a)\sqrt{npq}$. By the result we have shown,

$$p_n(x_{nj}) \sim \frac{1}{\sqrt{2\pi npq}} e^{-x_{nj}^2/2} \text{ uniformly in } j \in N_n.$$

Hence

$$P\left(a \leqslant \frac{\xi_n - np}{\sqrt{npq}} \leqslant b\right) = \sum_{j \in N_n} p_n(x_{nj})$$

$$= \frac{1}{\sqrt{2\pi}} \sum_{j \in N_n} \frac{1}{\sqrt{npq}} e^{-x_{nj}^2/2}(1 + o(1))$$

$$= \frac{1}{\sqrt{2\pi}} \sum_{j \in N_n} e^{-x_{nj}^2/2}(x_{n,j} - x_{n,j} - 1) + o(1)$$

$$\to \frac{1}{\sqrt{2\pi}}\int_a^b e^{-x^2/2}\,dx.$$

The proof is therefore completed.

3. Why do random variables, like the number of telephone calls, the number of passengers, the number of radioactive particles, obey the Poisson law? What these random variables represent are the numbers of events occurring in a given interval of time or space. Therefore, they have the following common features: stationarity, independent increments and rarity. To illustrate, consider the number ξ of calls received by a phone operator in a given interval of time. It has

(1) stationarity, which means the probability of $\xi = k$ in $[t_0, t_0 + t)$ depends only on t but does not depend on the starting point t_0. Write simply $P_k(t)$ for the probability. Thus we can regard the numbers of calls in distinct intervals as repeated experiments when the interval length t are identical.

(2) independent increments (memoryless property), indicating that the event $\xi = k$ in $[t_0, t_0 + t)$ is independent of all events before t_0. Thus the numbers of calls in disjoint intervals are independent.

(3) rarity, indicating that there is at most one call in a small enough interval of time. More precisely, as $t \to 0$,

$$1 - P_0(t) - P_1(t) = o(t). \tag{2.90}$$

Now we compute $P_k(t)$. Divide the interval $[t_0, t_0 + 1)$ into n equal subintervals, and then $[t_0, t_0 + t)$ is divided into nt subintervals. Take n to be large enough and think of nt as an integer such that, roughly speaking, there is at most one call in each small interval $[t_0 + \frac{r-1}{n}, t_0 + \frac{r}{n})$, where $\Delta t = \frac{1}{n}$, $r = 1, 2, \cdots, nt$. In this way, whether there is one call or not in a small interval is viewed as a Bernoulli trial. Taking into account the properties (1) and (2), the study of the number of calls in $[t_0, t_0 + t)$ is equivalent to that of nt repeated Bernoulli trials. Hence the event $\{\xi = k\}$ approximately obeys the binomial distribution:

$$P(\xi = k) \approx b(k; nt, p_n), \tag{2.91}$$

where p_n is the probability that there is one call in a small interval, which is proportional to the length Δt of interval, that is $p_n = \lambda \Delta t$.

Formula (2.91) is approximately equal. The reason is that we ignore an infinite small amount. When $\Delta t \to 0$ $(n \to \infty)$, (2.91) becomes an identity. It follows from the Poisson theorem that

$$P(\xi = k) = \lim_{n \to \infty} b(k; nt, p_n) = \frac{(\lambda t)^k}{k!} e^{-\lambda t},$$

where $k = 0, 1, 2, \cdots$. This shows ξ obeys the Poisson distribution with parameter λt.

If we write $\xi(t)$ for ξ, that is, define a random variable $\xi(t)$ for $t \geqslant 0$, then we obtain a family of random variables $\{\xi(t); t \geqslant 0\}$, called a stochastic process. In the theory of stochastic processes, we shall study $\{\xi(t); t \geqslant 0\}$ in a precise way and generalize the above results.

4. We turn to a brief description of the Gamma and the exponential distributions. Denote by $\xi(t)$ the number of customers served within t. According to what we explained in the paragraph above, $\xi(t)$ obeys a Poisson distribution with parameter λt. Let τ_r be the time the r-th customer is served, and let $F(t)$ be its distribution function.

When $t \leqslant 0$, $F(t) = 0$. When $t > 0$, since $\{\tau_r \leqslant t\} = \{\xi(t) \geqslant r\}$, then

$$\begin{aligned} F(t) &= P(\tau_r \leqslant t) = P(\xi(t) \geqslant r) \\ &= 1 - P(\xi(t) < r) = 1 - \sum_{k=0}^{r-1} \frac{(\lambda t)^k}{k!} e^{-\lambda t}, \end{aligned}$$

from which we derive the density function

$$\begin{aligned} p(t) = F'(t) &= -\sum_{k=0}^{r-1} \frac{k(\lambda t)^{k-1} \lambda e^{-\lambda t}}{k!} + \sum_{k=0}^{r-1} \frac{(\lambda t)^k \lambda e^{-\lambda t}}{k!} \\ &= \frac{(\lambda t)^{r-1} \lambda e^{-\lambda t}}{(r-1)!} = \frac{\lambda^r}{\Gamma(r)} t^{r-1} e^{-\lambda t}, \qquad t > 0. \end{aligned}$$

This is just the density function of the Gamma distribution. It shows that the time of serving the r-th customer obeys a Gamma distribution. Analogously, consider the replacement of machine parts, and then the time τ_r of replacing the r-th same part obeys a Gamma distribution. In particular, the time τ (the life time) of replacing the first part obeys an exponential distribution. Sometimes we call an exponential distribution the life time distribution.

5. Probability density functions of discrete random variables

Define a pulse function, that is, a distribution function $\delta(x)$ which satisfies the following integration relations: for any functions φ is continuous at 0,

$$\int_{-\infty}^{+\infty} \varphi(x)\delta(x)\,dx = \varphi(0).$$

Applying translation transformation, it is easy to see that

$$\int_{-\infty}^{+\infty} \varphi(x)\delta(x - x_0)\,dx = \varphi(x_0).$$

Given a function $F(x)$, if x_0 is a point of discontinuity, letting $k = F(x_{0+}) - F(x_{0-})$, then the value of $k\delta(x - x_0)$ at x_0 can be regarded as the derivative of $F(x)$ at x_0. Assume that X is a discrete random variable with a distribution sequence as follows:

$$\begin{pmatrix} x_1 & x_2 & \cdots & x_n & \cdots \\ p_1 & p_2 & \cdots & p_n & \cdots \end{pmatrix}, \tag{2.92}$$

then its density function can be written as

$$p(x) = \sum_i p_i \delta(x - x_i),$$

that is

$$\frac{dF(x)}{dx} = [F(x_i) - F(x_{i-})]\delta(x - x_i).$$

Likewise, we can use $p(x)$ to express probabilities of events, for instance,

$$P(x_1 < X \leqslant x_2) = \int_{x_1+}^{x_2+} p(x)\,dx,$$

and

$$P(x_1 \leqslant X \leqslant x_2) = \int_{x_1-}^{x_2+} p(x)\,dx.$$

Example 40 If X is degenerate and takes only a certain constant c, then its density function is $p(x) = \delta(x-c)$. If X is a two-point random variable and takes 0 and 1 with probabilities $1-p$ and p respectively, then its density function is:

$$p(x) = (1-p)\delta(x) + p\delta(x-1).$$

6. Markov chains

A Markov process $\{X_t\}$ is a stochastic process with the property that, given the value of X_t, the values of X_s, for $s > t$ are not influenced by the values of X_u for $u < t$. In words, the probability of any particular future behavior of the process, when its current state is known exactly, is not altered by additional knowledge concerning its past behavior. A discrete Markov chain is a Markov process whose state space (all possible values) is a finite or countable set, and whose (time) index set is $T = (0,1,2,\cdots)$. In formal terms, the Markov property is that

$$\begin{aligned} &P(X_{n+1} = j \mid X_0 = i_0, \cdots, X_{n-1} = i_{n-1}, X_n = i) \\ &= P(X_{n+1} = j \mid X_n = i) \end{aligned} \tag{2.93}$$

for all time points n and all states $i_0, \cdots, i_{n-1}, i, j$.

The probability of X_{n+1} being in state j given that X_n is in state i is called the one step transition probability and is denoted by $p_{ij}^{n,n+1}$. That is,

$$p_{ij}^{n,n+1} = P(X_{n+1} = j \mid X_n = i).$$

When the one step transition probabilities are independent of the time variable n, we say that the Markov chain has stationary transition probabilities. Then $p_{ij}^{n,n+1} = p_{ij}$ is the conditional probability that the state value undergoes a transition from i to j in one trial. It is customary to arrange these numbers p_{ij} in a matrix (p_{ij}), and refer to $\boldsymbol{P} = (p_{ij})$ the transition probability matrix. Clearly the quantities p_{ij} satisfy the conditions

$$p_{ij} \geqslant 0, \quad \text{for } i,j = 0,1,2,\cdots, \tag{2.94}$$

$$\sum_{j=0}^{\infty} p_{ij} = 1, \quad \text{for } i = 0,1,2,\cdots. \tag{2.95}$$

The condition (2.95) merely expresses the fact some transition occurs at each trial.

A Markov chain is completely defined once its transition probability matrix and initial state X_0 are specified. In fact, by the definition of conditional probabilities and (2.93) we obtain

$$P(X_0 = i_0, X_1 = i_1, \cdots, X_n = i_n) = p_{i_0} p_{i_0 i_1} \cdots p_{i_{n-1} i_n}. \tag{2.96}$$

The importance of Markov chains lies in that a large number of natural physical, biological, and economic phenomena can be described by them and it's enhanced by the amenability of Markov chains to quantitative manipulation.

Example 41 (The Ehrenfest urn model) A classical mathematical description of diffusion through a membrane is the famous Ehrenfest urn model. Imagine two containers containing a total of $2a$ balls (molecules). Suppose the first container, labeled A, holds k balls and the second container, B, holds the remaining $2a-k$ balls. A ball is selected at random (all selections are equally likely) from the totality of the $2a$ balls and moved to the other container. (A molecule diffuses at random through the membrane.) Each selection generates a transition of the process. Clearly the balls fluctuate between the two containers with average drift from the urn with the excess numbers to the one with the smaller concentration.

Let Y_n be the number of balls in Urn A at the n-th stage and define $X_n = Y_n - a$. Then $\{X_n\}$ is a Markov chain on the states $i = -a, -a+1, \cdots, -1, 0, +1, \cdots, a$ with transition probabilities

$$p_{ij} = \begin{cases} \dfrac{a-i}{2a}, & \text{if } j = i+1; \\ \dfrac{a+i}{2a}, & \text{if } j = i-1; \\ 0, & \text{otherwise.} \end{cases}$$

An important quantity in the Ehrenfest urn model is the long time equilibrium distribution of the number of balls in each urn.

7. The joint probability density functions of mixed random vectors

Assume that X is a discrete random variable with the distribution sequence, and Y is a continuous random variable with the density $p_Y(y)$. In turn, the joint distribution of (X,Y) is completely dependent on a family of lines $x=x_k$. The probability that (X,Y) lies in the line segment $\{x_k\}\times(y, y+\mathrm{d}y]$ is

$$P(X = x_k, y < Y \leqslant y + \mathrm{d}y).$$

So, the joint distribution function $F(x,y)$ is continuous in y, but discontinuous in x, where the discontinuity points are $x = x_k$, $k = 1, 2, \cdots$. If X, Y are independent, then applying the previous expression concerning density functions for discrete random variables, the joint probability density function of (X,Y) can be written as

$$p(x,y) = \sum_{k=1}^{\infty} p_k \delta(x - x_k) p_Y(y),$$

that is,

$$p(x,y) = \begin{cases} p_k p_Y(y), & x = x_k, k = 1, 2, \cdots; \\ 0, & \text{otherwise.} \end{cases}$$

Example 42 Assume that z is a uniform random variable over $(-\pi, \pi)$. Let $x = \cos z$, $y = \sin z$, and then the joint distribution function of (x,y) is completely concentrated in $x^2 + y^2 = 1$. As for its probability density function, letting $w = 1$, then w and z are independent and the joint density function is:

$$p(w,z) = \delta(w-1)p(z) = \begin{cases} \frac{1}{2\pi}, & w=1, -\pi < z < \pi; \\ 0, & \text{otherwise.} \end{cases}$$

Furthermore,

$$\begin{cases} x = w\cos z, \\ y = w\sin z \end{cases}$$

has a unique solution and $|J| = 1$. Hence, the joint density function of (X,Y) is

$$p(x,y) = \begin{cases} \frac{1}{2\pi}, & x^2+y^2=1; \\ 0, & \text{otherwise.} \end{cases}$$

8. Existence theorems

Although many commonly used random variables and distribution functions have their own practical applications, having something to do with real life experiments (observations), we usually assume, for convenience of theoretical study, that these random variables obey some specific kind of distribution or have particular density functions. In fact, the following existence theorem shows that once given a distribution function $G(x)$, we can always construct a proper experiment and a random variable such that it has G as its distribution function.

Theorem 10 *Assume that $G(x)$ is a distribution function, then there exists a probability space $(\Omega, \mathscr{F}, P)$ and a random variable X such that the distribution function $F_X(x)$ of X under P is just equal to $G(x)$.*

Proof We first regard all real numbers as outcomes of an experiment, that is, $\Omega = \mathbf{R}$, and define its σ-field $\mathscr{F}$ to be a Borel field, $\mathscr{B}$. Next we define the probability P as follows: for any x,

$$P((-\infty, x]) = G(x).$$

Thus it follows from measure theory that the probabilities of all events in $\mathscr{F}$ are well defined. Finally, we define a random variable X as follows:

$$X(\omega) = \omega, \ \omega \in \Omega.$$

This is again well defined since ω is a real number. Furthermore, we have for any x,

$$F_X(x) = P(X(\omega) \leqslant x) = P((-\infty, x]) = G(x).$$

9. Compound distributions

We have already studied some commonly used distribution functions, but sometimes people need to think of other types of distribution functions or combinations of those typical distribution functions. For instance, in the theory and application of insurance actuarial processes, the following risk model is often investigated. Let N be the number of claims in a given interval of time, X_i the amount of the i-th claim, and then the total claim amount S_N

during this interval is equal to $\sum_{i=1}^{N} X_i$. The defining feature of this model is that the number N of claims is a random variable, so the distribution of S_N is combination of the X_i and N. For simplicity, we make the following hypotheses:

(1) $X_n, n \geqslant 1$ are identically distributed with a common distribution function $F(x)$;

(2) N and $\{X_n, n \geqslant 1\}$ are independent.

In the above setting, the distribution $F_{S_N}(x)$ of S_N is given by the total probability formula

$$F_{S_N}(x) = P(S_N \leqslant x) = \sum_{n=0}^{\infty} P(S_N \leqslant x \mid N = n) P(N = n).$$

When N obeys a Poisson distribution, we call $F_{S_N}(x)$ a compound Poisson distribution.

Example 43 Assume that N is geometrically distributed as follows:

$$P(N = n) = pq^n, \ n = 0, 1, 2, \cdots,$$

where $0 < q < 1$, $p = 1 - q$. Assume that X_i is exponentially distributed with the distribution function $F_{X_i}(x) = 1 - \mathrm{e}^{-x}$, $x > 0$. Calculate $F_{S_N}(x)$.

Solution It easily follows that

$$P(S_N = 0) = P(N = 0) = p.$$

Then for any $x > 0$,

$$\begin{aligned} P(S_N \leqslant x) &= \sum_{n=0}^{\infty} P(S_N \leqslant x \mid N = n) P(N = n) \\ &= \sum_{n=0}^{\infty} pq^n P(S_N \leqslant x) \\ &= \sum_{n=0}^{\infty} pq^n \int_0^x \frac{1}{(n-1)!} z^{n-1} \mathrm{e}^{-z} \mathrm{d}z \\ &= pq \int_0^x \mathrm{e}^{-pz} \mathrm{d}z. \end{aligned}$$

This implies that S_N is a mixed type of distribution: S_N takes 0 with probability p and is exponentially distributed on $(0, \infty)$ with parameter p with probability q.

10. Densities of t-distribution and F-distribution

(1) The $t(n)$-distribution

Assume that $\xi \sim N(0, 1)$ and $\eta \sim \chi_n^2$, and that ξ and η are independent. We want to calculate the density function of $t = \xi / \sqrt{\eta / n}$.

Let us first compute the density function of $\theta = \sqrt{\eta / n}$. Note that $y = \sqrt{x / n}$ is strictly increasing in $x > 0$, whose inverse $x = ny^2$ has a continuous derivative. For $y > 0$,

$$\begin{aligned} p_\theta(x) &= p_\eta[x(y)] \mid x'(y) \mid \\ &= \frac{(1/2)^{n/2}}{\Gamma(n/2)} (ny^2)^{\frac{n}{2}-1} \mathrm{e}^{-\frac{ny^2}{2}} 2ny \\ &= \frac{2(n/2)^{n/2}}{\Gamma(n/2)} y^{n-1} \mathrm{e}^{-\frac{ny^2}{2}}. \end{aligned}$$

Trivially, $p_\theta(y)=0$ when $y\leqslant 0$.

Second, calculate the density function of $t=\xi/\theta$. Since ξ and η are independent, so are ξ and θ. By the quotient density formula, it follows that

$$\begin{aligned}p_t(z)&=\int_{-\infty}^{+\infty}p(zy,y)\mid y\mid \mathrm{d}y\\&=\int_{-\infty}^{+\infty}p_\xi(zy)p_\theta(y)\mid y\mid \mathrm{d}y\\&=\int_0^{+\infty}\frac{1}{\sqrt{2\pi}}\mathrm{e}^{-\frac{(zy)^2}{2}}\frac{2(n/2)^{n/2}}{\Gamma(n/2)}y^{n-1}\mathrm{e}^{-\frac{ny^2}{2}}\mathrm{d}y\\&=\frac{1}{\sqrt{\pi}}\frac{(n/2)^{n/2}}{\Gamma(n/2)}\frac{2^{n/2}}{(n+z^2)^{(n+1)/2}}\int_0^{+\infty}u^{(n+1)/2-1}\mathrm{e}^{-u}\mathrm{d}u\\&=\frac{\Gamma((n+1)/2)}{\sqrt{n\pi}\Gamma(n/2)}\left(1+\frac{z^2}{n}\right)^{-(n+1)/2},\end{aligned}$$

where $-\infty<z<\infty$.

(2) The $F(m,n)$-distribution

Assume that $\xi\sim\chi_m^2$, $\eta\sim\chi_n^2$, and ξ and η are independent. We want to calculate the density function of $F=\dfrac{\xi/m}{\eta/n}$.

Let us first compute the density function of $\xi_1=\xi/m$. Let $y=x/m$, and then $x=my$. So, the density of ξ_1 is:

$$\begin{aligned}p_1(y)=p_\xi(my)(my)'&=\frac{(1/2)^{m/2}}{\Gamma(m/2)}(my)^{\frac{m}{2}-1}\mathrm{e}^{-\frac{my}{2}}m\\&=\frac{(m/2)^{m/2}}{\Gamma(m/2)}y^{\frac{m}{2}-1}\mathrm{e}^{-\frac{my}{2}},\qquad y>0.\end{aligned}$$

Similarly, $\eta_1=\eta/n$ has density

$$p_2(x)=\frac{(n/2)^{n/2}}{\Gamma(n/2)}x^{\frac{n}{2}-1}\mathrm{e}^{-\frac{nx}{2}},\qquad x>0.$$

Hence the density of $F=\dfrac{\xi_1}{\eta_1}$ is

$$\begin{aligned}p_F(z)&=\int_{-\infty}^{+\infty}p_1(zx)p_2(x)\mid x\mid \mathrm{d}x\\&=\frac{(m/2)^{m/2}}{\Gamma(m/2)}\frac{(n/2)^{n/2}}{\Gamma(n/2)}z^{\frac{m}{2}-1}\int_0^{+\infty}x^{\frac{m+n}{2}-1}\mathrm{e}^{-\frac{(mz+n)x}{2}}\mathrm{d}x\\&=\frac{(m/2)^{m/2}}{\Gamma(m/2)}\frac{(n/2)^{m/2}}{\Gamma(n/2)}z^{\frac{n}{2}-1}\left(\frac{2}{mz+n}\right)^{\frac{m+n}{2}}\int_0^{+\infty}u^{\frac{m+n}{2}-1}\mathrm{e}^{-u}\mathrm{d}u\\&=\frac{\Gamma(m+n)/2}{\Gamma(m/2)}\frac{\Gamma(m+n)/2}{\Gamma(n/2)}m^{\frac{m}{2}}n^{\frac{n}{2}}\frac{z^{m/2-1}}{(mz+n)^{(m+n)/2}}\end{aligned}$$

when $z>0$; $p_F(z)=0$ when $z\leqslant 0$.

11. Simulation

To simulate is to reproduce the results of a real phenomenon by artificial means. We are interested here in phenomena that have a random element. What we really simulate are the outcome of a sample space and/or the states of a random variable in accordance with the distribution that we are assuming. Computers give a fast and convenient way of carrying out these sorts of simulations.

There are several reasons to use simulation in problems. Sometimes we do not yet have a good understanding of a phenomenon, and observing it repeatedly may give us intuition about a question related to it. In such case, analytic results of close form might be derivable, and the simulation mostly serves the purpose of suggesting what to try to prove. But many times the system under study is complex enough that analysis is either very difficult or impossible. Then simulation is the only means of getting approximate information, and the tools of probability and statistics also allow us to measure our confidence in our conclusions.

12. Calculate common probability distribution values and quantiles using Matlab

Probability distribution functions $F(x)=P(X\leqslant x)$ and quantile function $x=F^{-1}(y)$ in Matlab are (see Table 2-5):

Table 2-5

Distribution	Disrtibution Function	Quantile Function
Binomial	binocdf(x, n, p)	binoinv(y, n, p)
Poisson	poisscdf(x, lambda)	poissinv(y, lambda)
Uniform	unifcdf(x, a, b)	unifinv(y, a, b)
Exponential	expcdf(x, mu)	expinv(y, mu)
Normal	normcdf(x, mu, sigma)	norminv(y, mu, sigma)
Standard normal	normcdf (x)	norminv(y)
x^2	chi2cdf(x, n)	chi2inv(y, n)
t	tcdf(x, n)	tinv(y, n)
F	fcdf(x, m, n)	finv(y, m, n)

See the introduction in the section Matlab Help of software for many other distribution functions.

Example 44 Assume $X\sim B(10;0.2)$. Find $y=F(3)=P(x\leqslant 3)$.

(1) Open Matlab, and pull down menu File⇒New⇒M-File. See Figure 2-9.

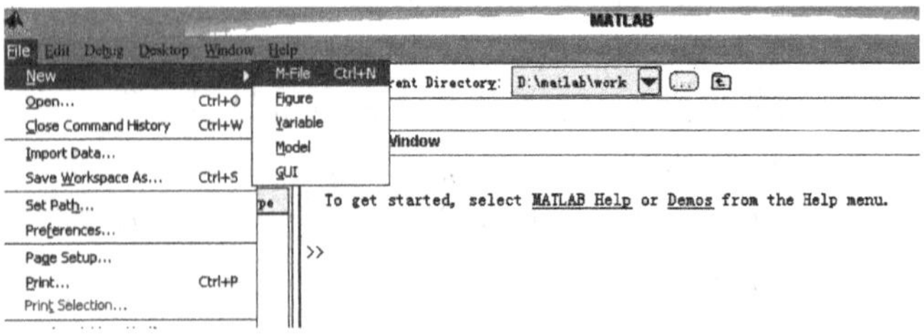

Figure 2-9

Write in programs in program editing window (Edit -Untiled). See Figure 2-10.

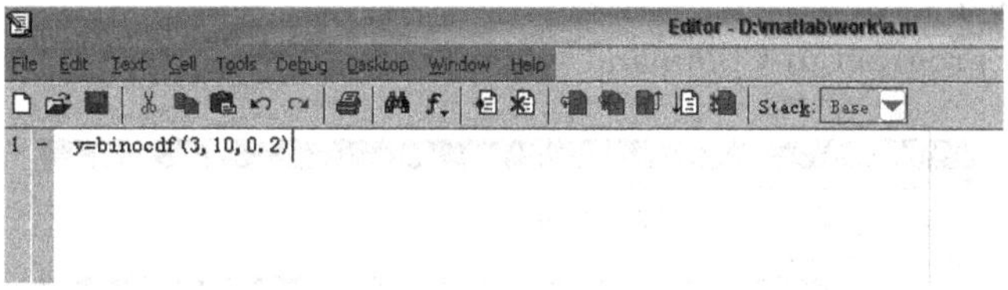

Figure 2-10

(2) Save a copy of the above programs to "work"fold of the Matlab system, and name it as, say, al.

(3) Down Menu window⇒Command window, and open Command window and input the file name a 1, to run the program. See Figure 2-11.

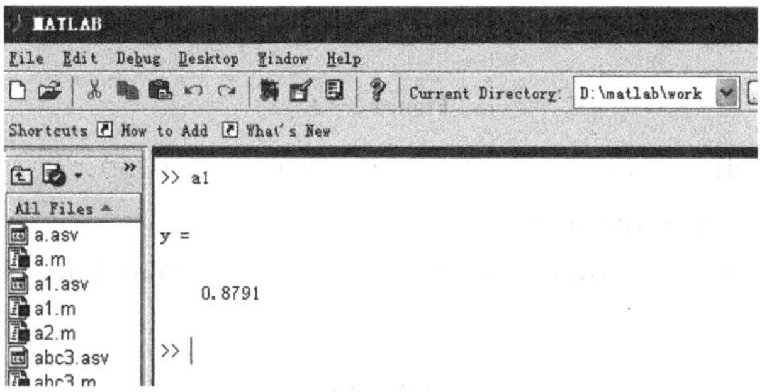

Figure 2-11

Output the result $y=0.8791$.

Example 45 Assume $X \sim N(0,1)$. Find $y=\Phi(1.5)=P(X \leqslant 1.5)$.

The steps are the same as Example 1 except that the program should be changed into $y=$normcdf(1.5) and the file should be renamed, say, a2, and then output the result $y=0.9332$.

Example 46 Find the quantiles $\chi^2_{0.90}(10)$, $t_{0.05}(5)$ and $F_{0.95}(8,10)$.

Remart: Matlab only gives the left quantile.

(1) Open Matlab, and pull dowm menu File⇒New⇒M-File(see Example 1). Write programs in Edit window. See Figure 2-12.

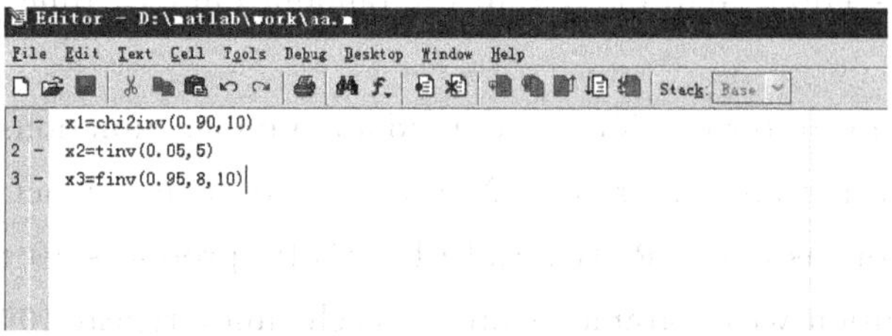

Figure 2-12

(2) Save a copy of the above programs to "work" fold of the Matlab system, and name it as. say, a3.

(3) Run this program in Command window. See Figure 2-13.

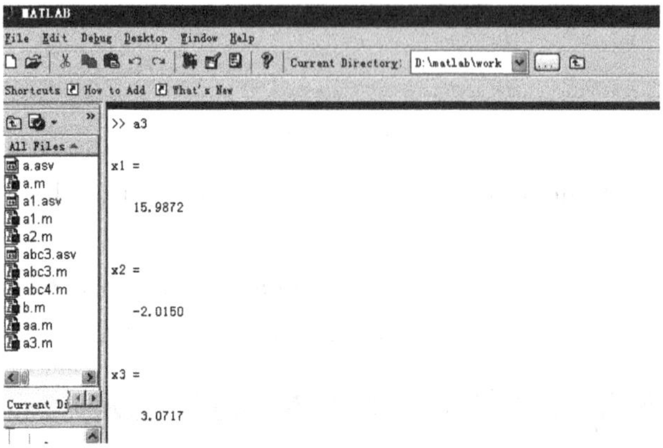

Figure 2-13

Output the result $\chi^2_{0.90}(10)=15.9872$, $t_{0.05}(5)=-2.0150$, $F_{0.95}(8,10)=3.0717$.

13. Generate random numbers

Among random number functions of commonly used distributions in Matlab softwae are(see Table 2-6):

Table 2-6

Distribution	Random Number Function
Binomial	binornd(n, p)
Poisson	poissrnd(lambda)
Uniform	unifrnd(a, b)
Exponential	exprnd(mu)
Uniform in (0,1)	rand
Normal	normrnd(mu, sigma)

Also, see the section Help of Matlab for random number functions of many other distributions.

If add the optional term (N, k) to random number functions, then generate k random numbers each time and repeat N times, and so construct a random number matrix of $N\times k$. For instance, $\mathbf{R}$=exprnd (1,100,10)produces 10 random numbers of exponential distribution with parameter mu=1 each time, repeat 100 times, so yields a matris of 100×10.

Example 47 Generate 10 random numbers of $U(0,1)$.

(1) Write programs in Edit window (Edit-Untitled). See Figure 2-14.

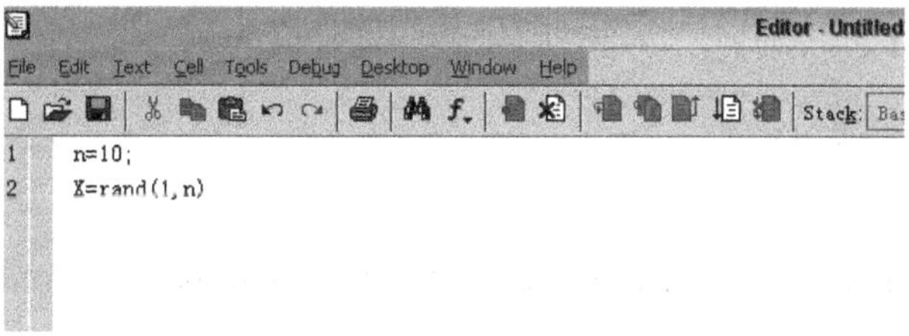

Figure 2-14

(2) Save the above program in the "work" fold of Matlab system and name it as, b1.

(3) Run this program in Command window. See Figure 2-15.

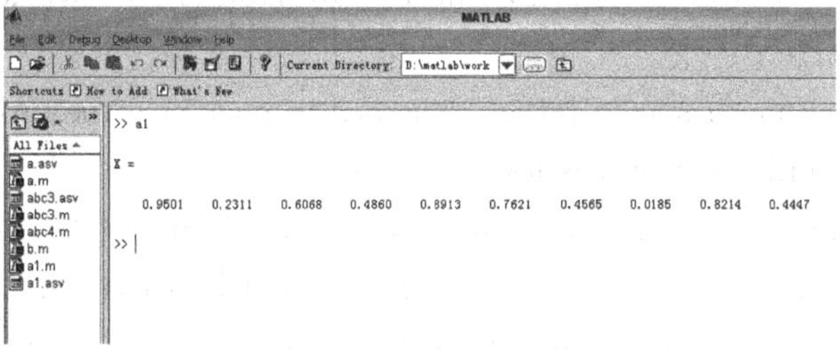

Figure 2-15

Then generate 10 random numbers of $U(0,1)$ as follows:

0.9501 0.2311 0.6068 0.4860 0.8913 0.7621 0.4565 0.0185 0.8214 0.4447.

Exercise 2

1. What values can c be in order that the following sequences become a distribution sequence?

(1) $P(\xi = k) = c/n,\ k = 1,2,\cdots,\ n;$

(2) $P(\xi = k) = c\lambda^k/k!,\ k = 1,2,\cdots,\lambda > 0.$

2. Suppose that ξ is the number of the first consecutive success or consecutive failure in a Bernoulli trial, calculate the distribution of ξ.

3. A particle starts from the origin at time 0 on the real line and moves one step toward the left or right with respective probability p or $1-p$ in every one unit of time, independently. Denote by ξ_n the number of the particles moving toward the right before time n, and S_n the displacement of the particle at time n. Find the distribution sequences of ξ_n and S_n.

4. There are five balls labelled 1,2,3,4,5 in a pocket. Now draw three balls and denote by ξ the maximum of the three numbers. Find the distribution sequence of ξ.

5. Suppose that ξ has the distribution sequence $P(\xi = k) = k/15$, $k = 1,2,3,4,5$. Calculate

(1) $P(\xi = 1 \text{ or } \xi = 2)$;

(2) $P\left(\frac{1}{2} < \xi < \frac{5}{2}\right)$;

(3) $P(1 \leqslant \xi \leqslant 2)$.

6. A computer has 20 terminals, each of which is operated independently by different departments. Each department uses its computer with a probability of 0.7. Find the probability that 10 or more terminals are in use.

7. There are 12 machines in a factory, each working independently. Suppose that the time each machine works occupies 2/3 of the total work time and that each uses 1 unit of power in operation.

(1) If the supply power is 9 units, find the probability that work is stopped because of insufficient power.

(2) At least how much supply power is needed to guarantee the probability that work is stopped because of insufficient power is less than 1%?

8. Select at random 10 seeds, each with a probability of 0.8 to grow. Find the probability that the number of seeds growing up is not less than 8.

9. A book of 500 pages contains 500 mistakes, each mistake appearing equally likely in each page. Find the probability that a particular page contains at least 3 mistakes.

10. Determine how many screws a box should contain in order to guarantee the probability that more than 100 screws is of good quality is not less than 80%.

11. The random variable ξ follows a Poisson distribution with $P(\xi = 1) = P(\xi = 2)$. Find $P(\xi = 4)$.

12. The number of a certain product sold in a particular department store every month follows a Poisson distribution with parameter 6. Determine how many goods the store should purchase in order to guarantee the probability that the goods will not be out of order during a month's time is greater than 0.999.

13. A factory needs some repairers in order to guarantee machines being in operating condition. Assume that each machine breaks down independently with probability 0.01, find the probability that machines break down but cannot be repaired immediately in the following two cases:

(1) one repairer is responsible for 20 machines;

(2) 3 repairers are responsible for 90 machines.

14. The probabilities that claims received by a certain insurance company are equal to 0, 1, 2, 3 within a given duration of time are 0.1, 0.3, 0.4, 0.2 respectively. The probabilities that the individual claim amount is 1, 2 and 3 are 0.5, 0.4 and 0.1 respectively. Find the probability distribution of the total claim amount.

15. The number that a certain kind of vaccine contains bacterium follows the Poisson distribution, and each milliliter contains one on average. Now put this kind of vaccine in 5 tubes, each 2 milliliters. Find

(1) the probability that each of the five tubes contains bacterium;

(2) at least three tubes contain bacterium.

16. Assume that $\xi \sim P(\lambda)$, find the best possible number (that is, what k is such that $p(k,\lambda)$ attains the maximum).

17. Can the following function become the distribution function of a random variable? If so, please complete the definition.

(1) $F(x) = 1/(1+x^2)$, $-\infty < x < \infty$;

(2) $F(x) = 1/(1+x^2)$ when $x > 0$, $F(x)$ is to be determined when $x \leqslant 0$;

(3) $F(x) = 1/(1+x^2)$ when $x < 0$, $F(x)$ is to be determined when $x \geqslant 0$.

18. Pick at random a point P in a disk with radius R centered at O.

(1) Find the distribution function of $\xi = OP$;

(2) Find $P(-2 < \xi < R/2)$.

19. Assume that $F_1(x)$ and $F_2(x)$ are distribution functions. Prove $F(x) = aF_1(x) + bF_2(x)$ is still a distribution function, where the constants $a, b > 0$ and $a + b = 1$.

20. Suppose that ξ has the distribution function $F(x) = A + B \arctan x$, find A, B.

21. Prove the ξ of Exercise 20 is a continuous random variable, and calculate its density function.

22. Determine the constant A in the following functions such that they become density functions.

(1) $p(x) = Ae^{-|x|}$.

(2)

$$p(x) = \begin{cases} A\cos x, & -\pi/2 \leqslant x \leqslant \pi/2; \\ 0, & \text{otherwise.} \end{cases}$$

(3)

$$p(x) = \begin{cases} Ax^2, & 1 \leqslant x < 2; \\ Ax, & 2 \leqslant x < 3; \\ 0, & \text{otherwise.} \end{cases}$$

23. Find the distribution functions corresponding to the densities of Exercise 22.

24. Suppose that the ξ has density function

$$p(x) = \begin{cases} x, & 0 \leqslant x < 1; \\ 2 - x, & 1 \leqslant x < 2; \\ 0, & \text{otherwise.} \end{cases}$$

Find

(1) the distribution function $F(x)$ of ξ;

(2) $P(\xi < 0.5)$, $P(\xi > 1.3)$ and $P(0.2 < \xi < 1.2)$.

25. The electric power used in a certain city each day is less than 1 million kilowatts. Denote the rate of consuming electricity by ξ, its density function is $p(x) = 12x(1-x)^2, 0 < x < 1$. What is the shortage probability when the amount supplied each day is 0.8 million kilowatts? What is it about if 0.9 million kilowatts?

26. Suppose $\xi \sim N(10,4)$. Find

(1) $P(6 < \xi < 9)$;

(2) $P(7 < \xi < 12)$;

(3) $P(13 \leqslant \xi \leqslant 15)$.

27. Suppose $\xi \sim N(5,4)$. Find a such that

(1) $P(\xi < a) = 0.90$;

(2) $P(|\xi - 5| > a) = 0.01$.

28. Suppose $\xi \sim U[0,5]$. Find the probability that equation $4x^2 + 4\xi x + \xi + 2 = 0$ has real roots.

29. In a generalized Bernoulli trial there are three outcomes A_1, A_2, A_3, each with probability p_1, p_2, p_3 respectively. Now repeat this experiment n times, denote by ξ, η respectively the number of times A_1, A_2 appears. Find the joint distribution sequence and marginal distributions of (ξ, η).

30. Show that the bivariate function

$$F(x,y) = \begin{cases} 1, & x + y \geqslant 0; \\ 0, & x + y < 0 \end{cases}$$

is monotonic non-decreasing, right continuous in each argument, and $F(-\infty, y) = F(x, -\infty) = 0, F(+\infty, +\infty) = 1$, but that $F(x,y)$ is not a distribution function.

31. Try to express the following probabilities using the distribution function $F(x, y)$ of (ξ, η)

(1) $P(a \leqslant \xi \leqslant b, \eta \leqslant y)$;

(2) $P(\xi = a, \eta < y)$;

(3) $P(\xi < -\infty, \eta < +\infty)$.

32. Assume the density function of (ξ, η) is

$$p(x,y) = \begin{cases} Ae^{-(2x+y)}, & x, y > 0; \\ 0, & \text{otherwise.} \end{cases}$$

Find

(1) the constant A;

(2) the distribution function $F(x,y)$;

(3) the marginal density of ξ;

(4) $P(\xi < 2, 0 < \eta < 1)$;

(5) $P(\xi + \eta < 2)$;

(6) $P(\xi = \eta)$.

33. Suppose that (ξ, η) is uniformly distributed over rectangular $D = \{0 < x < 1,$

$0 < y < 2\}$.

(1) Write down the joint density;

(2) Find the marginal densities;

(3) Find the joint distribution function;

(4) Find $P(\xi + \eta < 1)$.

34. Assume that a coin lands on head with probability p when rolling. Roll independently this coin n times, and denote by Y the total number of times of appearing heads in these n rollings. Assume that the first head appears at the X-th rolling, find the conditional distribution of X given that only a head appears in n rollings.

35. Assume that the joint density $p(x, y)$ is as in Exercise 32, find the conditional density $p_{\eta|\xi}(y \mid x)$.

36. Consider the bivariate normal density

$$p(x,y) = \frac{1}{2\pi}\exp\left\{-\frac{1}{2}(2x^2 + y^2 + 2xy - 22x - 14y + 65)\right\}.$$

(1) Express it in a standard form, and point out $a, b, \sigma_1^2, \sigma_2^2, r$;

(2) Find the marginal density $p_\xi(x)$;

(3) Find the conditional density $p_{\eta|\xi}(y \mid x)$.

37. In the following distribution diagram of (ξ, η), what are a, b and c so that ξ and η are independent(see Table 2-7, Table 2-8)?

Table 2-7

ξ	η		
	0	1	3
1	1/6	1/9	c
2	1/3	a	b

Table 2-8

ξ	η		
	0	1	3
0	c	a	1/5
1	b	1/10	1/5

38. Assume that ξ and η are independent and $P(\xi = 1) = P(\eta = 1) = p > 0, P(\xi = 0) = P(\eta = 0) = 1 - p > 0$. Define

$$\zeta = \begin{cases} 1, & \xi + \eta \quad \text{even;} \\ 0, & \xi + \eta \quad \text{odd.} \end{cases}$$

What is p so that ξ and ζ are independent?

39. Determine whether the ξ and η in Exercise 32 are independent or not.

40. Assume that (ξ, η) is uniformly distributed over the disk $x^2 + y^2 \leqslant r^2$.

(1) Find the densities of ξ and η;

(2) Determine whether the ξ and η are independent or not.

41. Assume that (ξ, η) has density $p(x, y)$, show that a sufficient and necessary condition for ξ and η to be independent is that $p(x, y)$ can be written as $p(x, y) = g(x)h(y)$. Then, how are $g(x)$ and $h(y)$ related to the marginal densities?

42. Apply the sufficient and necessary condition in Exercise 41 to check the

independence of ξ and η, whose densities are as follows.

(1) $p(x,y) = 4xy$ if $0 \leqslant x \leqslant 1$, $0 \leqslant y \leqslant 1$; 0 otherwise.

(2) $p(x,y) = 8xy$ if $0 \leqslant x \leqslant y \leqslant 1$; 0 otherwise.

43. Assume (ξ,η,ζ) has the joint density

$$p(x,y,z) = \begin{cases} \dfrac{1-\sin x \sin y \sin z}{8\pi^3}, & 0 \leqslant x,y,z \leqslant 2\pi; \\ 0, & \text{otherwise.} \end{cases}$$

Show ξ,η,ζ are pairwise independent, but not independent.

44. Assume ξ has the distribution sequence as follows:

$$\begin{pmatrix} 0 & \pi/2 & \pi \\ 1/4 & 1/2 & 1/4 \end{pmatrix},$$

calculate the distribution sequences of the following random variables:

(1) $\eta = 2\xi + \pi/2$;

(2) $\zeta = \sin \xi$.

45. Assume that ξ follows a Poisson distribution with parameter λ, find the distribution sequences of the following random variables:

(1) $\eta = a + b\xi$;

(2) $\zeta = \xi^2$.

46. The digits 0, 1, 1, 2 are respectively written on four pieces of paper. Now draw with replacement two pieces, denote by ξ,η the corresponding numbers. Find the distribution sequences of ξ and η and $\theta = \xi\eta$.

47. Assume that ξ and η are independent Poisson random variables with respective parameters λ_1, λ_2. Directly show that

(1) $\xi + \eta$ follows the Poisson distribution with parameter $\lambda_1 + \lambda_2$;

(2) $P(\xi = k \mid \xi + \eta = n) = \binom{n}{k}\left(\dfrac{\lambda_1}{\lambda_1+\lambda_2}\right)^k \left(\dfrac{\lambda_1}{\lambda_1+\lambda_2}\right)^{n-k}$,

where $k = 0, 1, \cdots, n$.

48. Let Y_{nk}, $k = 1,\cdots,k_n$, Z_{nk}, $k = 1,\cdots,k_n$, be array of independent random variables with Y_{nk} having a Poisson distribution $P(p_{nk})(0 \leqslant p_{nk} \leqslant 1)$, and Z_{nk} having a Bernoulli distribution $B(1, 1-(1-p_{nk})e^{p_{nk}}$, $k=1,\cdots,k_n$. Let

$$X_{nk} = \begin{cases} 0, & \text{if } Y_{nk} = 0 \text{ and } Z_{nk} = 0; \\ 1, & \text{otherwise.} \end{cases}$$

(1) Prove that X_{nk}, $k = 1,\cdots,k_n$ are independent, and X_{nk} has a Bernoulli distribution $B(1, p_{nk})$.

(2) Prove that $P(X_{nk} \neq Y_{nk}) \leqslant p_{nk}^2$.

(3) Denote $X_n = \sum_{k=1}^{k_n} X_{nk}$ and $Y_n = \sum_{k=1}^{k_n} Y_{nk}$. Prove that

$$|P(X_n \in A) - P(Y_n \in A)| \leqslant \sum_{k=1}^{k_n} p_{nk}^2, \text{ for all } A \in \mathscr{B}.$$

(4) Use the above conclusion to prove the Poisson theorem.

49. Assume that ξ has density $p(x)$, find the densities of the following random variables:

(1) $\eta = 1/\xi$, where $P(\xi = 0) = 0$;

(2) $\eta = |\xi|$;

(3) $\eta = \tan \xi$.

50. Assume that X is a symmetric random variable, i. e., $-X$ and X have the same distribution, I is a random variable independent of X with $P(I=1) = P(I=-1) = 1/2$, and $M>0$ is a constant. Let $Y = XI$ and

$$Z = \begin{cases} X, & \text{if } |X| \leqslant M; \\ -X, & \text{elsewhere.} \end{cases}$$

Prove that X, Y and Z are identically distributed.

51. Assume that $F(x)$ is the distribution function of random variable ξ, and $F(x) < 1$ for an arbitrary real number x. Define

$$R(x) = \ln \frac{1}{1-F(x)}, \qquad -\infty < x < \infty.$$

Show that $\eta = R(\xi)$ is exponentially distributed with parameter 1.

52. Measure approximately the diameter D of a circle. Assume that its value is uniformly distributed over $[a,b]$, find the density of the circle area S.

53. Assume $\xi \sim N(a,\sigma^2)$, find the density function of e^{ξ} (called a logarithmic normal distribution).

54. Assume that θ is uniformly distributed over $[-\frac{\pi}{2}, \frac{\pi}{2}]$, and let $\varphi = \tan\theta$, find the density of φ (called a Cauchy distribution).

55. Assume that ξ, η are independent and uniformly distributed over $[0,1]$, find the density of $\zeta = \xi + \eta$.

56. Assume that ξ, η are independent and normally distributed as $N(0,1)$, find the density of $\zeta = \xi/\eta$.

57. Let $f(x)$ and $g(x)$ be two density functions such that for a constant $C \geqslant 1$,

$$\frac{f(x)}{g(x)} \leqslant C \text{ for all } x.$$

Suppose that $Y_1, U_1, Y_2, U_2, \cdots$ are independent random variables such that the Y_i has the density function $g(x)$ and U_i is uniformly distributed on $(0,1)$, $i = 1,2,\cdots$. Define the random variables X and N as follows. If $U_1 \leqslant \frac{f(Y_1)}{Cg(Y_1)}$, then set $X = Y_1$ and $N = 1$, otherwise, we compare U_2 with $\frac{f(Y_2)}{Cg(Y_2)}$. If $U_2 \leqslant \frac{f(Y_2)}{Cg(Y_2)}$, then set $X = Y_2$ and $N = 2$, and so on.

(1) Show that X has density function $f(x)$.

(2) Find the distribution of N.

58. Assume $\xi_1, \cdots, \xi_n$ are independent and exponentially distributed with respective parameters $\lambda_1, \cdots, \lambda_n$, find the density of $\eta = \min(\xi_1, \cdots, \xi_n)$.

59. Assume that the system L consists of two subsystems L_1, L_2, whose lifetimes X, Y are exponentially distributed with parameter a, b $(a \neq b)$ respectively. Find the lifetime Z of L in the following three ways:

(1) L_1 and L_2 are series connected;

(2) L_1 and L_2 are parallel connected;

(3) L_2 is spare (L_2 works immediately when L_1 is broken).

60. The demands for a certain goods during a week is a random variable with density function:

$$p(x) = \begin{cases} xe^{-x}, & x \geqslant 0; \\ 0, & x < 0. \end{cases}$$

Assume that the demands in different weeks are independent, find the density of demands over certain two weeks.

61. Suppose that ξ and η are independent and exponentially distributed with mean λ and μ respectively. Find the density function of $\xi - \eta$.

62. Draw at random two points in the interval $(0, a)$, find the density function of their distance.

63. Assume that the position (ξ, η) obeys a two dimensional normal distribution $N(0, 0, \sigma^2, \sigma^2, 0)$, find the density function of the distance $\rho = \sqrt{\xi^2 + \eta^2}$.

64. Assume that each coordinate of velocity $V = (X, Y, Z)$ of gas molecule is independent and distributed as $N(0, \sigma^2)$. Show that $S = \sqrt{X^2 + Y^2 + Z^2}$ obeys a Maxwell distribution with the density:

$$p(s) = \begin{cases} \sqrt{\dfrac{2}{\pi}} \dfrac{s^2}{\sigma^3} \exp\left(-\dfrac{s^2}{2\sigma^2}\right), & s \geqslant 0; \\ 0, & s < 0. \end{cases}$$

65. Assume that ξ and η are independent and exponentially distributed with mean 1. Determine whether $(\xi + \eta)$ and $\xi/(\xi + \eta)$ are independent.

66. Assume that (ξ, η) has the joint density function

$$p(x, y) = \begin{cases} 4xy, & 0 < x, y < 1; \\ 0, & \text{otherwise}. \end{cases}$$

Find the joint density function of (ξ^2, η^2).

67. Assume that ξ and η are independent and exponentially distributed with mean 1. Find the joint density function and marginal density functions of $U = \xi + \eta$ and $V = \xi - \eta$.

68. Assume that $(\xi, \eta) \sim N(0, 0, \sigma_1^2, \sigma_2^2, r)$. Give a sufficient and necessary condition

for $\xi+\eta$ and $\xi-\eta$ to be independent.

69. Consider the transform of random vectors of Section 2.6. How do we apply (2.84) to compute the density of random vectors after transformation if the inverse functions are not necessarily unique?

70. Assume that the joint density function $p(x, y)$ of (ξ, η) is as follows:

$$p(x,y)=\begin{cases}\dfrac{1+xy}{4}, & |x|, |y|<1;\\ 0, & \text{otherwise.}\end{cases}$$

Show that ξ and η are not independent, but ξ^2 and η^2 are independent.

71. Assume that (ξ,η) has the joint density $p(x,y)$. Let $U=\xi$, $V=\xi+\eta$, find the joint density of (U,V), and marginal densities, and then compare them with the result via (2.74) of Section 2.6.

72. Assume that (ξ,η) has the joint density $p(x,y)$. Let $U=\xi$, $V=\xi/\eta$, calculate the joint density of (U,V), and its marginal densities, and then compare them with the result via (2.78) of Section 2.6.

73. Let X and Y be independent, identically distributed, positive random variables with continuous density function $f(x)$ with $f(0)>0$. Assume, further, that $U=X-Y$ and $V=\min(X,Y)$ are independent random variables. Prove that

$$f(x)=\begin{cases}\lambda e^{-\lambda x}, & \text{for } x\geqslant 0;\\ 0, & \text{elsewhere}\end{cases}$$

for some $\lambda>0$.

74. Assume X and Y have a joint density function

$$g(x^2+y^2),\ -\infty<x,y<\infty,$$

where $g(t), t\geqslant 0$, is a continuous positive function. Assume, further, X and Y are independent. Prove that X and Y are identically distributed normal random variables.

75. Assume that (ξ,η,ζ) has the joint density

$$p(x,y,z)=\begin{cases}\dfrac{6}{(1+x+y+z)^4}, & x,y,z>0;\\ 0, & \text{otherwise.}\end{cases}$$

Find the density of $U=\xi+\eta+\zeta$. (Hint: find two good functions $V=V(\xi,\eta,\zeta)$, $W=W(\xi,\eta,\zeta)$, and calculate the joint density of (U,V,W) and then the marginal density $p_U(u)$).

Chapter 3

Numerical Characteristics and Characteristic Functions

As we have known, distribution functions completely describe a random phenomenon. However, one does not readily determine the distribution function of a random variable so that a complete description is very hard in practical work. Thus it is necessary to introduce some numerical characteristics to reflect major features of random variables. On the other hand, from the empirical viewpoint, a random variable related to a random phenomenon follows usually some kind of distributions, some of whose parameters are determined by numerical characteristics. For these random phenomena, numerical characteristics seem more interesting. Among numerical characteristics, we shall introduce mathematical expectation and variance below. The former is just the mean of the values of a random variable, while the latter expresses the extent to which a random variable deviates from the mean. In addition, there is a concept of correlation coefficient which describes the extent two random variables have towards linear relation. In this chapter we shall study these concepts and their properties and methods of computing them. Moreover, we shall introduce another powerful tool to analyze random variables — characteristic function, and use it to deal with multivariate normal distributions which play a major role in multivariate analysis.

3.1 Mathematical expectations

3.1.1 Expectations of discrete random variables

Example 1 In order to evaluate A's shooting level, randomly observe his ten shootings and record the circular ring number he hits each time and the frequency as in Table 3-1 below.

Table 3-1

x_k	8	9	10
v_k	2	5	3
$f_k = v_k/N$	0.2	0.5	0.3

Find the average number of circular rings (here $N = \sum v_k = 10$).

Solution This is just a problem of finding the mean $\overline{x}$ as usual. In fact,

$$\overline{x} = \frac{1}{10}(8 \times 2 + 9 \times 5 + 10 \times 3)$$

$$= \frac{8 \times 2}{10} + \frac{9 \times 5}{10} + \frac{10 \times 3}{10} = 9.1.$$

Generally speaking, we have

$$\bar{x} = \frac{1}{N} \sum_k x_k v_k = \sum_k \frac{x_k v_k}{N} = \sum_k x_k f_k.$$

If we want to fairly evaluate A's shooting level, it is insufficient to depend only on the ten shootings above. This is because the frequency $v_k / N = f_k$ is related to the outcomes in these ten shootings (experiments). But as the number of shootings increases, the sum $\sum_k x_k f_k$ tends to $\sum_k x_k p_k$ since frequency tends to probability. It is a constant, not depending on particular trials, so it reflects better A's shooting level. We give the following definition by generalizing the example above.

Definition 1 *Suppose that a discrete random variable ξ has the distribution sequence*

$$\begin{pmatrix} x_1 & x_2 & \cdots & x_k & \cdots \\ p_1 & p_2 & \cdots & p_k & \cdots \end{pmatrix}.$$

If the series $\sum_k x_k p_k$ converges absolutely, that is, $\sum_k |x_k| p_k < \infty$, the sum is called mathematical expectation or mean of ξ, written as

$$E\xi = \sum_k x_k p_k. \tag{3.1}$$

Remark The condition $\sum_k |x_k| p_k < \infty$ implies the sum in formula (3.1) is not affected by the ordering of sum, so $E\xi$ is a constant. When this condition is not satisfied, we say that mathematical expectation of ξ does not exist.

Example 2 The degenerate distribution $P(\xi = a) = 1$ has mathematical expectation $E\xi = a$. In other words, the expectation of a constant is just itself.

Example 3 Calculate the mathematical expectation of the binomial distribution

$$P(\xi = k) = \binom{n}{k} p^k q^{n-k}, \quad k = 0, 1, 2, \cdots, n.$$

Solution By definition, we have

$$\begin{aligned} E\xi &= \sum_{k=0}^{n} k p_k = \sum_{k=0}^{n} k \frac{n!}{k!(n-k)!} p^k q^{n-k} \\ &= np \sum_{k=1}^{n} \frac{(n-1)!}{(k-1)![(n-1)-(k-1)]!} p^{k-1} q^{n-1-(k-1)} \\ &= np \sum_{r=0}^{n-1} \frac{(n-1)!}{r![(n-1)-r]!} p^r q^{n-1-r} \\ &= np(p+q)^{n-1} = np. \end{aligned}$$

In particular, when $n = 1$ we get the mathematical expectation of two points distribution (0-1 distribution).

Example 4 Calculate the mathematical expectation of the Poisson distribution

$$P(\xi = k) = \frac{\lambda^k}{k!}e^{-\lambda},\ k = 0,1,2,\cdots.$$

Solution By definition, we have

$$E\xi = \sum_{k=0}^{\infty} kp_k = \sum_{k=0}^{\infty} k\frac{\lambda^k}{k!}e^{-\lambda} = \lambda e^{-\lambda}\sum_{k=0}^{\infty}\frac{\lambda^{k-1}}{(k-1)!}$$
$$=\lambda.$$

Hence the parameter in the Poisson distribution is just its mathematical expectation.

Example 5 Calculate the mathematical expectation of the geometric distribution

$$P(\xi = k) = pq^{k-1},\ k = 1,2,\cdots,\ 0 < p < 1.$$

Solution By definition, we have

$$E\xi = \sum_{k=0}^{\infty} kpq^{k-1} = p\sum_{k=0}^{\infty}(x^k)'\big|_{x=q} = p\Big(\sum_{k=1}^{\infty}x^k\Big)'\big|_{x=q}$$
$$=p\left(\frac{x}{1-x}\right)'\Big|_{x=q} = p\,\frac{1}{(1-x)^2}\Big|_{x=q} = \frac{1}{p}.$$

In the above examples either there are finitely many summands or each x_i is nonnegative, so we need not check the absolute convergence. When ξ takes infinitely many values, some of which are possibly negative, we are required to verify the absolute convergence.

Example 6 Suppose that

$$P\left(\xi = (-1)^k\frac{2^k}{k}\right) = \frac{1}{2^k},\quad k = 1,2,\cdots.$$

Note that

$$\sum_{k=1}^{\infty} |x_k|\, p_k = \sum_{k=1}^{\infty}\frac{1}{k} = \infty.$$

We say that $E\xi$ does not exist, although $\sum_{k=1}^{\infty} x_k p_k < 1$.

3.1.2 Expectations of continuous random variables

Suppose that ξ is a continuous random variable with the density function $p(x)$, then its mathematical expectation can be defined by combining the concept of mathematical expectation in the discrete case and the process of defining the Riemann integral. First, assume that ξ takes its values only on a finite interval $[a,b]$. Now partition $[a,b]$ into smaller intervals: $a = x_0 < x_1 < \cdots < x_n = b$, the probability ξ lies inside each interval is

$$P(x_k < \xi \leqslant x_{k+1}) = \int_{x_k}^{x_{k+1}} p(x)\,dx \approx p(x_k)\Delta x_k,$$

where $\Delta x_k = x_{k+1} - x_k$, $k = 0,1,\cdots,n$. This is approximately thought as the probability ξ at a point; the expectation of the corresponding discrete random variable is $\sum_{k=0}^{n-1} x_k p(x_k)\Delta x_k$. Then letting $n \to \infty$, we obtain the integral $\int_a^b xp(x)\,dx$ as a limit of sums. It is natural to view it as expectation of ξ. If ξ takes its values on the real line $(-\infty,\infty)$, letting $a\to-\infty, b\to\infty$, we get the following definition.

Definition 2 *Suppose that ξ is a continuous random variable with density $p(x)$, and*

$$\int_{-\infty}^{+\infty}|x|\,p(x)\,dx<+\infty,$$

then we call

$$E\xi=\int_{-\infty}^{+\infty}xp(x)\,dx \tag{3.2}$$

the mathematical expectation of ξ. If $\int_{-\infty}^{+\infty}|x|\,p(x)dx=\infty$, we say that the expectation of ξ does not exist.

Example 7 Calculate the expectation of the uniform random variable ξ over $[a,b]$.

Solution Recall that a uniform random variable over $[a,b]$ has the density function

$$p(x)=\begin{cases}\dfrac{1}{b-a}, & a\leqslant x\leqslant b;\\ 0, & \text{otherwise.}\end{cases}$$

It is easy to see by definition

$$E\xi=\int_a^b\frac{x}{b-a}dx=\frac{a+b}{2}.$$

Example 8 Calculate the expectation of the exponential random variable ξ with parameter λ.

Solution Since ξ has the density function

$$p(x)=\begin{cases}\lambda e^{-\lambda x}, & x>0;\\ 0, & x\leqslant 0,\end{cases}$$

then we have

$$E\xi=\int_0^{+\infty}x\lambda e^{-\lambda x}\,dx=\frac{1}{\lambda}.$$

Remark The above result is readily understood from the relation between the exponential distribution and the geometric distribution.

Example 9 Calculate the expectation of the normal random variable $\xi\sim N(a,\sigma^2)$.

Solution First we note

$$\int_{-\infty}^{+\infty}|x|\frac{1}{\sqrt{2\pi}\sigma}e^{-\frac{(x-a)^2}{2\sigma^2}}\,dx<\infty,$$

which implies that ξ has expectation. Also, a computation leads to

$$\begin{aligned}E\xi&=\int_{-\infty}^{+\infty}x\frac{1}{\sqrt{2\pi}\sigma}e^{-\frac{(x-a)^2}{2\sigma^2}}\,dx\\&=\frac{a}{\sqrt{2\pi}}\int_{-\infty}^{+\infty}e^{-\frac{z^2}{2}}\,dz+\frac{\sigma}{\sqrt{2\pi}}\int_{-\infty}^{+\infty}ze^{-\frac{z^2}{2}}\,dz\\&=a,\end{aligned}$$

where the second equation follows from the substitution of variable $z=(x-a)/\sigma$. Thus the parameter a in $N(a,\sigma^2)$ stands for its mean, which can be seen from the shape of the density function.

Example 10 Show that the Cauchy distribution does not have expectation.

Proof The Cauchy distribution has the density

$$p(x)=\frac{1}{\pi(1+x^2)},\ -\infty<x<\infty.$$

Since

$$\int_{-\infty}^{+\infty}|x|\,p(x)\mathrm{d}x=2\int_0^{+\infty}\frac{x}{\pi(1+x^2)}\mathrm{d}x=\infty,$$

so the expectation does not exist.

3.1.3 General definition

In the previous subsection we have defined the expectation of a continuous random variable to be the limit $\int_{-\infty}^{+\infty}xp(x)\,\mathrm{d}x$ of $\sum_{k=0}^{n-1}x_kp(x_k)\Delta x_k$. If ξ has the distribution function $F(x)$, then

$$\begin{aligned}\sum_{k=0}^{n-1}x_kp(x_k)\Delta x_k&=\sum_{k=0}^{n-1}x_k\Delta F(x_k)\\&=\sum_{k=0}^{n-1}x_k(F(x_k)-F(x_{k-1})).\end{aligned}$$

So we introduce a new integral, called the Stieltjes integral, and write $\int_{-\infty}^{+\infty}x\mathrm{d}F(x)$ for the limit of the right hand side in the sum above. See Supplements and Remarks 1 at the end of this chapter for properties of the Stieltjes integral.

Definition 3 *Suppose that ξ has distribution function $F(x)$. If $\int_{-\infty}^{+\infty}|x|\,\mathrm{d}F(x)<\infty$, then we call*

$$E\xi=\int_{-\infty}^{+\infty}x\mathrm{d}F(x)\tag{3.3}$$

the mathematical expectation of ξ. When $\int_{-\infty}^{+\infty}|x|\,\mathrm{d}F(x)=\infty$, we say that the expectation of ξ does not exist.

Let $\xi^+=\max\{\xi,0\}$ and $\xi^-=\max\{-\xi,0\}$ be the positive and negative parts of ξ, F_{ξ^+} and F_{ξ^-} be their distribution functions respectively. It can be verified that $\int_0^{\infty}x\mathrm{d}F(x)=\int_0^{\infty}x\mathrm{d}F_{\xi^+}(x)$ and

$$\int_{-\infty}^{0}x\mathrm{d}F(x)=\int_{-\infty}^{0}x\mathrm{d}F(x-0)=-\int_0^{\infty}x\mathrm{d}(1-F(-x-0))=-\int_0^{\infty}x\mathrm{d}F_{\xi^-}(x).$$

So, $E\xi$ exists if and only if $E\xi^+$ and $E\xi^-$ exist. Further,

$$E\xi = E\xi^+ - E\xi^-, \quad E|\xi| = E\xi^+ + E\xi^-.$$

On the other hand,

$$\int_0^\infty x\mathrm{d}F(x) = \int_0^\infty \int_0^x \mathrm{d}t\mathrm{d}F(x) = \int_0^\infty \int_t^\infty \mathrm{d}F(x)\mathrm{d}t = \int_0^\infty P(\xi > t)\mathrm{d}t.$$

Similarly,

$$\int_{-\infty}^0 x\mathrm{d}F(x) = -\int_{-\infty}^0 P(\xi \leqslant t)\mathrm{d}t.$$

Hence, we can write the expectation as integrals of the probabilities $P(\xi > t)$ and $P(\xi \leqslant t)$ as follows

$$E\xi = \int_0^\infty P(\xi > t)\mathrm{d}t - \int_{-\infty}^0 P(\xi \leqslant t)\mathrm{d}t.$$

We remark that $F(x) = \int_{-\infty}^x \mathrm{d}F(t)$. It can be showed that for any random variable ξ, $P(\xi \in B)$ can be written as the Stieltjes integral

$$P(\xi \in B) = \int_{x \in B} \mathrm{d}F(x). \tag{3.4}$$

3.1.4 Expectations of functions of random variables

Theorem 1 *Suppose that ξ is a random variable with the distribution $F_\xi(x)$, $f(x)$ a Borel function on the real line. Let $\eta = f(\xi)$. Then*

$$Ef(\xi) = \int_{-\infty}^{+\infty} f(x)\mathrm{d}F_\xi(x). \tag{3.5}$$

The rigorous proof of this theorem needs the properties of integrals in the measure theory. Here, we only give the proof when ξ is a discrete random variable or continuous random variable. When ξ is a discrete random variable, we suppose that its distribution sequence is

$$\begin{bmatrix} x_1 & x_2 & \cdots & x_k & \cdots \\ p_1 & p_2 & \cdots & p_k & \cdots \end{bmatrix}.$$

Then all possible values of $\eta = f(\xi)$ are $f(x_i)$, $i = 1, 2, \cdots$. Merging those x_i' s for which $f(x_i)$' s have the same value, we obtain the distribution sequence of η as

$$\begin{bmatrix} y_1 & y_2 & \cdots & y_k & \cdots \\ p_1^* & p_2^* & \cdots & p_k^* & \cdots \end{bmatrix}.$$

where $p_i^* = \sum_{j: f(x_j) = y_i} p_j$. Hence, $E\eta$ exists if and only if

$$\sum_i |y_i| p_i^* = \sum_i \sum_{j: f(x_j) = y_i} |f(x_j)| p_j = \sum_k |f(x_k)| p_k$$
$$= \int_{-\infty}^\infty |f(x)| \mathrm{d}F_\xi(x) < \infty.$$

Further,

$$E\eta = \sum_i y_i p_i^* = \sum_i \sum_{j: f(x_j) = y_i} f(x_j) p_j = \sum_k f(x_k) p_k.$$

Then

$$Ef(\xi)=\sum_k f(x_k)p_k=\sum_k f(x_k)P(\xi=x_k)=\int_{-\infty}^{\infty} f(x)\mathrm{d}F_\xi(x).$$

(3.5) follows.

When ξ is a continuous random variable, we denote its density function by $p(x)$. Then the expectation of $\eta=f(\xi)$ exists if and only if

$$E\mid\eta\mid=\int_0^{\infty} P(\mid f(\xi)\mid>t)\mathrm{d}t=\int_0^{\infty}\int_{x:\mid f(x)\mid>t} p(x)\mathrm{d}x\mathrm{d}t$$

$$=\int_{-\infty}^{\infty}\left[\int_0^{\mid f(x)\mid}\mathrm{d}t\right]p(x)\mathrm{d}x=\int_{-\infty}^{\infty}\mid f(x)\mid p(x)\mathrm{d}x<\infty.$$

Further,

$$E\mid\eta\mid=\int_0^{\infty} P(\mid f(\xi)\mid>t)\mathrm{d}t-\int_{-\infty}^{0} P(\mid f(\xi)\mid>t)\mathrm{d}t$$

$$=\int_0^{\infty}\int_{x:\mid f(x)\mid>t} p(x)\mathrm{d}x\mathrm{d}t-\int_{-\infty}^{0}\int_{x:\mid f(x)\mid>t} p(x)\mathrm{d}x\mathrm{d}t.$$

$$=\int_{-\infty}^{\infty}\left[\int_{0\leqslant t<f(x)}\mathrm{d}t-\int_{f(x)\leqslant t\leqslant 0}\mathrm{d}t\right]p(x)\mathrm{d}x$$

$$=\int_{-\infty}^{\infty} f(x)p(x)\mathrm{d}x.$$

(3.5) also follows.

By (3.5), if ξ and η have the same distribution, then

$$Ef(\xi)=Ef(\eta).$$

Conversely, if the above equation holds for any bounded continuous function f, then ξ and η will have the same distribution. In fact, for any fixed z and $\varepsilon>0$, if we choose a continuous function $f(x)$ such that $f(x)=1$, $0\leqslant f(x)\leqslant 1$ and $f(x)=0$ on the intervals $(-\infty,z]$, $(z,z+\varepsilon]$ and $(z+\varepsilon,\infty)$ respectively, then

$$F_\xi(z)=\int_{-\infty}^{z} f(x)\mathrm{d}F_\xi(x)\leqslant\int_{-\infty}^{\infty} f(x)\mathrm{d}F_\xi(x)=\int_{-\infty}^{\infty} f(x)\mathrm{d}F_\eta(x)$$

$$=\int_{-\infty}^{z+\varepsilon} f(x)\mathrm{d}F_\eta(x)\leqslant\int_{-\infty}^{z+\varepsilon}\mathrm{d}F_\eta(x)=F_\eta(z+\varepsilon).$$

Letting $\varepsilon\to 0$ yields $F_\xi(z)\leqslant F_\eta(z)$. Similarly, $F_\eta(z)\leqslant F_{(}z)$. Therefore, ξ and η have the same distribution.

Formula (3.5) is very useful to compute expectations of functions of random variables. In fact, one can compute $E\eta=Ef(\xi)$ from $F_\xi(x)$ the distribution function of ξ directly by (3.5), while one needs first to calculate the distribution function of η according to the approach of Section 2.6 if he/she wants to apply (3.3). The latter is often quite complicated. In particular, when ξ has a density $p(x)$, formula (3.5) becomes

$$Ef(\xi)=\int_{-\infty}^{\infty} f(x)p(x)\mathrm{d}x. \tag{3.6}$$

Example 11 (Stein lemma) Suppose that $\xi\sim N(0,1)$ and g is a continuous and differentiable function with $E[\xi g(\xi)]<\infty$, $E\mid g'(\xi)\mid<\infty$.

(1) Prove the following equality

$$E[\xi g(\xi)] = Eg'(\xi).$$

(2) Suppose that the above equality holds for any bounded continuous function $g(x)$ with bounded continuous derivatives. Prove $\xi \sim N(0,1)$.

Proof (1) Suppose $\xi \sim N(0,1)$. By the formula of integral by parts, we obtian

$$E[\xi g(\xi)] = \frac{1}{\sqrt{2\pi}}\int_{-\infty}^{\infty} xg(x)\mathrm{e}^{-\frac{x^2}{2}}\mathrm{d}x = \frac{1}{\sqrt{2\pi}}\int_{-\infty}^{\infty} g(x)\mathrm{d}[-\mathrm{e}^{-\frac{x^2}{2}}]$$
$$= \frac{1}{\sqrt{2\pi}}\left[-g(x)\mathrm{e}^{-\frac{x^2}{2}}\Big|_{-\infty}^{\infty} + \int_{-\infty}^{\infty} g'(x)\mathrm{e}^{-\frac{x^2}{2}}\mathrm{d}x\right] = Eg'(\xi).$$

(2) Let $\eta \sim N(0,1)$. It is sufficient to show that $Eh(\xi)=Eh(\eta)$ for any a bounded continuous function $h(x)$. Without loss of generality, we assume $0 \leqslant h(x) \leqslant 1$. Let $g(x)$ satisfy

$$h(x) - Eh(\eta) = g'(x) - xg(x).$$

The above equation is called the Stein equation. It can be verified that the following function

$$g(x) = \mathrm{e}^{\frac{x^2}{2}}\int_{-\infty}^{x} [h(u) - \mathrm{E}h(\eta)]\mathrm{e}^{-\frac{u^2}{2}}\mathrm{d}u$$

is a solution of the Stein equation, and $g(x)$ and $g'(x)$ are both bounded continuous functions. Then

$$Eh(\xi) = \int_{-\infty}^{\infty} [g'(x) - xg(x) + Eh(\eta)]\mathrm{d}F_\xi(x)$$
$$= \int_{-\infty}^{\infty} g'(x)\mathrm{d}F_\xi(x) - \int_{-\infty}^{\infty} xg(x)\mathrm{d}F_\xi(x) + Eh(\eta)\int_{-\infty}^{\infty}\mathrm{d}F_\xi(x)$$
$$= Eg'(\xi) - E[\xi g(\xi)] + Eh(\eta) = Eh(\eta).$$

The conclusion follows. In the second equality above, the linearity property of the Stieltjes integral was used(c. f. subsection 2 of Supplements and Remarks).

Theorem 1 can be generalized such that the conclusion remains true for a function of a random vector. Let $F(x_1, x_2, \cdots, x_n)$ be the distribution of a random vector $(\xi_1, \xi_2, \cdots, \xi_n)$, and $g(x_1, x_2, \cdots, x_n)$ be an n-dimensional Borel function. Then

$$\mathrm{E}g(\xi_1, \xi_2, \cdots, \xi_n) = \int_{-\infty}^{\infty}\cdots\int_{-\infty}^{\infty} g(x_1, x_2, \cdots, x_n)\mathrm{d}F(x_1, x_2, \cdots, x_n). \quad (3.7)$$

Here, the definition and properties of the multivariate Stieltjies integral are similar to those of the univariate Stieltjes integral. For example,

$$\int_{a_1}^{b_1}\cdots\int_{a_n}^{b_n} g(x_1, x_2, \cdots, x_n)\mathrm{d}F(x_1, x_2, \cdots, x_n)$$
$$= \lim \sum_{k_1, k_2, \cdots, k_n} g(x_1(k_1), x_2(k_2), \cdots, x_n(k_n))\Delta F(x_1(k_1), x_2(k_2), \cdots, x_n(k_n)),$$

where $x_i(1), x_i(2), \cdots$ are partitions of the interval $(a_i, b_i]$, $\Delta F(x_1(k_1), x_2(k_2), \cdots, x_n(k_n))$ is the probability that $(\xi_1, \xi_2, \cdots, \xi_n)$ falls into the rectangle $(x_1(k_1), x_1(k_1+1)] \times (x_2(k_2), x_2(k_2+1)] \times \cdots \times (x_n(k_n), x_n(k_n+1)]$, and the limit is taking over all kinds of partitions such that the largest size of partition intervals $\delta = \max_{i,k_i}(x_i(k_i+1)-x_i(k_i))$ goes to 0.

If $(\xi_1, \xi_2, \cdots, \xi_n)$ is a discrete random vector with the distribution sequence $P(\xi_1 = x_1(i_1), \xi_2 = x_2(i_2), \cdots, \xi_n = x_n(i_n)) = p_{i_1 i_2 \cdots i_n}$, then

$$Eg(\xi_1, \xi_2, \cdots, \xi_n) = \sum_{i_1, i_2, \cdots, i_n} g(x_1(i_1), x_2(i_2), \cdots, x_n(i_n)) p_{i_1 i_2 \cdots i_n}.$$

If $(\xi_1, \xi_2, \cdots, \xi_n)$ is a continuous random vector with a density function $p(x_1, x_2, \cdots, x_n)$, then

$$Eg(\xi_1, \xi_2, \cdots, \xi_n) = \int_{-\infty}^{\infty} \cdots \int_{-\infty}^{\infty} g(x_1, x_2, \cdots, x_n) p(x_1, x_2, \cdots, x_n) \mathrm{d}x_1 \cdots \mathrm{d}x_n.$$

In particular, we have

$$E\xi_i = \int_{-\infty}^{+\infty} \cdots \int_{-\infty}^{+\infty} x_i \mathrm{d}F(x_1, \cdots, x_n) = \int_{-\infty}^{+\infty} x \mathrm{d}F_i(x),$$

where $F_i(x)$ is the distribution function of ξ_i, that is, the i-th 1-dimensional marginal distribution of $F(x_1, \cdots, x_n)$. For the 2-dimensinal random vector (ξ, η) with distribution function $F(x, y)$ it follows

$$E\xi\eta = \int_{-\infty}^{+\infty} \int_{-\infty}^{+\infty} xy \mathrm{d}F(x, y)$$

and

$$E\xi^2 = \int_{-\infty}^{+\infty} \int_{-\infty}^{+\infty} x^2 \mathrm{d}F(x, y).$$

3.1.5 Basic properties of expectations

Property 1 If $a \leqslant \xi \leqslant b$, then $E\xi$ exists and $a \leqslant E\xi \leqslant b$. In particular, if $\xi = c$, then $E\xi = Ec = c$.

Property 2 If each of $E\xi_1, \cdots, E\xi_n$ exists, then for arbitrary constants $c_1, \cdots, c_n$ and b, $E(\sum_{i=1}^{n} c_i \xi_i + b)$ exists and

$$E(\sum_{i=1}^{n} c_i \xi_i + b) = \sum_{i=1}^{n} c_i E\xi_i + b. \tag{3.8}$$

Particularly, we have

$$E(\sum_{i=1}^{n} \xi_i) = \sum_{i=1}^{n} E\xi_i, \quad E(c\xi) = cE\xi.$$

By Properties 1 and 2, the expectation has monotonicity, i. e., $E\xi \leqslant E\eta$ when the expectations of ξ and η exist and $\xi \leqslant \eta$. In general, we have the following property.

Property 1′ If $|\xi| \leqslant \eta$ and $E\eta$ exists, then $E\xi$ exists and $|E\xi| \leqslant E|\xi| \leqslant E\eta$.

In fact, $P(|\xi| > x) \leqslant P(\eta > x)$ for any $x > 0$. So, $\int_0^{\infty} P(|\xi| > x)\mathrm{d}x \leqslant \int_0^{\infty} P(|\eta| > x)\mathrm{d}x = E\eta > \infty$. Then $E|\xi|$ exists and $E|\xi| \leqslant E\eta$. Further, by noting $E\xi$ and $E|\xi|$ exist, and $-|\xi| \leqslant \xi \leqslant |\xi|$, we conclude $E|\xi| \leqslant E\xi$.

Property 3 If $\xi_1, \cdots, \xi_n$ are independent random variables with finite mathematical expectations, then

$$E(\xi_1 \cdots \xi_n) = E\xi_1 \cdots E\xi_n. \tag{3.9}$$

Proof Since

$$\begin{aligned} E \mid \xi_1 \cdots \xi_n \mid &= \int_{-\infty}^{+\infty} \cdots \int_{-\infty}^{+\infty} \mid x_1 \cdots x_n \mid \mathrm{d}F(x_1, x_2, \cdots, x_n) \\ &= \int_{-\infty}^{+\infty} \mid x_1 \mid \mathrm{d}F_1(x_1) \cdots \int_{-\infty}^{+\infty} \mid x_n \mid \mathrm{d}F_n(x_n) \\ &< \infty, \end{aligned}$$

$E(\xi_1 \cdots \xi_n)$ exists. Similarly, we can prove formula (3.9).

Property 4 (Bounded convergence theorem) Suppose that $\lim_{n\to\infty} \xi_n(\omega) = \xi(\omega)$ for any $\omega \in \Omega$, and $\mid \xi_n \mid \leqslant M$, where M is a constant. Then

$$\lim_{n\to\infty} E\xi_n = E\xi.$$

Proof It is obvious that $\mid \xi \mid \leqslant M$. By Property 1, the expectations $E\xi_n$ and $E\xi$ both exist, and

$$\mid E\xi_n - E\xi \mid = \mid E(\xi_n - \xi) \mid \leqslant E \mid \xi_n - \xi \mid.$$

For an arbitrary $\varepsilon > 0$, let $A_n = \{\omega: \mid \xi_n(\omega) - \xi(\omega) \mid > \varepsilon\}$. Then $\lim_{n\to\infty} A_n = \emptyset$. By the continuity of the probability, $P(A_n) \to 0$. On the other hand,

$$\mid \xi_n - \xi \mid \leqslant \varepsilon + 2MI_{A_n}.$$

where I_{A_n} is the indicator of A_n which takes value 1 if the event A_n occurs and zero otherwise. So

$$\lim_{n\to\infty} E \mid \xi_n - \xi \mid \leqslant \lim_{n\to\infty} (\varepsilon + 2MP(A_n)) = \varepsilon.$$

By the arbitrariness of $\varepsilon > 0$, $\lim_{n\to\infty} E \mid \xi_n - \xi \mid = 0$. The result follows.

Remark The conclusion of Property 4 remains true when the constant M is replaced by a non-negative random variable with finite expectation, and the theorem is referred to as dominated convergence theorem.

Example 12 Suppose that ξ obeys the binomial distribution $B(n, p)$, find $E\xi$.

Solution In Example 3, we already calculated out $E\xi = np$. Next we apply Property 2 to find $E\xi$. Consider a Bernoulli trial and set $p = P(A)$ and

$$\xi_i = \begin{cases} 1, & A \text{ occurs in the } i\text{-th trial}; \\ 0, & A \text{ does not occur in the } i\text{-th trial}. \end{cases}$$

Thus ξ_i obeys $0-1$ distribution, $E\xi_i = p$ and $\xi = \sum_{i=1}^{n} \xi_i$. Hence $\sum_{i=1}^{n} E\xi_i = np$.

Example 13 Suppose that ξ is a random variable obeying the hypergeometric distribution, that is,

$$P(\xi = m) = \frac{\binom{M}{m}\binom{N-M}{n-m}}{\binom{N}{n}}, \qquad m = 0, 1, \cdots, n.$$

Find $E\xi$.

Solution Design a sampling without replacement. Let ξ_i be the number of defective goods in the i-th draw. Then $\xi = \sum_{i=1}^{n} \xi_i$. It is known from Section 1.2 that $P(\xi_i = 1) = M/N$, so

$$E\xi = \sum_{i=1}^{n} E\xi_i = \frac{nM}{N}.$$

Example 14 Suppose that $\xi_1, \cdots, \xi_n$ are identically distributed positive random variables with a common density function $f(x)$. Show that for any $1 \leqslant k \leqslant n$,

$$E\left(\frac{\xi_1 + \cdots + \xi_k}{\xi_1 + \cdots + \xi_n}\right) = \frac{k}{n}.$$

Proof Since $\xi_i/(\xi_1 + \cdots + \xi_n)$, $i = 1, 2, \cdots, n$ have a common distribution function, for any $i = 1, 2, \cdots, n$,

$$E\left(\frac{\xi_i}{\xi_1 + \cdots + \xi_n}\right) = E\left(\frac{\xi_1}{\xi_1 + \cdots + \xi_n}\right),$$

from which it follows that

$$\begin{aligned} 1 &= E\left(\frac{\xi_1 + \cdots + \xi_n}{\xi_1 + \cdots + \xi_n}\right) \\ &= E\left(\frac{\xi_1}{\xi_1 + \cdots + \xi_n}\right) + \cdots + E\left(\frac{\xi_n}{\xi_1 + \cdots + \xi_n}\right) \\ &= nE\left(\frac{\xi_1}{\xi_1 + \cdots + \xi_n}\right). \end{aligned}$$

Thus we obtain

$$E\left(\frac{\xi_1 + \cdots + \xi_k}{\xi_1 + \cdots + \xi_n}\right) = \frac{k}{n}.$$

Let us look at some applications of mathematical expectation below.

Example 15 Make a census on some kind of disease in a community with large population. Now check blood for N citizens in two ways. (1) Each person each time, so need check N times. (2) Check the mixture of blood of a group of k people. If the outcome reports no virus, that means all these k people are not of this disease; while if the outcome reports virus, then each person from this group is checked again, so k people need check $k+1$ times in this way. Which way may decrease the number of checks?

Solution Denote by ξ the number of times each person needs check in a group of k people in the second way. If the mixture of blood is checked once and the outcome reports no virus, then each person needs check $\xi = 1/k$ times and its probability is

$$P\left(\xi = \frac{1}{k}\right) = P\,(\text{none of } k \text{ people is sick}) = (1-p)^k.$$

Otherwise k people need check $k+1$ times, and each person needs check $\xi = 1 + 1/k$ times; its probability is

$$P\left(\xi = 1 + \frac{1}{k}\right) = 1 - (1-p)^k.$$

Hence we have

$$E\xi = \frac{1}{k}(1-p)^k + \left(1+\frac{1}{k}\right)(1-(1-p)^k)$$
$$= 1-(1-p)^k + \frac{1}{k}.$$

When $(1-p)^k - 1/k > 0$, $E\xi < 1$, it means this way can decrease the number of times of checking blood from the average point. For instance, if $p = 0.1$, letting $k = 4$, then $(1-p)^k - \frac{1}{k} = 0.4$, which can decrease by 40%. If p is fixed, we can find k_0 such that $E\xi$ attains its minimum.

Example 16 The demand amount ξ of some Chinese commodity in the international market each year follows a uniform distribution over [2 000,4 000]. The boss will gain 30 000 yuan by selling one ton of this commodity; while he will lose 10 000 yuan each ton if the commodity is placed in the inventory. How much commodity should be reserved such that the gain attains its maximum?

Solution The gain η is a random variable, so by saying maximum gain we mean the average gain attains its maximum. Note that η does not only depend on the demand amount ξ, but also depends on the production amount y. More precisely,

$$\eta = H(\xi) = \begin{cases} 30\,000y, & y \leqslant \xi; \\ 30\,000\xi - 10\,000(y-\xi), & \xi < y. \end{cases}$$

Also, observe ξ has density

$$p(x) = \frac{1}{2\,000}, \qquad 2\,000 \leqslant x \leqslant 4\,000.$$

By (3.6) we obtain

$$E\eta = \int_{-\infty}^{+\infty} H(x)p(x)\mathrm{d}x = \int_{2\,000}^{4\,000} \frac{H(x)}{2\,000}\mathrm{d}x$$
$$= \frac{1}{2\,000}\left[\int_{2\,000}^{y}(30\,000x - 10\,000(y-x))\mathrm{d}x + \int_{y}^{4\,000} 30\,000y\mathrm{d}x\right]$$
$$= \frac{1}{1\,000}(-y^2 + 7\,000y - 4\times 10^6).$$

It is easy to see $E\eta$ attains its maximum 82 500 000 at $y = 3\,500$.

3.1.6 Conditional expectation

In Section 2.5 of Chapter 2, we have introduced the conditional distribution. The conditional distribution has all the properties of a distribution function, and we can define its mathematical expectation. Such an expectation is called conditional expectation.

Suppose that $F_{\eta|\xi}(y \mid x)$ is the conditional distribution function of η given $\xi = x$. Then the conditional expectation is defined as

$$E(\eta \mid \xi = x) = \int_{-\infty}^{\infty} y dF_{\eta|\xi}(y \mid x). \tag{3.10}$$

If given the condition $\xi = x$, η has a conditional probability sequence $p_{\eta|\xi}(y_j \mid x)$, then

$$E(\eta \mid \xi = x) = \sum_j y_j p_{\eta|\xi}(y \mid x).$$

If given the condition $\xi=x$, η has a conditional density function $p_{\eta|\xi}(y|x)$, then

$$E(\eta \mid \xi = x) = \int_{-\infty}^{\infty} y p_{\eta|\xi}(y \mid x) dy.$$

Obviously, $E(\eta \mid \xi = x) = E\eta$ when ξ and η are independent.

In Section 2.5 of Chapter 2, it has been shown that, if (ξ, η) has a bivariate normal distribution $N(a, b, \sigma_1^2, \sigma_2^2, r)$, then under the condition $\xi = x$, the conditional distribution of η is a normal distribution $N\left(b + r\frac{\sigma_2}{\sigma_1}(x-a), \sigma_2^2(1-r^2)\right)$. So, by the formula of the expectation of a univariate normal distribution we obtain

$$E(\eta \mid \xi = x) = b + r\frac{\sigma_2}{\sigma_1}(x-a).$$

The conditional expectation $E(\eta \mid \xi = x)$ is a function of x. We denote it by $m(x)$. Then $m(\xi)$ is a random variable and is called the conditional expectation of η given ξ.

Denition 4 *Suppose that (ξ, η) is a random vector with $E|\eta| < \infty$. If $m(x)$ is the conditional expectation of η given $\xi = x$ as*

$$m(x) = E(\eta \mid \xi = x),$$

then the random variable $m(\xi)$ is said to be the conditional expectation of η given ξ, and written as $E(\eta \mid \xi)$.

By taking the mathematical expectation of the random variable $E(\eta \mid \xi)$, we obtain an interesting result as follows:

$$E[E(\eta \mid \xi)] = E\eta. \tag{3.11}$$

Here we only give a proof of the above equation for continuous random vector. Suppose that (ξ, η) has a joint density function $p(x, y)$ and accordingly, ξ has a density function $p_\xi(x) = \int_{-\infty}^{\infty} p(x, y) dy$. Then

$$p_{\eta|\xi}(y \mid x) = \frac{p(x, y)}{p_\xi(x)} \quad (p_\xi(x) \neq 0).$$

$$E(\eta \mid \xi = x) = m(x) = \int_{-\infty}^{\infty} y p_{\eta|\xi}(y \mid x) dy = \int_{-\infty}^{\infty} y \frac{p(x, y)}{p_\xi(x)} dy.$$

By the formula (3.5) on the expectation of function of a random variable, we obtain the desired result as follows

$$\begin{aligned} E[E(\eta \mid \xi)] &= Em(\xi) = \int_{-\infty}^{\infty} m(x) p_\xi(x) dx \\ &= \int_{-\infty}^{\infty} \left(\int_{-\infty}^{\infty} y \frac{p(x, y)}{p_\xi(x)} dy \right) p_\xi(x) dx \end{aligned}$$

$$= \int_{-\infty}^{\infty}\int_{-\infty}^{\infty} yp(x,y)\mathrm{d}x\mathrm{d}y = E\eta.$$

When ξ is a discrete random variable with the distribution sequence $p_i = P(\xi = x_i)$, then (3.11) becomes

$$E\eta = \sum_i p_i E(\eta \mid \xi = x_i) = \sum_i E(\eta \mid \xi = x_i)P(\xi = x_i).$$

The above formula is very similar to the total probability formula, and so is called the total expectation formula.

Example 17 Suppose that ξ_1, ξ_2, ⋯ is a sequence of random variables with a common binomial distribution $B(n,p)$. Let ν be a Poisson $P(\lambda)$ random variable and is independent of ξ_1, ξ_2, ⋯. Find $\mathrm{E}(\sum_{k=1}^{\nu}\xi_k)$, where $\sum_{k=1}^{0}(\cdot)$ is defined to be 0.

Solution Let $\eta = \sum_{k=1}^{\nu}\xi_k$. Then

$$E(\eta \mid \nu = r) = E(\sum_{k=1}^{r}\xi_k \mid \nu = r) = E(\sum_{k=1}^{r}\xi_k) = \sum_{k=1}^{r} E\xi_k = rnp.$$

So, $E(\eta \mid \nu) = \nu np$. Hence

$$E\eta = E(\nu np) = npE\nu = np\lambda.$$

Properties of the conditional expectation The conditional expectation $E(\cdot \mid \xi = x)$ is a kind of mathematical expectation, so it has all the properties of mathematical expectation. The following are some basic properties of the conditional expectation.

(1) If the expectations of η_1 and η_2 exist, then

$$E(a_1\eta_1 + a_2\eta_2 \mid \xi = x) = a_1E(\eta_1 \mid \xi = x) + a_2E(\eta_2 \mid \xi = x).$$

(2) If the expectation of $g(\eta)$ exists, then

$$E(g(\eta) \mid \xi = x) = \int_{-\infty}^{\infty} g(y)\mathrm{d}F_{\eta|\xi}(y \mid x).$$

If under the condition $\xi = x$, η has a conditional distribution sequence $p_{\eta|\xi}(y_j|x)$, then

$$E(g(\eta) \mid \xi = x) = \sum_j g(y_j)p_{\eta|\xi}(y_j \mid x);$$

If under the condition $\xi = x$, η has a conditional density function $p_{\eta|\xi}(y|x)$, then

$$E(g(\eta) \mid \xi = x) = \int_{-\infty}^{\infty} g(y_j)p_{\eta|\xi}(y_j \mid x)\mathrm{d}y.$$

(3) If the expectations of η_1, η_2 exist, and $\eta_1 \leqslant \eta_2$, then

$$E(\eta_1 \mid \xi = x) \leqslant E(\eta_2 \mid \xi = x).$$

(4) $E(h(\xi)\eta \mid \xi) = h(\xi)E(\eta \mid \xi)$.

(5) (Cauchy-Schwarz's inequality) $| E(XY \mid Z) | \leqslant \sqrt{E(X^2 \mid Z)}\sqrt{\mathrm{E}(Y^2 \mid Z)}$.

Example 18 (Quick-sort algorithm) Suppose that we are presented with a set of n distinct values x_1, x_2, ⋯, x_n and that we desire to put them in an increasing order as $x_{(1)} < x_{(2)} < \cdots < x_{(n)}$, or as it is commonly stated, to sort them. To sort these n values, if we compare them pair by pair, and then we will have totally $\frac{n(n-1)}{2}$ different pairs to compare. A natural question is, is there any procedure for sorting n distinct values such that the number of comparisons is less than $\frac{n(n-1)}{2}$? An efficient procedure for accomplishing this task is the quick-sort algorithm, which is defined as follows: When $n=2$, the algorithm compares the two values and then puts them in the appropriate order. When $n>2$, one of the elements is randomly chosen from the set $\{x_1, x_2, \cdots, x_n\}$ —say it is x_J—and then all of the other values are compared with x_J. Those smaller than x_J are put in a bracket to the left of x_J and those larger than x_J are put in a bracket to the right of x_J. The algorithm then repeats itself on these brackets and continues until all values have been sorted. Let ξ denote the number of comparisons that it takes the quick-sort algorithm to sort n distinct numbers. Find $q_n = E\xi$.

Solution After x_J is randomly chosen and the other values are compared with x_J, let L (resp. R) be the set of those values which are put to the bracket to the left (resp. right) of x_J, and let ξ_L, ξ_R denote the number of comparisons that it takes the quick-sort algorithm to sort numbers in L *and* R respectively. Then

$$\xi = \xi_L + \xi_R + (n-1).$$

Given the condition $x_J = x_{(i)}$, there are $i-1$ and $n-i$ distinct numbers in L and R respectively. So,

$$E(\xi_L \mid x_J = x_{(i)}) = q_{i-1}, \ E(\xi_R \mid x_J = x_{(i)}) = q_{n-i}.$$

It follows that

$$\begin{aligned} E(\xi \mid x_J = x_{(i)}) &= E(\xi_L \mid x_J = x_{(i)}) + E(\xi_R \mid x_J = x_{(i)}) + (n-1) \\ &= q_{i-1} + q_{n-i} + (n-1). \end{aligned}$$

On the other hand, $P(x_J = x_{(i)}) = \frac{1}{n}$ since x_J is chosen randomly. We obtain

$$\begin{aligned} q_n = E\xi &= \sum_{i=1}^{n} E(\xi \mid x_J = x_{(i)}) P(x_J = x_{(i)}) \\ &= n - 1 + \frac{1}{n}\sum_{i=1}^{n}(q_{i-1} + q_{n-i}) = n - 1 + \frac{2}{n}\sum_{i=1}^{n} q_{i-1}, \end{aligned}$$

where $q_0 = q_1 = 0$, $q_2 = 1$. Therefore

$$nq_n = n(n-1) + 2\sum_{i=1}^{n} q_{i-1},$$

$$nq_n - (n-1)q_{n-1} = n(n-1) - (n-1)(n-2) + 2q_{n-1}.$$

Then $nq_n = 2(n-1) + (n+1)q_{n-1}$. It follows that

$$\begin{aligned} \frac{q_n}{n+1} &= \frac{q_{n-1}}{n} + \frac{2(n-1)}{n(n+1)} = \frac{q_{n-1}}{n} + \frac{2}{n} + 4\left(\frac{1}{n+1} - \frac{1}{n}\right) \\ &= \cdots = 2\sum_{k=1}^{n} \frac{1}{k} + 4\left(\frac{1}{n+1} - \frac{1}{n}\right) \\ &= 2\left(\log n + \gamma + \frac{1}{2n} + O(n^{-2})\right) - \frac{4n}{n+1}, \end{aligned}$$

where γ is the Euler constant. We conclude that

$$q_n = 2(n+1)\log n + n(2\gamma - 4) + 2\gamma + 1 + O(n^{-1}).$$

q_n is obviously much smaller than $\frac{n(n-1)}{2}$.

Example 19* (Best predictor) Suppose $E\eta^2 < \infty$. Let $m(\xi) = E(\eta \mid \xi)$. Prove that for any real function $g(x)$, we have

$$E(\eta - m(\xi))^2 \leqslant E(\eta - g(\xi))^2. \tag{3.12}$$

Proof When $Eg^2(\xi) = \infty$, by the inequality $b^2 \leqslant 2(a-b)^2 + 2a^2$ we have

$$\infty = Eg^2(\xi) \leqslant 2E(\eta - g(\xi))^2 + 2E\eta^2$$

So, $E(\eta - g(\xi))^2 = \infty$. (3.12) follows.

Next, we assume $Eg^2(\xi) < \infty$. By the Cauchy-Schwarz's inequality,

$$(E(\eta \mid \xi = x))^2 \leqslant E(\eta^2 \mid \xi = x),$$

which is $(E(\eta \mid \xi))^2 \leqslant E(\eta^2 \mid \xi)$. It follows that $Em^2(\xi) \leqslant E\eta^2$. Now, denote $h(\xi) = m(\xi) - g(\xi)$. Then $Eh^2(\xi) < \infty$. We have

$$\begin{aligned} E(\eta - g(\xi))^2 &= E(\eta - m(\xi) + m(\xi) - g(\xi))^2 \\ &= E(\eta - m(\xi))^2 + Eh^2(\xi) + 2E[h(\xi)(\eta - m(\xi))]. \end{aligned}$$

By the properties of the conditional expectation,

$$\begin{aligned} E[h(\xi)(\eta - m(\xi)) \mid \xi] &= h(\xi)E(\eta - m(\xi) \mid \xi) \\ &= h(\xi)(E(\eta \mid \xi) - m(\xi)) = 0. \end{aligned}$$

Hence

$$E[h(\xi)(\eta - m(\xi))] = E(E[h(\xi)(\eta - m(\xi)) \mid \xi]) = 0.$$

It follows that

$$\begin{aligned} E(\eta - g(\xi))^2 &= E(\eta - m(\xi))^2 + E(m(\xi) - g(\xi))^2 \\ &\geqslant E(\eta - m(\xi))^2. \end{aligned}$$

The result follows.

Sometime a situation arises where the value of a random variable ξ is observed and then, based on the observed value, an attempt is to predict the value of a second random variable η. Let $g(\xi)$ denote the predictor. That is, if $\xi = x$ is observed, then $g(x)$ is the prediction for the value η. Clearly, we would like to choose g so that $g(\xi)$ tends to be close to η. One possible criterion for the closeness is to choose g so that minimize $E(\eta - g(\xi))^2$. The inequality (3.12) shows that $m(\xi)$ is the best predictor of η under this criterion. In statistics, $m(x)$ is called the regression function of η with respect to ξ. Also, from the above proof we can find that, if $Eh^2(\xi) < \infty$, then

$$E[(\eta - m(\xi))h(\xi)] = 0.$$

So, $m(\xi) = E(\eta \mid \xi)$ is also an orthogonal projection of η onto the space $\mathscr{L}^2(\xi) = \{h(\xi): Eh^2(\xi) < \infty\}$.

3.2 Variances, covariances and correlation coefficients

3.2.1 Variances

Example 20 Let ξ and η denote the numbers of times two shooters, say A and B, hit the target respectively. Suppose that ξ and η obey respectively the following distributions

$$\begin{pmatrix} 7 & 8 & 9 \\ 0.1 & 0.8 & 0.1 \end{pmatrix}$$

and

$$\begin{pmatrix} 6 & 7 & 8 & 9 & 10 \\ 0.1 & 0.2 & 0.4 & 0.2 & 0.1 \end{pmatrix}.$$

Which one is better at shooting?

At first let us look at the average number of times these two people hitting the target. Obviously, $E\xi = E\eta$, it is impossible from means to decide who is better. But intuitively speaking, A hits most of time No. 8 cycle, while B sometimes hits No. 10 cycle and other times No. 6 cycle, so B seems instable. Thus it is reasonable to say that A is better at shooting than B.

The above example shows that we need to consider the deviation extent to which it takes values, besides its mean for a random variable.

We call $\xi - E\xi$ the deviation of ξ from its mean $E\xi$. It is still a random variable. To give a numerical constant describing the deviation extent, the first choice is to use $E(\xi - E\xi)$. But since $E(\xi - E\xi) = 0$ holds for all random variables with finite expectation, that is, the positive and negative deviations cancelling each other, it does not make sense to use $E(\xi - E\xi)$. Instead, we use $E(\xi - E\xi)^2$ to describe the deviation extent to which the ξ takes values, and it is just the variance.

Definition 5 *If $E(\xi - E\xi)^2$ exists and is a finite constant, then we call it the variance of ξ, and write $Var\xi$ or $D\xi$, i.e.,*

$$Var\xi = E(\xi - E\xi)^2. \tag{3.13}$$

But $\mathrm{Var}\xi$ and ξ have different dimensions. To unify, we sometimes use $\sqrt{\mathrm{Var}\xi}$, called the standard deviation of ξ.

Variance is just the expected value of the function $(\xi - E\xi)^2$, by formula (3.5) of Section 3.1 we easily obtain

$$\mathrm{Var}\xi = \int_{-\infty}^{+\infty} (x - E\xi)^2 \mathrm{d}F_\xi(x)$$

$$=\begin{cases}\sum_i (x_i - E\xi)^2 P(\xi = x_i), & \text{(discrete)};\\ \int_{-\infty}^{+\infty} (x - E\xi)^2 p_\xi(x)\mathrm{d}x, & \text{(continuous)}.\end{cases} \tag{3.14}$$

Moreover, note that

$$\begin{aligned} E(\xi - E\xi)^2 &= E[\xi^2 - 2\xi E\xi + (E\xi)^2] \\ &= E\xi^2 - (E\xi)^2, \end{aligned} \tag{3.15}$$

i.e.,

$$\mathrm{Var}\xi = E\xi^2 - (E\xi)^2. \tag{3.16}$$

In many cases, it is more convenient to compute variances by (3.16).

Example 21 Find $\mathrm{Var}\xi$ and $\mathrm{Var}\eta$ of ξ and η in Example 20.

Solution Applying (3.16), we have

$$E\xi^2 = \sum_i x_i^2 P(\xi = x_i) = 64.2,$$

so

$$\mathrm{Var}\xi = E\xi^2 - (E\xi)^2 = 0.2.$$

Similarly, $\mathrm{Var}\eta = E\eta^2 - (E\eta)^2 = 65.2 - 64 = 1.2$. Thus $\mathrm{Var}\eta > \mathrm{Var}\xi$ and so η takes its values more dispersedly, which implies A shoots better.

Example 22 Find the variance of the Poisson distribution $P(\lambda)$.

Solution We have

$$\begin{aligned} E\xi^2 &= \sum_{k=0}^{\infty} k^2 \frac{\lambda^k}{k!}\mathrm{e}^{-\lambda} = \sum_{k=1}^{\infty} k \frac{\lambda^k}{(k-1)!}\mathrm{e}^{-\lambda} \\ &= \sum_{k=1}^{\infty} (k-1) \frac{\lambda^k}{(k-1)!}\mathrm{e}^{-\lambda} + \sum_{k=1}^{\infty} \frac{\lambda^k}{(k-1)!}\mathrm{e}^{-\lambda} \\ &= \lambda \sum_{j=0}^{\infty} j \frac{\lambda^j}{j!}\mathrm{e}^{-\lambda} + \lambda \sum_{j=0}^{\infty} \frac{\lambda^j}{j!}\mathrm{e}^{-\lambda} = \lambda^2 + \lambda. \end{aligned}$$

So, $\mathrm{Var}\xi = \lambda^2 + \lambda - \lambda^2 = \lambda$.

Example 23 Suppose that ξ is uniform over $[a, b]$, find $\mathrm{Var}\xi$.

Solution It is easy to see

$$E\xi^2 = \int_a^b x^2 \frac{1}{b-a}\mathrm{d}x = \frac{1}{3}(a^2 + ab + b^2).$$

Applying (3.16), we have

$$\mathrm{Var}\xi = \frac{1}{3}(a^2 + ab + b^2) - \left(\frac{a+b}{2}\right)^2 = \frac{1}{12}(b-a)^2.$$

Example 24 Assume that ξ is normally distributed as $N(a, \sigma^2)$, find $\mathrm{Var}\xi$.

Solution Recall $E\xi = a$. Applying (3.14), we have

$$\begin{aligned} \mathrm{Var}\xi = E(\xi - a)^2 &= \frac{1}{\sqrt{2\pi}\sigma}\int_{-\infty}^{+\infty} (x-a)^2 \frac{\mathrm{e}^{-(x-a)^2/2\sigma^2}}{\sqrt{2\pi}\sigma}\mathrm{d}x \\ &= \frac{\sigma^2}{\sqrt{2\pi}}\int_{-\infty}^{+\infty} z^2 \mathrm{e}^{-\frac{z^2}{2}}\mathrm{d}z \end{aligned}$$

$$= \frac{\sigma^2}{\sqrt{2\pi}}\left(-ze^{-\frac{z^2}{2}}\Big|_{-\infty}^{+\infty} + \int_{-\infty}^{+\infty} e^{-\frac{z^2}{2}}\mathrm{d}z\right)$$
$$= \sigma^2.$$

This shows that the parameter σ^2 is just its variance, σ its standard deviation.

Variances have a few simple and important properties. Before preceding, we introduce an important inequality.

Chebyshev's inequality. *Assume that the random variable ξ has variance, then it follows for any given $\varepsilon>0$ that*

$$P(|\xi - E\xi| \geqslant \varepsilon) \leqslant \frac{\mathrm{Var}\xi}{\varepsilon^2}. \tag{3.17}$$

Proof Let $F(x)$ be the distribution function of ξ, then

$$\begin{aligned} P(|\xi - E\xi| \geqslant \varepsilon) &= \int_{|x-E\xi|\geqslant\varepsilon} \mathrm{d}F(x) \\ &\leqslant \int_{|x-E\xi|\geqslant\varepsilon} \frac{(x-E\xi)^2}{\varepsilon^2}\mathrm{d}F(x) \\ &\leqslant \frac{1}{\varepsilon^2}\int_{-\infty}^{+\infty}(x-E\xi)^2\mathrm{d}F(x) \\ &= \frac{\mathrm{Var}\xi}{\varepsilon^2}, \end{aligned}$$

which is the desired inequality.

In probability theory, the estimation of the probability $P(|\xi - E\xi| \geqslant \varepsilon)$ is quite useful. We frequently employ a moment, say a variance, as its upper estimation.

Property 1 $\mathrm{Var}\xi = 0$ if and only if $P(\xi = c) = 1$, where c is a constant.

Proof The sufficiency is trivial. Conversely, if $\mathrm{Var}\xi = 0$, letting $E\xi = c$, by Chebyshev's inequality, it follows that

$$P(|\xi - E\xi| \geqslant \varepsilon) = 0$$

for any $\varepsilon > 0$. Thus

$$\begin{aligned} P(\xi = c) &= 1 - P(|\xi - c| > 0) \\ &= 1 - \lim_{n\to\infty} P\left(|\xi - c| \geqslant \frac{1}{n}\right) = 1. \end{aligned}$$

Property 2 Let c, b be constants. Then

$$\mathrm{Var}(c\xi + b) = c^2\mathrm{Var}\xi. \tag{3.18}$$

Proof It follows immediately from the definition of a variance

$$\begin{aligned} \mathrm{Var}(c\xi + b) &= E(c\xi + b - E(c\xi + b))^2 \\ &= c^2E(\xi - E\xi)^2 = c^2\mathrm{Var}\xi. \end{aligned}$$

Property 3 If $c \neq E\xi$, then $\mathrm{Var}\xi < E(\xi - c)^2$.

Proof Recall $\mathrm{Var}\xi = E\xi^2 - (E\xi)^2$. Also note that

$$E(\xi - c)^2 = E\xi^2 - 2cE\xi + c^2.$$

A simple computation completes the proof. This shows the deviation extent of the ξ around $E\xi$ is minimum.

Property 4 $\operatorname{Var}(\sum_{i=1}^{n}\xi_i)=\sum_{i=1}^{n}\operatorname{Var}\xi_i+2\sum_{1\leqslant i<j\leqslant n}E(\xi_i-E\xi_i)(\xi_j-E\xi_j)$. (3.19)

In particular, if $\xi_1,\cdots,\xi_n$ are pairwise independent, then

$$\operatorname{Var}(\sum_{i=1}^{n}\xi_i)=\sum_{i=1}^{n}\operatorname{Var}\xi_i. \tag{3.20}$$

Proof

$$\begin{aligned}
&\operatorname{Var}(\sum_{i=1}^{n}\xi_i)\\
&=E(\sum_{i=1}^{n}\xi_i-E\sum_{i=1}^{n}\xi_i)^2=E(\sum_{i=1}^{n}(\xi_i-E\xi_i))^2\\
&=E\Big[\sum_{i=1}^{n}(\xi_i-E\xi_i)^2+2\sum_{1\leqslant i<j\leqslant n}(\xi_i-E\xi_i)(\xi_j-E\xi_j)\Big]\\
&=\sum_{i=1}^{n}\operatorname{Var}\xi_i+2\sum_{1\leqslant i<j\leqslant n}E(\xi_i-E\xi_i)(\xi_j-E\xi_j).
\end{aligned}$$

This completes the proof of (3.19).

When $\xi_1,\cdots,\xi_n$ are pairwise independent, $\xi_1-E\xi_1,\cdots,\xi_n-E\xi_n$ are also pairwise independent by Example 39 in Chapter 2. So,

$$\begin{aligned}
E(\xi_i-E\xi_i)(\xi_j-E\xi_j)&=E(\xi_i-E\xi_i)E(\xi_i-E\xi_j)\\
&=0.
\end{aligned}$$

This proves (3.20).

Applying these properties simplifies computation of variances of some random variables.

Example 25 Assume that ξ follows the binomial distribution $B(n, p)$, find $\operatorname{Var}\xi$.

Solution Construct ξ_i, $i=1,\cdots,n$, as in Example 12. These are independent and identically distributed random variables. Then

$$\operatorname{Var}\xi_i=E\xi_i^2-(E\xi_i)^2=1^2\cdot p+0^2\cdot q-p^2=pq.$$

By Property 4 we have

$$\operatorname{Var}\xi=\operatorname{Var}(\sum_{i=1}^{n}\xi_i)=\sum_{i=1}^{n}\operatorname{Var}(\xi_i)=npq.$$

Example 26 Suppose that $\xi_1,\cdots,\xi_n$ are independent identically distributed random variables, and $E\xi_i=a$, $\operatorname{Var}\xi_i=\sigma^2$. Let $\bar{\xi}=\frac{1}{n}\sum_{i=1}^{n}\xi_i$, and find $E\bar{\xi}$ and $\operatorname{Var}\bar{\xi}$.

Solution Combining Property 2 in Section 3.1 and Properties 2 and 4 in this section together, we have

$$E\bar{\xi}=\frac{1}{n}\sum_{i=1}^{n}E\xi_i=a,$$

$$\operatorname{Var}\bar{\xi}=\frac{1}{n^2}\sum_{i=1}^{n}\operatorname{Var}\xi_i=\frac{1}{n^2}n\sigma^2=\frac{\sigma^2}{n}.$$

This shows that, in the case of independent identically distributed random variables, as the arithmetic mean of the ξ_i, $\bar{\xi}$ has the same expected value as each ξ_i, but the variance is as small as $1/n$ times $\text{Var}\xi$. This fact plays an important role in mathematical statistics.

Example 27 Suppose that the random variable ξ has finite expectation and positive variance. Let

$$\xi^* = \frac{\xi - E\xi}{\sqrt{\text{Var}\xi}}.$$

We call ξ^* the standardized random variable of ξ. Find $E\xi^*$ and $\text{Var}\xi^*$.

Solution By the above properties on expectations and variances, it easily follows that

$$E\xi^* = \frac{E(\xi - E\xi)}{\sqrt{\text{Var}\xi}} = 0$$

and

$$\text{Var}\xi^* = \frac{\text{Var}(\xi - E\xi)}{\text{Var}\xi} = \frac{\text{Var}\xi}{\text{Var}\xi} = 1.$$

3.2.2 Covariances

We have seen that expectation and variance describe distribution characteristics of random variables. For random vectors, say, $(\xi_1, \xi_2, \cdots, \xi_n)'$, besides expectation and variance of each coordinate, there is another numerical characteristic, called covariance, which expresses the connection between coordinate random variables.

Definition 6 *Let $F_{ij}(x, y)$ be the joint distribution of ξ_i and ξ_j. If $E|(\xi_i - E\xi_i)(\xi_j - E\xi_j)| < \infty$, we call*

$$E(\xi_i - E\xi_i)(\xi_j - E\xi_j) = \int_{-\infty}^{+\infty}\int_{-\infty}^{+\infty}(x - E\xi_i)(y - E\xi_j)\mathrm{d}F_{ij}(x, y) \tag{3.21}$$

the covariance of ξ_i and ξ_j, written as $\text{Cov}(\xi_i, \xi_j)$.

It is obvious that $\text{Cov}(\xi_i, \xi_i) = \text{Var}\xi_i$. Also, formula (3.19) can be written as

$$\text{Var}\left(\sum_{i=1}^n \xi_i\right) = \sum_{i=1}^n \text{Var}\xi_i + 2\sum_{1\leqslant i<j\leqslant n}\text{Cov}(\xi_i, \xi_j). \tag{3.22}$$

One readily verifies the following properties.

Property 1 $\text{Cov}(\xi, \eta) = \text{Cov}(\eta, \xi) = E\xi\eta - E\xi E\eta.$

Property 2 Assume that a, b are constants, then

$$\text{Cov}(a\xi, b\eta) = ab\text{Cov}(\xi, \eta).$$

Property 3 $\text{Cov}\left(\sum_{i=1}^n \xi_i, \eta\right) = \sum_{i=1}^n \text{Cov}(\xi_i, \eta).$

Given an n-dimensional random vector $\boldsymbol{\xi} = (\xi_1, \cdots, \xi_n)'$, one can write its covariance matrix as

$$B = E(\xi - E\xi)(\xi - E\xi)' = \begin{pmatrix} b_{11} & b_{12} & \cdots & b_{1n} \\ b_{21} & b_{22} & \cdots & b_{2n} \\ \vdots & \vdots & & \vdots \\ b_{n1} & b_{n2} & \cdots & b_{nn} \end{pmatrix}, \tag{3.23}$$

where $b_{ij} = \mathrm{Cov}(\xi_i, \xi_j)$.

From the definition and Properties 2,3, it follows that $\boldsymbol{B}$ is a symmetric matrix and for any real numbers t_j, $j = 1, 2, \cdots, n$,

$$\sum_{j,k=1}^{n} t_j t_k b_{jk} = \sum_{j,k=1}^{n} t_j t_k E(\xi_j - E\xi_j)(\xi_k - E\xi_k)$$
$$= E\left(\sum_{j=1}^{n} t_j (\xi_j - E\xi_j)\right)^2 \geqslant 0.$$

That is, the covariance matrix $\boldsymbol{B}$ is nonnegative definite.

Property 4 Let

$$\xi = (\xi_1, \xi_2, \cdots, \xi_n)', \quad C = \begin{pmatrix} C_{11} & C_{12} & \cdots & C_{1n} \\ C_{21} & C_{22} & \cdots & C_{2n} \\ \vdots & \vdots & & \vdots \\ C_{n1} & C_{n2} & \cdots & C_{nn} \end{pmatrix},$$

then $\boldsymbol{C\xi}$ has covariance matrix $\boldsymbol{CBC'}$, where $\boldsymbol{B}$ is covariance matrix of $\boldsymbol{\xi}$.

Indeed, it is easy to see

$$E(C\xi - E(C\xi))(C\xi - E(C\xi))' = CBC'.$$

Hence the (i, j)-th entry in $\boldsymbol{CBC'}$ is just the covariance of the i-th entry and the j-th entry in $\boldsymbol{C\xi}$.

Property 5 Let $\boldsymbol{\xi} = (\xi_1, \cdots, \xi_n)'$ have a multi-normal distribution $N(\boldsymbol{\mu}, \boldsymbol{B})$, where $\boldsymbol{\mu}$ is an n-dimensional real vector, and $\boldsymbol{B}$ is an $n \times n$ positive definite symmetric matrix. Then the expectation of $\boldsymbol{\eta}$ is $\boldsymbol{\mu}$, and covariance matrix of ξ is $\boldsymbol{B}$.

In fact, in the special case that $\boldsymbol{\mu} = \boldsymbol{0}$ and $\boldsymbol{B} = \boldsymbol{I}$, $\xi_1, \cdots, \xi_n$ are independent standard normal random variables, and so $E\xi_i = 0$, $\mathrm{Var}(\xi_i) = 1$, $\mathrm{Cov}(\xi_i, \xi_j) = 0$, $i \neq j$. That is, the random vector $\boldsymbol{\xi}$ has expectation $\boldsymbol{0}$ and covariance matrix $\boldsymbol{I}$. In the general case, we can write $\boldsymbol{B}$ as $\boldsymbol{B} = \boldsymbol{LL'}$, where $\boldsymbol{L}$ is a positive definite symmetric matrix. Then $\boldsymbol{\xi}$ can be written as $\boldsymbol{\xi} = \boldsymbol{L\eta} + \boldsymbol{\mu}$, where $\boldsymbol{\eta} = \boldsymbol{L}\ {-1}\ (\boldsymbol{\xi} - \boldsymbol{\mu}) \sim N(\boldsymbol{0}, \boldsymbol{I})$. Hence, $\boldsymbol{\eta}$ has expectation $\boldsymbol{0}$ and covariance matrix $\boldsymbol{I}$. It follows that $\mathrm{E}\boldsymbol{\xi} = \boldsymbol{L}(\mathrm{E}\boldsymbol{\eta}) + \boldsymbol{\mu} = \boldsymbol{\mu}$, and by Property 4, the covariance matrix of $\boldsymbol{\xi}$ is $\boldsymbol{LIL'} = \boldsymbol{B}$ as desired.

3.2.3 Correlation coefficients

Covariance describes the extent of the connection between two random variables, but the size of $\mathrm{Cov}(\xi, \eta)$ depends on the scale of ξ and η. To avoid this kind of dependence, we shall use standardized random variables (see Example 27) in the argument below.

Definition 7 *We call*

$$r_{\xi\eta} = \mathrm{Cov}(\xi^*, \eta^*) = \frac{E(\xi - E\xi)(\eta - E\eta)}{\sqrt{\mathrm{Var}\xi \mathrm{Var}\eta}} \tag{3.24}$$

the correlation coefficient of ξ *and* η.

Before proceeding, let us give an important inequality.

Cauchy-Schwarz's inequality. For any pair of random variables ξ and η, it holds that

$$| E\xi\eta |^2 \leqslant E\xi^2 E\eta^2, \tag{3.25}$$

and the equation is valid if and only if there is a constant t_0 such that

$$P(\eta = t_0\xi) = 1. \tag{3.26}$$

Proof Define for an arbitrary real number t

$$u(t) = E(t\xi - \eta)^2 = t^2 E\xi^2 - 2tE\xi\eta + E\eta^2.$$

This can be viewed as a nonnegative quadratic function in t, so its discriminant is

$$(E\xi\eta)^2 - E\xi^2 E\eta^2 \leqslant 0,$$

which is the desired result. It is not hard to see that the equation (3.25) is valid if and only if $u(t)$ has a double root t_0. In other words,

$$u(t_0) = E(t_0\xi - \eta)^2 = 0,$$

which implies from the proof of Property 1 of variance that $P(t_0\xi - \eta = 0)$, as desired.

From the above we obtain an important property on correlation coefficients.

Property 1 Let $r_{\xi\eta}$ be the correlation coefficient, then

$$| r_{\xi\eta} | \leqslant 1. \tag{3.27}$$

Also, $r_{\xi\eta} = 1$ if and only if

$$P\left(\frac{\xi - E\xi}{\sqrt{\mathrm{Var}\xi}} = \frac{\eta - E\eta}{\sqrt{\mathrm{Var}\eta}}\right) = 1;$$

$r_{\xi\eta} = -1$ if and only if

$$P\left(\frac{\xi - E\xi}{\sqrt{\mathrm{Var}\xi}} = -\frac{\eta - E\eta}{\sqrt{\mathrm{Var}\eta}}\right) = 1.$$

Proof It follows from (3.25) that

$$| r_{\xi\eta} | = | E\xi^* \eta^* | \leqslant \sqrt{E\xi^{*2} E\eta^{*2}} = \sqrt{\mathrm{Var}\xi^* \mathrm{Var}\eta^*} = 1$$

as desired.

Next let us turn to the second conclusion. By the definition of correlation coefficient, $r_{\xi\eta} = r_{\xi^*\eta^*}$. Also, it is easy to see from the proof of Cauchy-Schwarz's inequality that $| r_{\xi\eta} | = 1$ is equivalent to saying

$$u(t) = t^2 E\xi^{*2} - 2tE\xi^* \eta^* + E\eta^{*2}$$

has a multi-root $t_0 = E\xi^* \eta^* / E\xi^{*2} = r_{\xi\eta}$. Therefore we know from (3.26) that $r_{\xi\eta} = 1$ if and only if $P(\xi^* = \eta^*) = 1$, similarly $r_{\xi\eta} = -1$ if and only if $P(\xi^* = -\eta^*) = 1$.

Remark Property 1 shows that when the correlation coefficient $r_{\xi\eta} = \pm 1$, there almost surely exists a kind of linear relation between ξ and η. When $r_{\xi\eta} = 0$, we say ξ and η are uncorrelated.

Property 2 The following statements are equivalent

(1) $\mathrm{Cov}(\xi,\eta)=0$;

(2) ξ and η are uncorrelated;

(3) $E\xi\eta=E\xi E\eta$;

(4) $\mathrm{Var}(\xi+\eta)=\mathrm{Var}\xi+\mathrm{Var}\eta$.

Proof Obviously, (1) and (2) are equivalent. Also, by Property 1 of covariances one easily proves (1) and (3) are equivalent. Finally, by (3. 22), (1) and (4) are equivalent.

Property 3 If ξ and η are independent, then ξ and η are uncorrelated.

The proof is trivial, but the reverse is not true.

Example 28 Suppose that θ is uniform over $[0,2\pi]$. Let $\xi=\cos\theta,\eta=\sin\theta$. Since $\xi^2+\eta^2=1$, ξ and η are not independent. However ξ and η are uncorrelated. Indeed, a simple computation shows

$$E\xi=E\cos\theta=\int_0^{2\pi}\frac{1}{2\pi}\cos\varphi\mathrm{d}\varphi=0,$$

$$E\eta=E\sin\theta=\int_0^{2\pi}\frac{1}{2\pi}\sin\varphi\mathrm{d}\varphi=0,$$

and

$$E\xi\eta=E\sin\theta\cos\theta=\int_0^{2\pi}\frac{1}{2\pi}\sin\varphi\cos\varphi\mathrm{d}\varphi=0.$$

Thus $\mathrm{Cov}(\xi,\eta)=E\xi\eta-E\xi E\eta=0$.

Remark Property 2 of covariances cannot be extended to the case of $n(n\geqslant 3)$ random variables. In fact, the condition that $\xi_1,\cdots,\xi_n$ are pairwise uncorrelated implies only $\mathrm{Var}\left(\sum_{i=1}^n\xi_i\right)=\sum_{i=1}^n\mathrm{Var}\xi_i$, but does not lead to $E(\xi_1\cdots\xi_n)=E\xi_1\cdots E\xi_n$. As for Property 3, a useful observation is that the independence is consistent with the uncorrelated property in the case of normal distribution. This will be further discussed below.

Example 29 Assume that ξ,η are jointly normally distributed as $N(a,b,\sigma_1^2,\sigma_2^2,r)$, find $\mathrm{Cov}(\xi,\eta)$ and $r_{\xi\eta}$.

Solution

$$\begin{aligned}\mathrm{Cov}(\xi,\eta)&=\int_{-\infty}^{+\infty}\int_{-\infty}^{+\infty}(x-a)(y-b)p(x,y)\mathrm{d}x\mathrm{d}y\\&=\frac{1}{2\pi\sigma_1\sigma_2\sqrt{1-r^2}}\int_{-\infty}^{+\infty}\int_{-\infty}^{+\infty}(x-a)(y-b)\\&\quad\cdot\exp\left(-\frac{1}{2(1-r^2)}\left(\frac{x-a}{\sigma_1}-r\frac{y-b}{\sigma_2}\right)^2-\frac{(y-b)^2}{2\sigma_2^2}\right)\mathrm{d}x\mathrm{d}y.\end{aligned}$$

Let

$$z=\frac{x-a}{\sigma_1}-r\frac{y-b}{\sigma_2},\quad t=\frac{y-b}{\sigma_2},$$

then

$$\frac{x-a}{\sigma_1}=z+rt,\quad J=\left|\frac{\partial(x,y)}{\partial(z,t)}\right|=\sigma_1\sigma_2.$$

Thus we have

$$\begin{aligned}&\operatorname{Cov}(\xi,\eta)\\&=\frac{\sigma_1\sigma_2}{2\pi\sqrt{1-r^2}}\int_{-\infty}^{+\infty}\int_{-\infty}^{+\infty}(zt+rt^2)\mathrm{e}^{-\frac{z^2}{2(1-r^2)}}\mathrm{e}^{-\frac{t^2}{2}}\mathrm{d}z\mathrm{d}t\\&=\sigma_1\sigma_2\frac{1}{\sqrt{2\pi}}\int_{-\infty}^{+\infty}t\mathrm{e}^{-\frac{t^2}{2}}\mathrm{d}t\cdot\frac{1}{\sqrt{2\pi}\sqrt{1-r^2}}\int_{-\infty}^{+\infty}z\mathrm{e}^{-\frac{z^2}{2(1-r^2)}}\mathrm{d}z\\&\quad+\frac{r\sigma_1\sigma_2}{\sqrt{2\pi}}\int_{-\infty}^{+\infty}t^2\mathrm{e}^{-\frac{t^2}{2}}\mathrm{d}t\cdot\frac{1}{\sqrt{2\pi}\sqrt{1-r^2}}\int_{-\infty}^{+\infty}\mathrm{e}^{-\frac{z^2}{2(1-r^2)}}\mathrm{d}z\\&=r\sigma_1\sigma_2.\end{aligned}$$

Therefore

$$r_{\xi\eta}=\frac{\operatorname{Cov}(\xi,\eta)}{\sqrt{\operatorname{Var}\xi\operatorname{Var}\eta}}=r.$$

The above procedure involves double integrals. Conditional expectation can be used to avoid the computation of double integrals. In fact, by the total expectation formula we have

$$\operatorname{Cov}(\xi,\eta)=E[(\xi-a)(\eta-b)]=E\,[E((\xi-a)(\eta-b)\mid\xi)].$$

Since the conditional distribution of η given $\xi=x$ is a normal distribution $N\left(b+r\frac{\sigma_2}{\sigma_1}(x-a),(1-r^2)\sigma_2^2\right)$, we obtain

$$E\,((\xi-a)(\eta-b)\mid\xi=x)=(x-a)E(\eta-b\mid\xi=x)=r\frac{\sigma_2}{\sigma_1}(x-a)^2.$$

Therefore

$$\operatorname{Cov}(\xi,\eta)=E\left[r\frac{\sigma_2}{\sigma_1}(\xi-a)^2\right]=r\frac{\sigma_2}{\sigma_1}E(\xi-a)^2=r\sigma_1\sigma_2.$$

Hence $r_{\xi\eta}=r$.

It shows that the parameter r in the normal distribution is the correlation coefficient between ξ and η. So, for a bivariate normal distribution, the uncorrelated property is equivalent to $r=0$. On the other hand, we have proven the independence of ξ and η is equivalent to $r=0$. Consequently, we have

Property 4 For a bivariate normal distribution, the uncorrelated property is equivalent to the independence.

3.2.4 Moments

Moments are one of most commonly used numerical characteristics. There are two kinds of typical moments, one of which is the moment around the origin. For a positive integer k, we call $m_k=E\xi^k$ the origin moment of order k. Mathematical expectation is the origin moment of the first order.

The other is the moment around the center (mathematical expectation). For a positive integer k, we call $c_k = E(\xi - E\xi)^k$ the center moment of order k. Variance is the center moment of the second order. Besides, the center moments of the third and fourth orders are frequently used to describe the behaviors of random variables. In most cases, people like to use their ratios. We call $c_3/c_2^{3/2}$ the skewness, when it is positive. Meanwhile we call $c_4/c_2^2 - 3$ the kurtosis coefficient; when it is positive it shows the density function is sharper than the standard normal distribution.

Example 30 Assume that ξ is a normal random variable $N(0, \sigma^2)$, then $E\xi = 0$ and

$$m_n = c_n = \frac{1}{\sqrt{2\pi}\sigma}\int_{-\infty}^{+\infty} x^n e^{-x^2/2\sigma^2}\,dx = \begin{cases} 0, & n = 2k+1; \\ 1 \cdot 3 \cdot \cdots \cdot (n-1)\sigma^n, & n = 2k. \end{cases}$$

In particular, $m_4 = c_4 = 3\sigma^4$. Hence for an arbitrary σ, both skewness and kurtosis of the standard normal distribution are 0.

We can use the origin moments to express the center moments:

$$c_k = \sum_{r=0}^{k} (-1)^r \binom{k}{r} m_1^r m_{k-r}.$$

Conversely, we can also use center moments to express origin moments:

$$m_k = \sum_{r=0}^{k} (-1)^r \binom{k}{r} m_1^r c_{k-r}.$$

The absolute moment of order α is defined by $M_\alpha = E|\xi|^\alpha$, where α is a real number. Consider the random variable ξ in Example 28,

$$E|\xi|^n = \begin{cases} \sqrt{\frac{2}{\pi}}\, 2^k k!\,\sigma^{2k+1}, & n = 2k+1; \\ 1 \cdot 3 \cdot \cdots \cdot (n-1)\sigma^n, & n = 2k. \end{cases}$$

Applying the above results, one can calculate moments of other distributions. For instance, the Rayleigh distribution has the density function

$$R_\sigma(x) = \frac{x}{\sigma^2} e^{-x^2/2\sigma^2}, \quad x > 0,$$

and then

$$\begin{aligned} E\xi^n &= \int_0^{+\infty} x^n \frac{x}{\sigma^2} e^{-x^2/2\sigma^2}\,dx \\ &= \frac{1}{2\sigma^2}\int_{-\infty}^{+\infty} |x|^{n+1} e^{-x^2/2\sigma^2}\,dx \\ &= \begin{cases} \sqrt{\frac{\pi}{2}} \cdot 1 \cdot 3 \cdot \cdots \cdot n\sigma^n, & n = 2k+1; \\ 2^k k!\,\sigma^{2k}, & n = 2k. \end{cases} \end{aligned}$$

In particular, $E\xi = \sigma\sqrt{\pi/2}$, $E\xi^2 = 2\sigma^2$, and so $\sigma_\xi^2 = (2 - \pi/2)\sigma^2$. Still, the Maxwell distribution has the density function

$$p(x) = \begin{cases} \dfrac{\sqrt{2}}{\sqrt{\pi}\sigma^3} x^2 e^{-x^2/2\sigma^2}, & x > 0; \\ 0, & x \leqslant 0. \end{cases}$$

Then

$$\begin{aligned} E\xi^n &= \frac{\sqrt{2}}{\sqrt{\pi}\sigma^3}\int_0^{+\infty} x^{n+2} e^{-x^2/2\sigma^2}\, dx \\ &= \frac{1}{\sqrt{2\pi}\sigma^3}\int_{-\infty}^{+\infty} |x|^{n+2} e^{-x^2/2\sigma^2}\, dx \\ &= \begin{cases} \sqrt{\dfrac{2}{\pi}}\, 2^{k+1}(k+1)!\sigma^{2k+1}, & n = 2k+1; \\ 1 \cdot 3 \cdot \cdots \cdot (n+1)\sigma^n, & n = 2k. \end{cases} \end{aligned}$$

In particular, $E\xi = 2\sigma\sqrt{\dfrac{2}{\pi}}$, $E\xi^2 = 3\sigma^2$.

Example 31 If ξ follows the exponential distribution with parameter λ, then we have for $k \geqslant 1$,

$$E\xi^k = \int_0^{+\infty} x^k \lambda e^{-\lambda x}\, dx = \frac{k}{\lambda} E\xi^{k-1},$$

from which we further recursively derive

$$E\xi^k = \frac{k!}{\lambda^k}.$$

This implies that the exponential distribution has moments of any order.

3.3 Characteristic functions

Generally speaking, the numerical characteristic can not completely determine the distribution function of a random variable. In this section we shall introduce the concept of characteristic function which can completely determine the distribution function and possess nice properties as well, so will become a powerful tool in the study of laws of random variables.

3.3.1 Definitions

Definition 8 *Suppose that ξ and η are real random variables, we call $\zeta = \xi + i\eta$ a complex random variable, where $i^2 = -1$. We call $E\zeta = E\xi + iE\eta$ the expectation of ζ.*

A complex random variable is in essence a two dimensional random vector, and many relevant notions and properties can be directly obtained from real random variables. For instance, $E\zeta$ possesses properties similar to that of a real mathematical expectation.

Definition 9 *Suppose ξ is a real random variable, we call*

$$f(t) = Ee^{it\xi}, \qquad -\infty < t < \infty \tag{3.28}$$

the characteristic function of ξ.

Since $E \mid e^{it\xi} \mid = 1$, (3.28) makes sense for any ξ and all $t \in (-\infty, \infty)$. In other words, each random variable ξ (or each distribution function $F(x)$) corresponds to a characteristic function $f(t)$, which is a complex function in real argument defined on $(-\infty, \infty)$.

By definition, a characteristic function is the expected value of $e^{it\xi}$, and then we have from Section 3.1

$$f(t) = \int_{-\infty}^{\infty} e^{itx}\,dF(x).$$

In particular, if ξ is a discrete random variable with $P(\xi = x_n) = p_n$, then its characteristic function is

$$f(t) = \sum_{n=1}^{\infty} p_n e^{itx_n}, \qquad -\infty < t < \infty.$$

If ξ is a continuous random variable with the density function $p(x)$, then

$$f(t) = \int_{-\infty}^{\infty} e^{itx} p(x)\,dx, \qquad -\infty < t < \infty, \tag{3.29}$$

which is just the Fourier transformation of $p(x)$.

Example 32 The characteristic function of the degenerate distribution $P(\xi = c) = 1$ is

$$f(t) = e^{ict}, \qquad -\infty < t < \infty.$$

Example 33 The characteristic function of the binomial distribution $B(n, p)$ is

$$\begin{aligned} f(t) &= \sum_{k=0}^{n} \binom{n}{k} p^k q^{n-k} e^{itk} = \sum_{k=0}^{n} \binom{n}{k} (pe^{it})^k q^{n-k} \\ &= (pe^{it} + q)^n, \quad p + q = 1, \quad -\infty < t < \infty. \end{aligned}$$

Example 34 The characteristic function of the Poisson distribution $P(\lambda)$ is

$$\begin{aligned} f(t) &= \sum_{k=0}^{\infty} \frac{\lambda^k}{k!} e^{-\lambda} e^{itk} = \sum_{k=0}^{\infty} \frac{(\lambda e^{it})^k}{k!} e^{-\lambda} \\ &= e^{\lambda(e^{it}-1)}, \qquad -\infty < t < \infty. \end{aligned}$$

Example 35 The characteristic function of the uniform distribution $U[a, b]$ is

$$f(t) = \int_a^b \frac{1}{b-a} e^{itx}\,dx = \frac{e^{itb} - e^{ita}}{i(b-a)t}, \qquad -\infty < t < \infty.$$

Example 36 The characteristic function of the normal distribution $N(a, \sigma^2)$ is

$$f(t) = \frac{1}{\sigma\sqrt{2\pi}} \int_{-\infty}^{\infty} e^{itx - \frac{(x-a)^2}{2\sigma^2}}\,dx = e^{iat - \frac{\sigma^2 t^2}{2}}, \qquad -\infty < t < \infty.$$

To verify this conclusion, we let $\eta = (\xi - a)/\sigma$. Then $\eta \sim N(0,1)$ and $f(t) = Ee^{it(a+\sigma\eta)} = e^{ita} f_\eta(\sigma t)$. So it is sufficient to show $f_\eta(t) = e^{-\frac{t^2}{2}}$. Obviously,

$$f_\eta(t) = E\mathrm{e}^{\mathrm{i}t\eta} = \frac{1}{\sqrt{2\pi}}\int_{-\infty}^{\infty} \mathrm{e}^{\mathrm{i}tx}\mathrm{e}^{-\frac{x^2}{2}}\mathrm{d}x$$
$$= \frac{1}{\sqrt{2\pi}}\int_{-\infty}^{\infty} \cos(tx)\mathrm{e}^{-\frac{x^2}{2}}\mathrm{d}x + \mathrm{i}\frac{1}{\sqrt{2\pi}}\int_{-\infty}^{\infty} \sin(tx)\mathrm{e}^{-\frac{x^2}{2}}\mathrm{d}x$$
$$= \frac{1}{\sqrt{2\pi}}\int_{-\infty}^{\infty} \cos(tx)\mathrm{e}^{-\frac{x^2}{2}}\mathrm{d}x.$$

Taking the derivative with respect to t under the integral and applying the formula of integral by parts, we obtain

$$f'\eta(t) = -\frac{1}{\sqrt{2\pi}}\int_{-\infty}^{\infty} x\sin(tx)\mathrm{e}^{-\frac{x^2}{2}}\mathrm{d}x = \frac{1}{\sqrt{2\pi}}\int_{-\infty}^{\infty} \sin(tx)\mathrm{d}\mathrm{e}^{-\frac{x^2}{2}}$$
$$= -t\frac{1}{\sqrt{2\pi}}\int_{-\infty}^{\infty} \cos(tx)\mathrm{e}^{-\frac{x^2}{2}}\mathrm{d}x = -tf_\eta(t).$$

That is, the function $f_\eta(t)$ satisfies the differential equation $f_\eta{}'(t) + tf_\eta(t) = 0$. It follows that

$$\frac{\mathrm{d}}{\mathrm{d}t}(f_\eta(t)\mathrm{e}^{\frac{t^2}{2}}) = f_\eta{}'(t)\mathrm{e}^{\frac{t^2}{2}} + tf_\eta(t)\mathrm{e}^{\frac{t^2}{2}} = 0.$$

Hence $f_\eta(t)\mathrm{e}^{\frac{t^2}{2}} = f_\eta(0)\mathrm{e}^{-\frac{0^2}{2}} = 1$. We obtain $f_\eta(t) = \mathrm{e}^{-\frac{t^2}{2}}$ as desired.

Example 37 The characteristic function of the Cauchy distribution is

$$f(t) = \int_{-\infty}^{\infty} \mathrm{e}^{\mathrm{i}tx}\frac{1}{\pi(1+x^2)}\mathrm{d}x = \mathrm{e}^{-|t|}, \quad -\infty < t < \infty.$$

Here the computations involve contour integrals in complex function theory, so are omitted (see Supplements and Remarks 6 at the end of this chapter).

3.3.2 Properties

Assume that $f(t)$ is a characteristic function.

Property 1

$$|f(t)| \leqslant f(0) = 1, \tag{3.30}$$
$$f(-t) = \overline{f(t)}. \tag{3.31}$$

Proof Obviously, by definition, we have

$$|f(t)| = \left|\int_{-\infty}^{+\infty} \mathrm{e}^{\mathrm{i}tx}\mathrm{d}F(x)\right| \leqslant \int_{-\infty}^{+\infty} |\mathrm{e}^{\mathrm{i}tx}|\mathrm{d}F(x) = 1$$

and

$$f(0) = \int_{-\infty}^{+\infty} \mathrm{e}^{\mathrm{i}0x}\mathrm{d}F(x) = 1,$$

so (3.30) follows.

Also,

$$f(-t) = \int_{-\infty}^{+\infty} \mathrm{e}^{-\mathrm{i}tx}\mathrm{d}F(x) = \overline{\int_{-\infty}^{+\infty} \mathrm{e}^{\mathrm{i}tx}\mathrm{d}F(x)} = \overline{f(t)},$$

as desired.

Property 2 $f(t)$ is uniformly continuous on $(-\infty,\infty)$.

Proof For any $t \in (-\infty, \infty)$ and $\varepsilon > 0$, it follows that

$$\begin{aligned}|f(t+h)-f(t)| &= \left|\int_{-\infty}^{+\infty}(e^{itx}e^{ihx}-e^{itx})dF(x)\right| \\ &\leqslant \left(\int_{|x|\geqslant A}+\int_{|x|<A}\right)|e^{ihx}-1|\,dF(x) \\ &\equiv I_1+I_2.\end{aligned}$$

Since $\int_{-\infty}^{+\infty}dF(x)=1$, there exists an A so large that $\int_{|x|\geqslant A}dF(x)<\varepsilon/4$. So,

$$I_1 \leqslant \int_{|x|\geqslant A}(|e^{ihx}|+1)dF(x) = 2\int_{|x|\geqslant A}dF(x) < \frac{\varepsilon}{2}.$$

Also, note that

$$|e^{ihx}-1| = |e^{i\frac{h}{2}x}|\,|e^{i\frac{h}{2}x}-e^{-i\frac{h}{2}x}| = 2\left|\sin\frac{hx}{2}\right|.$$

For A determined above, take $\delta=\varepsilon/(2A)$. Thus when $|x|<A$ and $0<h<\delta$, $\sin(hx/2)<|hx/2|<\varepsilon/4$. We obtain

$$I_2 \leqslant \frac{\varepsilon}{2}\int_{-A}^{A}dF(x) \leqslant \frac{\varepsilon}{2}.$$

Consequently, $|f(t+h)-f(t)|<\varepsilon$. It is easily seen from the proof that the choice of δ is independent of t.

Property 3 $f(t)$ is nonnegative definite, i.e., for an arbitrary integer n, any real numbers $t_1, \cdots, t_n$ and complex numbers $\lambda_1, \cdots, \lambda_n$ it follows

$$\sum_{k=1}^{n}\sum_{j=1}^{n}f(t_k-t_j)\lambda_k\overline{\lambda_j} \geqslant 0. \tag{3.32}$$

Proof

$$\begin{aligned}\sum_{k=1}^{n}\sum_{j=1}^{n}f(t_k-t_j)\lambda_k\overline{\lambda_j} &= \sum_{k=1}^{n}\sum_{j=1}^{n}\int_{-\infty}^{+\infty}e^{i(t_k-t_j)x}dF(x)\lambda_k\overline{\lambda_j} \\ &= \int_{-\infty}^{+\infty}\left(\sum_{k=1}^{n}e^{it_kx}\lambda_k\right)\overline{\left(\sum_{j=1}^{n}e^{it_jx}\lambda_j\right)}dF(x) \\ &= \int_{-\infty}^{+\infty}\left|\sum_{k=1}^{n}e^{it_kx}\lambda_k\right|^2dF(x) \geqslant 0.\end{aligned}$$

This property is one of the most basic properties of characteristic functions. Indeed, we have

Theorem 2 *(Bochner-Khinchine) The function $f(t)$ is a characteristic function if and only if $f(t)$ is nonnegative definite, continuous and $f(0)=1$.*

The proof is rather tedious, so omitted. It tells us a criterion for characteristic functions in principle, but it is not easy to decide whether a function is nonnegative definite. Hence the theorem is of no great use in practice. One usually uses other ways to check whether a function is a characteristic function in many specific cases (see the end of this section and Section 4.1 below).

Property 4 Assume that $\xi_1, \cdots, \xi_n$ are independent with characteristic functions

$f_1(t), \cdots, f_n(t)$ respectively. Let $\eta = \xi_1 + \cdots + \xi_n$, and then

$$f_\eta(t) = f_1(t) f_2(t) \cdots f_n(t). \tag{3.33}$$

This is because the independence of $\xi_1, \cdots, \xi_n$ implies $e^{it\xi_1}, \cdots, e^{it\xi_n}$ are independent. Hence

$$Ee^{it\eta} = E(e^{it\xi_1} \cdots e^{it\xi_n}) = Ee^{it\xi_1} \cdots Ee^{it\xi_n}.$$

The above property plays an important role in the study of sums of independent random variables.

Property 5 If $E\xi^n$ exists, then $f(t)$ is differentiable of n orders, and when $k \leqslant n$

$$f^{(k)}(t) = i^k \int_{-\infty}^{\infty} x^k e^{itx} dF(x), f^{(k)}(0) = i^k E\xi^k.$$

In particular, when $n = 2$, $E\xi = -if'(0)$, $E\xi^2 = -f''(0)$, $\text{Var}\xi = -f''(0) + [f'(0)]^2$.

Conversely, if $f^{(n)}(0)$ exists for an even integer n, then $E\xi^n$ exists.

Proof We first prove the first part. We have

$$\int_{-\infty}^{\infty} \left| \frac{d^k}{dt^k} e^{itx} \right| dF(x) = \int_{-\infty}^{\infty} |i^k x^k e^{itx}| dF(x)$$

$$= \int_{-\infty}^{\infty} |x^k| dF(x) = E|\xi|^k < \infty.$$

Hence, $\int_{-\infty}^{\infty} \frac{d^k}{dt^k} e^{itx} dF(x)$ uniformly converges in t. So $f^{(k)}(t)$ exists and

$$f^{(k)}(t) = \int_{-\infty}^{\infty} \frac{d^k}{dt^k} e^{itx} dF(x) = i^k \int_{-\infty}^{\infty} x^k e^{itx} dF(x),$$

$$f^{(k)}(0) = i^k \int_{-\infty}^{\infty} x^k dF(x) = i^k E\xi^k.$$

Next, we prove the second part by induction. If $n=2$, then

$$f''(0) = \lim_{h\to 0} \frac{f(h) - 2f(0) + f(-h)}{h^2} = \lim_{h\to 0} \int_{-\infty}^{\infty} \frac{e^{ihx} - 2 + e^{-ihx}}{h^2} dF(x)$$

$$= -\lim_{h\to 0} \int_{-\infty}^{\infty} 2\frac{1 - \cos hx}{h^2} dF(x).$$

Note that $0 \leqslant 2\frac{1-\cos hx}{h^2} \leqslant x^2$, and $\lim_{h\to 0} 2\frac{1-\cos hx}{h^2} = x^2$ uniformly in x on any a finite interval. It follows that for any $a > 0$,

$$-f''(0) \geqslant \lim_{h\to 0} \int_{-a}^{a} 2\frac{1-\cos hx}{h^2} dF(x)$$

$$= \int_{-a}^{a} \lim_{h\to 0} 2\frac{1-\cos hx}{h^2} dF(x) = \int_{-a}^{a} x^2 dF(x).$$

Letting $a \to \infty$ yields $\int_{-\infty}^{\infty} x^2 dF(x) \leqslant -f''(0)$. So, $E\xi^2$ exists as desired.

Now, suppose that $f^{(2k)}(0)$ exists and $E\xi^{2k-2}$ also exists. By the conclusion of the first part, $f(t)$ is differentiable up to $(2k-2)$-th order, and

$$f^{(2k-2)}(t) - i^{2k-2} \int_{-\infty}^{\infty} e^{itx} x^{2k-2} dF(x) = (-1)^{k-1} \int_{-\infty}^{\infty} e^{itx} x^{2k-2} dF(x).$$

Let $G(y) = \int_{-\infty}^{y} x^{2k-2} \mathrm{d}F(x)$, where $G(\infty) = \mathrm{E}\xi^{2k-2} < \infty$. Then $\frac{G(y)}{G(\infty)}$ is a distribution function with its characteristic function

$$g(t) = \frac{1}{G(\infty)}\int_{-\infty}^{\infty} \mathrm{e}^{ity} \mathrm{d}G(y) = \frac{1}{G(\infty)}\int_{-\infty}^{\infty} \mathrm{e}^{ity} y^{2k-2} \mathrm{d}F(y) = \frac{(-1)^{k-1}}{G(\infty)} f^{(2k-2)}(t).$$

Hence $g''(0) = \frac{(-1)^{k-1}}{G(\infty)} f^{(2k)}(0)$ exists. By the conclusion for $n = 2$ as we have shown, we conclude that

$$\frac{1}{G(\infty)}\int_{-\infty}^{\infty} x^{2k} \mathrm{d}F(x) = \frac{1}{G(\infty)}\int_{-\infty}^{\infty} y^2 \mathrm{d}G(y)$$

exists. So, $\mathrm{E}\xi^{2k}$ exists. The proof is completed.

Property 6 Let $\eta = a\xi + b$, where a, b are arbitrary constants. Then

$$f_\eta(t) = \mathrm{e}^{ibt} f(at). \tag{3.34}$$

To see this note that

$$E\mathrm{e}^{i(a\xi+b)t} = E\mathrm{e}^{iat\xi} \cdot \mathrm{e}^{ibt} = \mathrm{e}^{ibt} f(at).$$

Example 38 Suppose $\xi \sim N(a,\sigma^2)$, use the characteristic function method to calculate $E\xi$ and $\mathrm{Var}\xi$.

Solution It is known that $f(t) = \mathrm{e}^{iat-\frac{\sigma^2 t^2}{2}}$. Differentiating $f(t)$, we have

$$f'(t) = (ia - \sigma^2 t)\mathrm{e}^{iat-\frac{\sigma^2 t^2}{2}},$$

$$f''(t) = [-\sigma^2 + (ia - \sigma^2 t)^2]\mathrm{e}^{iat-\frac{\sigma^2 t^2}{2}}.$$

So $f'(0) = ia$, and by Property 5, $E\xi = a$. Similarly, $f''(0) = -a^2 - \sigma^2$, so $E\xi^2 = a^2 + \sigma^2$ and $\mathrm{Var}\xi = \sigma^2$.

Example 39 Are the following functions characteristic functions of random variables?

(1) $f(t) = \sin t$;

(2) $f(t) = \ln(\mathrm{e} + |t|)$;

(3) $f(t) = 0$ when $t < 0$, $f(t) = 1$ when $t \geqslant 0$.

Solution Note that (1) $f(0) = 0 \neq 1$; (2) $|f(t)| > \ln \mathrm{e} = 1$ when $|t| \neq 0$; (3) $f(t)$ is discontinuous at $t = 0$. Therefore, none of them is a characteristic function.

The distribution function of a random variable can completely determine its characteristic function. Conversely, can a characteristic function determine a unique distribution function? The inverse formula and uniqueness theorem below answer positively this question.

3.3.3 Inverse formula and uniqueness theorem

Theorem 3 *(Inverse formula) Suppose that $f(t)$ is a characteristic function corresponding to distribution function $F(x)$. Let x_1, x_2 be two continuity points of $F(x)$, and then*

$$F(x_2)-F(x_1)=\lim_{T\to\infty}\frac{1}{2\pi}\int_{-T}^{T}\frac{e^{-itx_1}-e^{-itx_2}}{it}f(t)\mathrm{d}t. \tag{3.35}$$

Proof Without loss of generality, we assume that $x_1<x_2$. The right hand side of (3.35) is

$$\lim_{T\to\infty}\frac{1}{2\pi}\int_{-T}^{T}\int_{-\infty}^{+\infty}\frac{e^{-itx_1}-e^{-itx_2}}{it}e^{itx}\,\mathrm{d}F(x)\mathrm{d}t.$$

Exchanging the order of integrations, dividing the interval $[-T, T]$ into two intervals $[-T, 0]$ and $[0, T]$, and setting $t=-u$ in $[-T, 0]$, the above is in turn equal to

$$\begin{aligned}&\lim_{T\to+\infty}\frac{1}{2\pi}\int_{-\infty}^{+\infty}\left\{\int_0^T\left[\frac{e^{it(x-x_1)}-e^{-it(x-x_1)}}{it}-\frac{e^{it(x-x_2)}-e^{-it(x-x_2)}}{it}\right]\mathrm{d}t\right\}\mathrm{d}F(x)\\&=\lim_{T\to+\infty}\frac{1}{\pi}\int_{-\infty}^{+\infty}\left\{\int_0^T\left[\frac{\sin t(x-x_1)}{t}-\frac{\sin t(x-x_2)}{t}\right]\mathrm{d}t\right\}\mathrm{d}F(x).\end{aligned} \tag{3.36}$$

Note that

$$\lim_{T\to+\infty}\int_0^T\frac{\sin at}{t}\mathrm{d}t=\int_0^{+\infty}\frac{\sin at}{t}\mathrm{d}t=\begin{cases}\frac{\pi}{2}, & a>0;\\ 0, & a=0;\\ -\frac{\pi}{2}, & a<0.\end{cases}$$

Put

$$g(T,x,x_1,x_2)=\frac{1}{\pi}\int_0^T\left(\frac{\sin t(x-x_1)}{t}-\frac{\sin t(x-x_2)}{t}\right)\mathrm{d}t,$$

and then

$$\lim_{T\to\infty}g(T,x,x_1,x_2)=\begin{cases}0, & x<x_1 \text{ or } x>x_2;\\ \frac{1}{2}, & x=x_1 \text{ or } x=x_2;\\ 1, & x_1<x<x_2.\end{cases} \tag{3.37}$$

Note that $\int_0^T\frac{\sin at}{t}\mathrm{d}t$ is a bounded continuous function of T. There is an $M>0$ such that $|g(T,x,x_1,x_2)|\leqslant M$ for all T, x, x_1 and x_2. Let the limit function in the right hand of (3.37) be denoted by $h(x,x_1,x_2)$, and $F(x)$ be the distribution function of a random variable ξ. Writing the integral with respect to $F(x)$ in (3.36) as the expectation of a function of ξ, we obtain

$$\text{Right hand of (3.35)}=\lim_{T\to\infty}E[g(T,\xi,x_1,x_2)]. \tag{3.38}$$

By the bounded convergence theorem, exchanging the order of taking the limit and integrating in the right hand of (3.38), we obtain

$$\begin{aligned}&\lim_{T\to\infty}\frac{1}{2\pi}\int_{T}^{T}\frac{e^{-itx_1}-e^{-itx_2}}{it}\mathrm{d}t=E[h(\xi, x_1, x_2)]\\&=0\cdot P(\xi<x_1 \text{ or } \xi>x_2)+\frac{1}{2}\cdot P(\xi=x_1 \text{ or } \xi=x_2)+1\cdot P(x_1<\xi<x_2)\\&=\frac{1}{2}(P(\xi=x_1)+P(\xi=x_2))+P(x_1<\xi<x_2).\end{aligned} \tag{3.39}$$

The second equality above is due to the fact that $h(\xi, x_1, x_2)$ is a discrete random variable with values $0, \frac{1}{2}$ and 1. Since x_1, x_2 are continuous points of $F(x)$, we have $P(\xi = x_1) = P(\xi = x_2) = 0, P(x_1 < \xi < x_2) = F(x_2) - F(x_1)$. (3.35) follows.

We remark that in the process of proof above, one needs to show that it is reasonable to exchange the order of integrations and to exchange the order of integration and limit. For example, in order to get the left hand side of (3.36) from (3.35), we have used the fact that

$$\left| \frac{e^{-itx_1} - e^{-itx_2}}{it} e^{itx} \right| = \left| \frac{e^{it(x_2 - x_1)} - 1}{it} \right| \leqslant | x_2 - x_1 |$$

is uniformly bounded in t, and hence, it is reasonable to exchange the order of integrations. The others are similar, so omitted.

Theorem 4 (*Uniqueness*) *A distribution function can be uniquely determined by its characteristic function.*

Proof In (3.35), letting $y = x_1$ tends to $-\infty$ along continuity points of $F(x)$, and letting $x = x_2$, we have

$$F(x) = \lim_{y \to -\infty} \lim_{T \to \infty} \frac{1}{2\pi} \int_{-T}^{T} \frac{e^{-ity} - e^{-itx}}{it} f(t) \mathrm{d}t. \tag{3.40}$$

Thus it is easy to see that $f(t)$ determines the value of $F(x)$ at its continuity points. As for the discontinuous points, in view of right continuity of $F(x)$, it suffices to take right limits along continuity points. The theorem is proved.

Theorem 5 (*Inverse Fourier transform*) *Suppose that $f(t)$ is a characteristic function and $\int_{-\infty}^{+\infty} | f(t) | \mathrm{d}t < \infty$, then the derivative of the corresponding distribution function $F(x)$ exists and is continuous. Moreover*

$$F'(x) = \frac{1}{2\pi} \int_{-\infty}^{+\infty} e^{-itx} f(t) \mathrm{d}t. \tag{3.41}$$

Remark Since $F'(x)$ is continuous, and $F(x) = \int_{-\infty}^{x} F'(t) \mathrm{d}t$, $p(x) = F'(x)$ is the density function of a continuous random variable ξ. This theorem illustrates that the random variable corresponding to $f(t)$ is of continuous type when $f(t)$ is absolutely integrable, and its density is given by (3.41). Formula (3.41) and (3.29) are just a pair of Fourier transforms.

Proof Let us first prove F is continuous at x. Choose $\delta > 0$ such that $x + \delta$ and $x - \delta$ are continuity points of F, and then by (3.35) and absolute integrability of $f(t)$,

$$F(x + \delta) - F(x - \delta) = \frac{1}{2\pi} \int_{-\infty}^{+\infty} \frac{e^{-it(x-\delta)} - e^{-it(x+\delta)}}{it} f(t) \mathrm{d}t.$$

Obviously, the right hand side is continuous in δ. Letting $\delta \to 0$ in the way such that $x + \delta$ and $x + \delta$ are continuity points of F, the right hand side of the above equation tends to 0, and so the left hand side tends to 0 as well. Since F is right continuous, then F must be continuous at x.

To prove (3.41), consider the following inverse formula

$$F(x_2) - F(x_1) = \frac{1}{2\pi}\int_{-\infty}^{+\infty} \frac{e^{-itx_1} - e^{-itx_2}}{it} f(t)\,dt. \tag{3.42}$$

Note that

$$\lim_{x_2 \to x_1} \frac{1 - e^{-it(x_2 - x_1)}}{it(x_2 - x_1)} = 1.$$

Dividing both sides of (3.42) by $x_2 - x_1$, letting $x_2 \to x_1$, and exchanging the order of the integration and the limit, we obtain (3.41).

A similar argument shows that $\lim\limits_{h \to 0} F'(x+h) = F'(x)$ holds for any x, so $F'(x)$ is continuous.

We have an analogue of discrete random variables. Assume that ξ is a nonnegative integer valued random variable with $P(\xi = k) = p_k$, $k = 0,1,2,\cdots$, then its characteristic function is

$$f(t) = \sum_{k=0}^{\infty} p_k e^{itk}.$$

If $f(t)$ is given, then we can multiply both sides by e^{-itk} and integrate. Noting that

$$\int_0^{2\pi} e^{int}\,dt = \begin{cases} 2\pi, & n = 0; \\ 0, & n \neq 0, \end{cases}$$

we have

$$p_k = \frac{1}{2\pi}\int_0^{2\pi} e^{-itk} f(t)\,dt.$$

The three theorems above enable us to calculate distribution functions or density functions from characteristic functions. But it is very hard to apply (3.40) directly to compute distribution functions. Its significance is mainly in theoretic aspect.

Example 40 Show that $f(t) = \cos t$ is a characteristic function of random variable, and find its distribution function.

Solution $f(t)$ can be written as

$$f(t) = \cos t = \frac{1}{2}(e^{it} + e^{-it}) = \frac{1}{2}e^{it} + \frac{1}{2}e^{-it}.$$

It is now easy to see that this is just the characteristic function of random variable with the distribution

$$\begin{pmatrix} 1 & -1 \\ \frac{1}{2} & \frac{1}{2} \end{pmatrix}.$$

In general, if $f(t)$ can be written as $\sum a_n e^{ix_n t}$, where $a_n > 0$ and $\sum a_n = 1$, then $f(t)$ is a characteristic function, whose corresponding random variable has distribution sequence $P(\xi = x_n) = a_n$, $n = 1,2,\cdots$.

Example 41 If $f(t)$ is a characteristic function of some random variable, so are $\overline{f(t)}$

and $|f(t)|^2$.

Solution By Properties 1 and 6, $f(-t)=\overline{f(t)}$ is characteristic function of $-\xi$. Also, assume that ξ_1 and ξ_2 are independent identically distributed with a common characteristic function $f(t)$, then $\eta=-\xi_2$ is independent of ξ_1 and has characteristic function $\overline{f(t)}$. Further, by Property 4, $\xi_1+\eta=\xi_1-\xi_2$ has characteristic function $f(t)\cdot\overline{f(t)}=|f(t)|^2$.

3.3.4 Additivity of distribution functions

Characteristic functions have a lot of important applications, but here we shall restrict ourselves to the study of additivity of distribution functions.

The additivity, also called regenerativity, means that if ξ and η are independent and follow a common type of distributions, then so do their sum $\xi+\eta$ and the parameter of $\xi+\eta$ is the sum of parameters of ξ and η. This property of distributions is called additivity. We have already known as applications of convolution formula of Section 2.5 that the binomial distribution, the Poisson distribution and the Gamma distribution all have additivity, but the proofs are a bit intricate. It will be much more convenient to apply characteristic functions and a direct discussion is feasible for sums of many random variables.

Example 42 If $\xi_j(j=1,2,\cdots,k)$ follows the binomial distribution $B(n_j,p)$ respectively and are independent of each other. Find the distribution of $\sum_{j=1}^{k}\xi_j$.

Solution It is easy to see that the characteristic function $f_j(t)$ of ξ_j is $(pe^{it}+q)^{n_j}$. By Property 4, the characteristic function of $\sum_{j=1}^{k}\xi_j$ is

$$\prod_{j=1}^{k}f_j(t)=(pe^{it}+q)^{\sum_{j=1}^{k}n_j}.$$

In turn, by the uniqueness theorem, $\sum_{j=1}^{k}\xi_j$ follows the binomial distribution $B(\sum_{j=1}^{k}n_j,p)$.

Example 43 Suppose that $\xi_1,\cdots,\xi_n$ are independent, and ξ_k is normally distributed as $N(a_k,\sigma_k^2)$, $k=1,\cdots,n$. Find the distribution of $\sum_{k=1}^{n}\xi_k$.

Solution It is known that the characteristic function of ξ_k is $e^{ia_kt-\sigma_k^2t^2/2}$, so the characteristic function of $\sum_{k=1}^{n}\xi_k$ is

$$\prod_{k=1}^{n}e^{ia_kt-\frac{\sigma_k^2t^2}{2}}=\exp\left\{i\left(\sum_{k=1}^{n}a_k\right)t-\frac{1}{2}\left(\sum_{k=1}^{n}\sigma_k^2\right)t^2\right\}.$$

Thus $\sum_{k=1}^{n}\xi_k$ has the normal distribution whose expectation is the sum of respective expectations, whose variance is the sum of respective variances.

3.3.5 Multivariate characteristic functions

Definition 10 *Suppose the random vector* $\xi=(\xi_1,\cdots,\xi_n)'$ *has distribution* *function* $F(x_1,\cdots,x_n)$, *then its characteristic function is defined by*

$$\begin{aligned} f(t_1,\cdots,t_n) &= E\mathrm{e}^{\mathrm{i}(t_1\xi_1+\cdots+t_n\xi_n)} \\ &= \int_{-\infty}^{+\infty}\cdots\int_{-\infty}^{+\infty} \mathrm{e}^{\mathrm{i}(t_1x_1+\cdots+t_nx_n)}\,\mathrm{d}F(x_1,\cdots,x_n). \end{aligned} \tag{3.43}$$

Let $\boldsymbol{t}=(t_1,\cdots,t_n)'$, $\boldsymbol{x}=(x_1,\cdots,x_n)'$. Then (3.44) can be written as

$$f(\boldsymbol{t}) = E\mathrm{e}^{\mathrm{i}\boldsymbol{t}'\boldsymbol{\xi}} = \int_{\mathbf{R}^n} \mathrm{e}^{\mathrm{i}\boldsymbol{t}'\boldsymbol{x}}\,\mathrm{d}F(\boldsymbol{x}), \quad \boldsymbol{t}\in\mathbf{R}^n. \tag{3.44}$$

This is very similar to (3.28). In fact, it is just (3.28) in the case $n=1$.

Many properties of multivariate characteristic functions are similar to that of univariate characteristic functions, for instance, uniform continuity and uniqueness. Here we restrict ourselves to some new properties.

Property 1′ The characteristic function of $\eta = a_1\xi_1+\cdots+a_n\xi_n$ is

$$\begin{aligned} f_\eta(t) &= E\mathrm{e}^{\mathrm{i}t\eta} = E\mathrm{e}^{\mathrm{i}t\sum_{k=1}^n a_k\xi_k} \\ &= E\mathrm{e}^{\mathrm{i}\sum_{k=1}^n (a_kt)\xi_k} = f(a_1t,\cdots,a_nt). \end{aligned} \tag{3.45}$$

Property 2′ If the characteristic function of $(\xi_1,\cdots,\xi_n)'$ is $f(t_1,\cdots,t_n)$, then k-dimensional subvector $(\xi_{l_1},\cdots,\xi_{l_k})'$ has characteristic function

$$f(0,\cdots,0,t_{l_1},0,\cdots,0,\ t_{l_k},0,\cdots,0).$$

Property 3′ Assume that f_ξ is the characteristic function of $\boldsymbol{\xi}=(\xi_1,\cdots,\xi_n)'$. Let L the an $m\times n$ matrix, $\boldsymbol{a}=(a_1,\cdots,a_m)'$, $\boldsymbol{\eta}=L\boldsymbol{\xi}+\boldsymbol{a}$. Then

$$f_\eta(t_1,\cdots,t_m) = f_\xi(\boldsymbol{t}'L)\mathrm{e}^{\mathrm{i}\boldsymbol{t}'\boldsymbol{a}},$$

where $\boldsymbol{t}=(t_1,\cdots,t_m)'$.

Applying the uniqueness theorem, we can prove the following two properties.

Property 4′ Assume that ξ_j has characteristic function $f_j(t)$, $j=1,\cdots,n$, then $\xi_1,\cdots,\xi_n$ are independent if and only if the characteristic function of $(\xi_1,\cdots,\xi_n)'$ is

$$f(t_1,\cdots,t_n) = f_1(t_1)\cdots f_n(t_n).$$

Property 5′ The random vectors $(\xi_1,\cdots,\xi_k)'$ and $(\xi_{k+1},\cdots,\xi_n)'$ are independent if and only if the product of their characteristic functions is just equal to the characteristic function of $(\xi_1,\cdots,\xi_n)'$.

Property 6′ Assume that $f(t_1,\cdots,t_n)$ is the characteristic function of $(\xi_1,\cdots,\xi_n)'$. If $E\prod_{l=1}^m \xi_{i_l}\ (1\leqslant m\leqslant n)$ exists, then

$$\left.\frac{\partial f(t_1,\cdots,t_n)}{\partial t_{i_1}\cdots\partial t_{i_m}}\right|_{t_1,\cdots,t_n=0} = \mathrm{i}^m E\prod_{l=1}^m \xi_{i_l}.$$

3.4 Multivariate normal distributions

Multivariate normal distributions are the most important in the theory of multivariate distributions. They are essential in the multivariate analysis. This section is devoted to concepts and properties of multivariate normal distributions with help of multivariate characteristic functions. For the bivariate case, we have derived most of results, and here we discuss the general case using matrix method.

3.4.1 Density functions and characteristic functions

In Section 2.3, we already gave the density function of the n-dimensional normal distribution $N(\boldsymbol{a},\boldsymbol{B})$

$$p(x) = \frac{1}{(2\pi)^{n/2} \mid \boldsymbol{B} \mid^{1/2}} \exp\left\{-\frac{1}{2}(\boldsymbol{x}-\boldsymbol{a})'\boldsymbol{B}^{-1}(\boldsymbol{x}-\boldsymbol{a})\right\}. \tag{3.46}$$

For deriving the characteristic function of $N(\boldsymbol{a},\boldsymbol{B})$, we write the positive definite symmetric matrix $\boldsymbol{B}$ as $\boldsymbol{B}=\boldsymbol{LL}'$, where $\boldsymbol{L}$ is a non—singular matrix (and it can be chosen to be also a positive definite symmetric matrix). Let $\boldsymbol{t} = (t_1,\cdots,t_n)'$, by (3.45) in Section 3.3, and it follows

$$\begin{aligned} f(\boldsymbol{t}) &= \int_{\mathbf{R}^n} e^{it'x} p(\boldsymbol{x})\mathrm{d}\boldsymbol{x} \\ &= \frac{1}{(2\pi)^{n/2} \mid \boldsymbol{B} \mid^{1/2}} \int_{\mathbf{R}^n} \exp\left\{it'\boldsymbol{x}-\frac{1}{2}(\boldsymbol{x}-\boldsymbol{a})'\boldsymbol{B}^{-1}(\boldsymbol{x}-\boldsymbol{a})\right\}\mathrm{d}\boldsymbol{x}. \end{aligned}$$

Let $\boldsymbol{y} = \boldsymbol{L}^{-1}(\boldsymbol{x}-\boldsymbol{a})$, and then $\boldsymbol{x} = \boldsymbol{Ly}+\boldsymbol{a}$. Let $\boldsymbol{s} = \boldsymbol{L}^{\mathrm{T}}\boldsymbol{t} = (s_1,\cdots,s_n)'$, and note that $\boldsymbol{t}'\boldsymbol{a} = \boldsymbol{a}'\boldsymbol{t}$, $\boldsymbol{s}'\boldsymbol{y} = \sum_k s_k y_k$, $\sum_k s_k^2 = \boldsymbol{s}'\boldsymbol{s} = \boldsymbol{t}'\boldsymbol{Bt}$, so we have

$$\begin{aligned} f(\boldsymbol{t}) &= \frac{1}{(2\pi)^{n/2} \mid \boldsymbol{B} \mid^{1/2}} \int_{\mathbf{R}^n} \exp\left\{i\boldsymbol{t}'\boldsymbol{a}+i\boldsymbol{t}'\boldsymbol{Ly}-\frac{\boldsymbol{y}'\boldsymbol{y}}{2}\right\} \mid \boldsymbol{B} \mid^{-1/2}\mathrm{d}\boldsymbol{y} \\ &= \frac{1}{(2\pi)^{n/2}} \exp(i\boldsymbol{a}'\boldsymbol{t}) \int_{-\infty}^{+\infty}\cdots\int_{-\infty}^{+\infty} \exp\left\{-\frac{1}{2}\sum_{k=1}^{n}(y_k^2-2is_k y_k)\right\}\mathrm{d}y_1\cdots\mathrm{d}y_n \quad (3.47) \\ &= \exp(i\boldsymbol{a}'\boldsymbol{t}) \exp\left(\frac{1}{2}\sum_k (i^2 s_k^2)\right) \prod_{k=1}^{n} \frac{1}{\sqrt{2\pi}} \int_{-\infty}^{+\infty} e^{-\frac{(y_k-is_k)^2}{2}}\mathrm{d}y_k \\ &= \exp\left(i\boldsymbol{a}'\boldsymbol{t}-\frac{1}{2}\boldsymbol{t}'\boldsymbol{Bt}\right), \end{aligned}$$

i.e.,

$$f(t_1,\cdots,t_n) = \exp\left\{i\sum_{k=1}^{n} a_k t_k - \frac{1}{2}\sum_{l=1}^{n}\sum_{s=1}^{n} b_{ls} t_l t_s\right\}. \tag{3.48}$$

This is a generalization of the characteristic function of 1-dimensional normal distribution. When $n=1$, it is just the characteristic function of 1-dimensional normal distribution.

In the above we define the multivariate normal distribution only in the case of a positive definite symmetric matrix. When $\boldsymbol{B}$ is nonnegative definite, (3.46) may not be well defined, but (3.47) still makes sense. In fact, it is a characteristic function of the following random variable. Let $\boldsymbol{B}=\boldsymbol{LL}'$, $\boldsymbol{\xi}=(\xi_1,\cdots\xi_n)'$ an n-dimensional standard normal random vector, i.e., $\xi_i, i=1,\cdots,n$, are n independent standard normal variables, and then by Property 3′ in Section 3.3, the characteristic function of $\boldsymbol{\eta}=\boldsymbol{L\xi}+\boldsymbol{a}$ is $\exp(i\boldsymbol{a}'\boldsymbol{t}-\boldsymbol{t}'\boldsymbol{Bt}/2)$. When the rank of $\boldsymbol{B}$ is r $(r<n)$, we call the corresponding distribution a singular normal distribution or a

degenerate normal distribution. It is actually only a distribution in r-dimensional subspace. Therefore, the characteristic function in (3.48) has wider applications than density function in (3.46) so that we usually define multivariate normal distributions by the characteristic function in (3.47).

3.4.2 Properties

Bivariate normal distributions possess many special properties. For instance, marginal distributions and conditional distributions are also normal distributions; parameters are just equal to mathematical expectation and variance and correlation coefficient; independence is equivalent to uncorrelated property. These properties remain valid in the case of n-dimensional normal distributions, which are easily seen from characteristic functions. The statements are as follows.

Property 1 Any sub-vector $(\xi_{l_1},\cdots,\xi_{l_k})'$ of ξ also follows normal distribution as $N(\tilde{\boldsymbol{a}},\tilde{\boldsymbol{B}})$, where $\tilde{\boldsymbol{a}}=(a_{l_1},\cdots,a_{l_k})'$, *and* $\tilde{\boldsymbol{B}}$ is a $k\times k$ matrix consisting of elements in both $l_1,\cdots,l_k$ rows and $l_1,\cdots,l_k$ columns in $\boldsymbol{B}$.

Proof We know from Property 2 of characteristic functions that the characteristic function of $(\xi_{l_1},\cdots,\xi_{l_k})'$ is $f(0,\cdots,0,t_{l_1},0,\cdots,0,\ t_{l_k},0,\cdots,0)$. Let all t_i be 0 except t_{lj}, and write $\tilde{\boldsymbol{t}}=(t_{l_1},\cdots,t_{l_k})$, then

$$\begin{aligned}\tilde{f}(\tilde{\boldsymbol{t}}) &= \exp\left(\mathrm{i}\sum_{j=1}^{k}a_{l_j}t_{l_j}-\frac{1}{2}\sum_{j=1}^{k}\sum_{s=1}^{k}b_{l_s}t_{l_j}t_{l_s}\right)\\ &= \exp\left(\mathrm{i}\,\tilde{\boldsymbol{t}}\,\tilde{\boldsymbol{a}}'-\frac{1}{2}\,\tilde{\boldsymbol{t}}'\tilde{\boldsymbol{B}}\tilde{\boldsymbol{t}}\right).\end{aligned}$$

This is just the characteristic function of $N(\tilde{\boldsymbol{a}},\tilde{\boldsymbol{B}})$. In particular, when $k=1$, $\xi_j\sim N(a_j,b_{jj})$. This shows that marginal distributions of a multivariate normal distributions are still normal distributions.

Property 2 $N(\boldsymbol{a},\boldsymbol{B})$ has expected value $\boldsymbol{a}$, covariance matrix $\boldsymbol{B}$.

Proof By Property 1, we have $E\xi_j=a_j$, $\mathrm{Var}\xi_j=b_{jj}$. So it suffices to prove that entries off diagonal are covariances. Since $E\xi_j^2=\mathrm{Var}\xi_j+(E\xi_j)^2$ exists, then $|E\xi_j\xi_k|\leqslant\sqrt{E\xi_j^2E\xi_k^2}$ exists. Also, it is easy to see

$$\mathrm{i}^2E\xi_j\xi_k=\frac{\partial^2 f}{\partial t_j\partial t_k}\Big|_{(0,\cdots,0)}.$$

Further, we derive from (3.49) that

$$\begin{aligned}\frac{\partial f}{\partial t_j} &= \left[\mathrm{i}a_j-\frac{1}{2}\left(\sum_{l=1}^{n}b_{lj}t_l+\sum_{s=1}^{n}b_{js}t_s\right)\right]\\ &\quad\cdot\exp\left(\mathrm{i}\sum_{k=1}^{n}a_kt_k-\frac{1}{2}\sum_{l=1}^{n}\sum_{s=1}^{n}b_{l_s}t_lt_s\right)\end{aligned}$$

and

$$\frac{\partial^2 f}{\partial t_j \partial t_k} = \left\{-\frac{1}{2}(b_{kj}+b_{jk}) + \left[\mathrm{i}a_j - \frac{1}{2}\left(\sum_{l=1}^{n} b_{lj}t_l + \sum_{s=1}^{n} b_{js}t_s\right)\right]\right.$$

$$\left.\cdot \left[\mathrm{i}a_k - \frac{1}{2}\left(\sum_{l=1}^{n} b_{lk}t_l + \sum_{s=1}^{n} b_{ks}t_s\right)\right]\right\}$$

$$\cdot \exp\left(\mathrm{i}\sum_{k=1}^{n} a_k t_k - \frac{1}{2}\sum_{l=1}^{n}\sum_{s=1}^{n} b_{l_s} t_l t_s\right).$$

Let $t_r = 0$, $r = 1, \cdots, n$, and note that $b_{jk} = b_{kj}$. We obtain

$$E\xi_j\xi_k = b_{jk} + a_j a_k, \quad \mathrm{Cov}(\xi_j, \xi_k) = b_{jk}.$$

Property 3 The random variables $\xi_1, \cdots, \xi_n$ with joint normal distribution are mutually independent if and only if they are pairwise uncorrelated.

Proof We know that $\xi_1, \cdots, \xi_n$ are independent if and only if $f(t_1, \cdots, t_n) = f_1(t_1)\cdots f_n(t_n)$. This is equivalent to saying $\boldsymbol{t}'\boldsymbol{B}\boldsymbol{t} = \sum_{j=1}^{n} b_{jj}t_j^2$. Solving the equation, we obtain

$$b_{ij} = 0, \quad i, j = 1, 2, \cdots, n, \quad i \neq j.$$

In turn, this holds if and only if ξ_i and ξ_j are uncorrelated for $i, j = 1, 2, \cdots, n$, $i \neq j$.

Applying operation of block matrices, we know Property 3 similarly holds for arbitrarily many sub-vectors of $\boldsymbol{\xi}$. For example, write $\boldsymbol{\xi} = (\boldsymbol{\eta}_1{}', \boldsymbol{\eta}_2{}')'$, where $\boldsymbol{\eta}_1$ and $\boldsymbol{\eta}_2$ are k and $n-k$-dimensional sub-vectors respectively. Let

$$\boldsymbol{B} = \begin{pmatrix} \boldsymbol{B}_{11} & \boldsymbol{B}_{12} \\ \boldsymbol{B}_{21} & \boldsymbol{B}_{22} \end{pmatrix},$$

and then $\boldsymbol{\eta}_1$ and $\boldsymbol{\eta}_2$ are mutually independent if and only if covariance matrices $\boldsymbol{B}_{12} = \boldsymbol{B}_{21}{}' = 0$. The proof is left to the reader.

Example 44 Let $\boldsymbol{\xi} = (\xi_1, \xi_2, \xi_3)' \sim N(\boldsymbol{a}, \boldsymbol{B})$, where

$$\boldsymbol{B} = \begin{pmatrix} 2 & 1 & 0 \\ 1 & 1 & 0 \\ 0 & 0 & 2 \end{pmatrix}.$$

Then one easily sees that ξ_1 and ξ_2 are not independent, but $(\xi_1, \xi_2)'$ is independent of ξ_3.

Property 4 $\boldsymbol{\xi}$ is normally distributed if and only if any linear combination of its components follows normal distributions. Specifically, let $\boldsymbol{l} = (l_1, \cdots, l_n)'$ be any n-dimensional real vector, and then

$$\boldsymbol{\xi} \sim N(\boldsymbol{a}, \boldsymbol{B}) \Longleftrightarrow \zeta = \boldsymbol{l}'\boldsymbol{\xi} \sim N(\boldsymbol{l}'\boldsymbol{a}, \boldsymbol{l}'\boldsymbol{B}\boldsymbol{l}) \tag{3.49}$$

$$\Longleftrightarrow \zeta = \sum_{j=1}^{n} l_j\xi_j \sim N\left(\sum_{j=1}^{n} l_j a_j, \sum_{j=1}^{n}\sum_{k=1}^{n} l_j l_k b_{jk}\right). \tag{3.50}$$

Proof Assume $\boldsymbol{\xi} \sim N(\boldsymbol{a}, \boldsymbol{B})$. For any real number u, write $u\boldsymbol{l} = \boldsymbol{t}$, and then the characteristic function of ζ is

$$f_\zeta(u) = E\mathrm{e}^{\mathrm{i}u\zeta} = E\mathrm{e}^{\mathrm{i}(u\boldsymbol{l}')\boldsymbol{\xi}} = f_{\boldsymbol{\xi}}(u\boldsymbol{l})$$

$$= \exp\left\{\mathrm{i}\boldsymbol{a}'u\boldsymbol{l} - \frac{(u\boldsymbol{l})'\boldsymbol{B}(u\boldsymbol{l})}{2}\right\}$$

$$= \exp\left\{i(\boldsymbol{a}'\boldsymbol{l})u - \frac{(\boldsymbol{l}'\boldsymbol{B}\boldsymbol{l})u^2}{2}\right\},$$

which implies that $\zeta \sim N(\boldsymbol{l}'\boldsymbol{a}, \boldsymbol{l}'\boldsymbol{B}\boldsymbol{l})$.

On the contrary, assume that $\zeta \sim N(\boldsymbol{l}'\boldsymbol{a}, \boldsymbol{l}'\boldsymbol{B}\boldsymbol{l})$, then for any real number u it follows that

$$f_\zeta(u) = Ee^{i(u\boldsymbol{l}')\boldsymbol{\xi}} = \exp\left\{i\boldsymbol{a}'u\boldsymbol{l} - \frac{(u\boldsymbol{l})'\boldsymbol{B}(u\boldsymbol{l})}{2}\right\}.$$

Take $u = 1$, and then

$$f_{\boldsymbol{\xi}}(\boldsymbol{l}) = Ee^{i\boldsymbol{l}'\boldsymbol{\xi}} = \exp\left\{i\boldsymbol{a}'\boldsymbol{l} - \frac{\boldsymbol{l}'\boldsymbol{B}\boldsymbol{l}}{2}\right\}.$$

So, $\boldsymbol{\xi} \sim N(\boldsymbol{a}, \boldsymbol{B})$.

In view of Property 4, we know that a multivariate random variable is normal if and only if any linear combination of its components follows normal distributions.

The necessary condition in Property 4 can be generalized to m dimensional linear transformations.

Property 5 Suppose $\boldsymbol{\xi} = (\xi_1, \cdots, \xi_n) \sim N(\boldsymbol{a}, \boldsymbol{B})$, $\boldsymbol{C} = (C_{ij})_{m\times n}$ is an $m \times n$ matrix, then

$$\boldsymbol{\eta} = \boldsymbol{C}\boldsymbol{\xi} \sim N(\boldsymbol{C}\boldsymbol{a}, \boldsymbol{C}\boldsymbol{B}\boldsymbol{C}') \tag{3.51}$$

is an m-dimensional normal distribution.

Proof It is easy to see

$$\begin{aligned} f_{\boldsymbol{\eta}}(\boldsymbol{t}) &= Ee^{i(\boldsymbol{t}'\boldsymbol{C})\boldsymbol{\xi}} = Ee^{i(\boldsymbol{C}'\boldsymbol{t})'\boldsymbol{\xi}} = f_{\boldsymbol{\xi}}(\boldsymbol{C}'\boldsymbol{t}) \\ &= \exp\left\{i\boldsymbol{a}'\boldsymbol{C}'\boldsymbol{t} - \frac{1}{2}(\boldsymbol{C}'\boldsymbol{t})'\boldsymbol{B}\boldsymbol{C}'\boldsymbol{t}\right\} \\ &= \exp\left\{i(\boldsymbol{C}\boldsymbol{a})'\boldsymbol{t} - \frac{1}{2}\boldsymbol{t}'(\boldsymbol{C}\boldsymbol{B}\boldsymbol{C}')\boldsymbol{t}\right\}. \end{aligned}$$

Corollary There exists an orthogonal matrix $\boldsymbol{U}$ such that, under the transform $\boldsymbol{\eta} = \boldsymbol{U}\boldsymbol{\xi}$, covariance matrix of $\boldsymbol{\eta}$ is diagonal matrix

$$\boldsymbol{U}\boldsymbol{B}\boldsymbol{U}' = \boldsymbol{D} = \begin{pmatrix} d_1 & & \\ & \ddots & \\ & & d_n \end{pmatrix}. \tag{3.52}$$

Since $\boldsymbol{B}$ is a real symmetric matrix, such an orthogonal matrix $\boldsymbol{U}$ must exist, $d_1, \cdots, d_n$ are eigenvalues of $\boldsymbol{B}$, and each row of $\boldsymbol{U}$ is corresponding eigenvector. If the rank of $\boldsymbol{B}$ is r, then $\boldsymbol{\eta}$ has r independent normal components.

Using block matrices, one can prove that Property 5 remains valid when ξ_i and ξ_j are subvectors of $\boldsymbol{\xi}$ and $\boldsymbol{\eta}$ respectively.

Property 6 Assume that $\boldsymbol{\xi} \sim N(\boldsymbol{a}, \boldsymbol{B})$, $\boldsymbol{\xi} = (\boldsymbol{\xi}'_1, \boldsymbol{\xi}'_2)'$, where $\boldsymbol{\xi}_1$, $\boldsymbol{\xi}_2$ are k and $n-k$-dimensional subvectors of $\boldsymbol{\xi}$ respectively,

$$\boldsymbol{B} = \begin{pmatrix} \boldsymbol{B}_{11} & \boldsymbol{B}_{12} \\ \boldsymbol{B}_{21} & \boldsymbol{B}_{22} \end{pmatrix} \tag{3.53}$$

is positive definite and $\boldsymbol{\xi}_1 \sim N(\boldsymbol{a}_1, \boldsymbol{B}_{11})$, $\boldsymbol{\xi}_2 \sim N(\boldsymbol{a}_2, \boldsymbol{B}_{22})$. Then conditioning on $\boldsymbol{\xi}_1 = \boldsymbol{x}_1$ the conditional distribution of $\boldsymbol{\xi}_2$ is a normal distribution

$$N(\boldsymbol{a}_2 + \boldsymbol{B}_{21}\boldsymbol{B}_{21}^{-1}\boldsymbol{B}_{11}^{-1}(\boldsymbol{x}_1 - \boldsymbol{a}_1), \boldsymbol{B}_{22} - \boldsymbol{B}_{21}\boldsymbol{B}_{11}^{-1}\boldsymbol{B}_{12}). \tag{3.54}$$

This is a generalization of Example 19 in Section 2.4, and the proof is omitted. (3.54) shows the conditional expectation is linear in $\boldsymbol{x}_1$, while the conditional variance is independent of $\boldsymbol{x}_1$.

The use of the above properties simplifies much of computation about a multivariate normal distribution.

Example 45 Assume that $\boldsymbol{\xi} = (\xi_1, \xi_2)' \sim N(a_1, a_2, \sigma^2, \sigma^2, r)$, then prove $\eta_1 = \xi_1 + \xi_2$ and $\eta_2 = \xi_1 - \xi_2$ are independent, and find respective distributions of η_1, η_2.

Solution This example can be solved by transform of random variables introduced in the previous chapter. But now we use Property 5. We have

$$\begin{pmatrix}\eta_1\\ \eta_2\end{pmatrix} = \begin{pmatrix}\xi_1 + \xi_2\\ \xi_1 - \xi_2\end{pmatrix} = \begin{pmatrix}1 & 1\\ 1 & -1\end{pmatrix}\begin{pmatrix}\xi_1\\ \xi_2\end{pmatrix} \equiv \boldsymbol{C\xi},$$

$$\boldsymbol{B} = \sigma^2\begin{pmatrix}1 & r\\ r & 1\end{pmatrix},$$

and

$$\boldsymbol{CBC}' = 2\sigma^2\begin{pmatrix}1+r & 0\\ 0 & 1-r\end{pmatrix}.$$

Hence $\mathrm{Cov}(\eta_1, \eta_2) = 0$, and η_1, η_2 are independent by Property 3. Also, since

$$\boldsymbol{C}\begin{pmatrix}a_1\\ a_2\end{pmatrix} = \begin{pmatrix}a_1 + a_2\\ a_1 - a_2\end{pmatrix},$$

we obtain

$$\eta_1 \sim N(a_1 + a_2, 2\sigma^2(1+r)), \quad \eta_2 \sim N(a_1 - a_2, 2\sigma^2(1-r)).$$

Supplements and Remarks

1. Properties of the Stieltjes integral

(1) If $g(x)$ is continuous on $[a, b]$, $F(x)$ has the first class of discontinuous points at finitely many points c_i $(i = 1, 2, \cdots, m)$. Besides, assume $F'(x)$ is absolutely integrable on $[a, b]$. Then

$$\int_a^b g(x)\mathrm{d}F(x) = \int_a^b g(x)F'(x)\mathrm{d}x + \sum_{i=0}^{m} g(c_i)[F(c_i + 0) - F(c_i - 0)].$$

If $F(x)$ is a distribution function of a continuous random variable, then the second part is equal to 0; while the first part is equal to 0 if F is a distribution function of a discrete random variable.

(2) $$\int_{-\infty}^{+\infty}[ag(x) + bf(x)]\mathrm{d}F(x)$$

$$= a\int_{-\infty}^{+\infty} g(x)\mathrm{d}F(x) + b\int_{-\infty}^{+\infty} f(x)\mathrm{d}F(x).$$

(3) $$\int_{-\infty}^{+\infty} g(x)\mathrm{d}[aF_1(x) + bF_2(x)]$$

$$= a\int_{-\infty}^{+\infty} g(x)\mathrm{d}F_1(x) + b\int_{-\infty}^{+\infty} g(x)\mathrm{d}F_2(x).$$

(4) When $a \leqslant c \leqslant b$

$$\int_a^b g(x)\mathrm{d}F(x) = \int_a^c g(x)\mathrm{d}F(x) + \int_c^b g(x)\mathrm{d}F(x).$$

(5) If $g(x) \geqslant 0$, $F(x)$ is nondecreasing. Let $b > a$, and then

$$\int_a^b g(x)\mathrm{d}F(x) \geqslant 0.$$

(6) $\int_a^b g(x)\mathrm{d}F(x) = g(x)F(x)\Big|_a^b - \int_a^b F(x)\mathrm{d}g(x).$

Equations (2) ~ (6) hold only under the condition that the integration in the left hand side makes sense. These formulae are very similar to those for the Riemann integration.

2. Moment inequalities

We give some useful inequalities, which compare moments of different orders for random variables.

(1) Hölder's inequality. Let ξ and η be two random variables, $1 < p < \infty$, $1/p + 1/q = 1$, and then

$$|E\xi\eta| \leqslant E|\xi\eta| \leqslant (E|\xi|^p)^{1/p}(E|\eta|^q)^{1/q}.$$

In particular, let $\eta = 1$, and we have $E|\xi| \leqslant (E|\xi|^p)^{1/p}$. When $p = 2$, it reduces to Cauchy-Schwarz's inequality.

Furthermore, replacing $|\xi|$ by $|\xi|^r$, we obtain

$$(E|\xi|^r)^{1/r} \leqslant (E|\xi|^{r'})^{1/r'}, \quad 1 < r < r' < \infty.$$

This is called Lyapunov's inequality.

(2) Minkowski's inequality. Let ξ and η be two random variables, $1 < p < \infty$, and then

$$(E|\xi + \eta|^p)^{1/p} \leqslant (E|\xi|^p)^{1/p} + (E|\eta|^p)^{1/p}.$$

3. Median and p-quantile

Suppose that X is a random variable, $0 < p < 1$. If α_p is such that

$$P(X \leqslant \alpha_p) \geqslant p, \ P(X \geqslant \alpha_p) \geqslant 1 - p,$$

then we call α_p a p-quantile of X. In particular, when $p = 1/2$ we call $\alpha_{1/2}$ a median of X. It is easy to see that for a continuous random variable X with distribution function $F(x)$ and a given value of p, the p-quantile α_p is the solution to equation $F(x) = p$; and either there is a unique value of α_p such that $F(\alpha_p) = p$, or this relationship is satisfied for a full interval.

Example 46 Suppose that a random variable ξ is such that $P(\xi = 0) = P(\xi = 1) =$

1/2. Then for an arbitrary $0<a<1$,

$$P(\xi \leqslant a) = P(\xi = 0) = \frac{1}{2}, \quad P(\xi \geqslant a) = P(\xi = 1) = \frac{1}{2}.$$

Hence any real number in the interval $(0, 1)$ is a median of ξ, which indicates that median maybe not unique.

Example 47 Suppose that a random variable ξ is such that $P(\xi=-1)=P(\xi=0)=P(\xi=1)=\frac{1}{3}$. Then

$$P(\xi \leqslant 0) = \frac{2}{3} > \frac{1}{2}, \quad P(\xi \geqslant 0) = \frac{2}{3} > \frac{1}{2},$$

so ξ has a median 0. In fact, 0 is its unique median, but neither $P(\xi \leqslant 0)$ nor $P(\xi \geqslant 0)$ is equal to 1/2.

4. Suppose that X is a random variable with normal distribution $N(0,1)$, then $Y=\mathrm{e}^X$ follows the log-normal distribution with the density function

$$p(x) = \frac{1}{\sqrt{2\pi}x}\mathrm{e}^{-(\log x)^2/2}, \quad x > 0.$$

Assume that $-1 \leqslant a \leqslant 1$, the define

$$p_a(x) = [1 + a\sin(2\pi\log x)]p(x).$$

It is not hard to verify $p_a(x)$ is a density function. Note that moments of any order of Y are finite, and for $k \geqslant 1$

$$\int_{-\infty}^{+\infty} x^k p_a(x)\mathrm{d}x = \int_{-\infty}^{+\infty} x^k p(x)\mathrm{d}x.$$

This shows $p_a(x)$, $-1\leqslant a\leqslant 1$ are a family of distinct density functions but have same moments.

We have the following theorem.

Theorem 6 *Assume that ξ and η are two random variables having common moments m_k of order k, $k\geqslant 1$. Suppose that one of the following three conditions is satisfied.*

(1) $\sum_{k=1}^{\infty} \frac{m_{2k}}{(2k)!} t^{2k} < \infty$ *for some* $t > 0$;

(2) $\sum_{k=1}^{\infty} m_{2k}^{-1/(2k)} = \infty$ (*the Carleman condition*);

(3) $\limsup_{k\to\infty} \frac{|m_k|^{1/k}}{k} < \infty$.

Then ξ and η have the same distribution function.

5. Generating function and moment generating function

For a discrete random variable ξ with non-negative integer values $0, 1, 2, \cdots$, and the distribution sequence $p_k = P(\xi = k)$, define its generating function by

$$g_\xi(s) = Es^\xi = \sum_{k=0}^{\infty} p_k s^k, \quad |s| \leqslant 1.$$

It is also called the generating function of the sequence $\{p_k\}$.

For example, if ξ has a Poisson distribution $P(\lambda)$, then its generating function is

$$g_\xi(s) = \sum_{k=0}^{\infty} s^k \frac{\lambda^k}{k!} e^{-\lambda} = e^{-\lambda(1-s)}.$$

The generating function $g_\xi(s)$ and the distribution sequence $\{p_k\}$ determine each other uniquely. In fact, $p_0 = g_\xi(0)$, $p_k = g_\xi^{(k)}(0)$, $k = 1, 2, \cdots$. It can be verified that the generating function has the following properties.

(1) Let $g_\xi(s)$ be the generating function of ξ. Then, the expectation $E\xi$ exists if and only if the derivative $g_\xi'(1)$ exists, and $E\xi = g_\xi'(1)$; the variance $\mathrm{Var}(\xi)$ exists if and only if the second derivative $g_\xi''(1)$ exists, and

$$\mathrm{Var}(\xi) = g_\xi''(1) + g_\xi'(1) - (g_\xi''(1))^2.$$

(2) If $g_\xi(s)$, $g_\eta(s)$ are respectively the generating functions of two independent random variables ξ and η who take non-negative integer values, the generating function of $\xi + \eta$ is $g_\xi(s) g_\eta(s)$.

(3) If $g_\xi(s)$ is the generating function of ξ, then its characteristic function is $g_\xi(e^{ix})$.

The above generating function only has definition for random variable with non-negative integer values. For a general random variable ξ, we usually consider its moment generating function as

$$M_\xi(t) = Ee^{t\xi}, \quad t \in T$$

for some $T \subseteq \mathbf{R}$ provided that the required expected values exist.

Moment generating function is an important tool in the study of random variables and distribution functions, but it does not necessarily exist for all t.

Example 48 If ξ obeys normal distribution $N(\mu, \sigma^2)$, then

$$M_\xi(t) = Ee^{t\xi} = e^{\mu t + \frac{1}{2}\sigma^2 t^2}, \quad t \in \mathbf{R}. \tag{3.55}$$

Example 49 If ξ obeys the exponential distribution with parameter λ, then when $t < \lambda$,

$$M_\xi(t) = \frac{\lambda}{\lambda - t}; \tag{3.56}$$

when $t \geqslant \lambda$, $M_\xi(t)$ does not exist.

Example 50 If ξ obeys the Cauchy distribution, then $M_\xi(t) = 1$ only when $t = 0$; for any other t, $M_\xi(t)$ does not exist.

We have known that the moment generating function of a random variable does not necessarily exist, but the characteristic function always exists and they have some common properties.

6. A method of computing characterstic functions

For readers not being familiar with contour integral of complex variable functions, we give a method of computing characteristic functions. Assume that ξ is a random variable, and there exists a $t > 0$ such that its moment generating function $M_\xi(s)$ is finite for all $|s| \leqslant t$, then its characteristic function $f_\xi(t) = M_\xi(it)$. Thus we can calculate the characteristic function by computing the moment generating function (the mathematical expectation of a real value random variable).

Example 51 If ξ follows the normal distribution $N(\mu,\sigma^2)$, then $f_\xi(t) = e^{i\mu t-\frac{1}{2}\sigma^2 t^2}$ by (3.55). If ξ follows the exponential distribution with parameter λ, then $f_\xi(t)=\lambda/(\lambda-it)$ by (3.56).

Similarly to Property 5 of a characteristic function in Section 3.3 we have

$$M_\xi^{(k)}(0) = E\xi^k,\ k=1,2,\cdots,$$

where $M_\xi^{(k)}(0)$ is the value of k-th derivative of $M_\xi(t)$ at 0. So, the function $M_\xi(t)$ generates all of the moments of ξ. Hence $M_\xi(t)$ is also called the moment generating function of ξ.

7. The general definition of conditional expectation

In Section 3.1, the conditional expectation $E(\eta \mid \xi = x)$ is defined as the mathematical expectation of the conditional distribution $F_{\eta|\xi}(y \mid x)$ of η given the condition $\xi = x$. When (ξ,η) is a discrete or continuous random vector, the conditional distribution defined as

$$F_{\eta|\xi}(y \mid x) = \lim_{\Delta x\to 0}\frac{P(\eta\leqslant y,\ x<\xi\leqslant x+\Delta x)}{P(x<\xi\leqslant x+\Delta x)}$$

is usually well-defined, and its distribution sequence (or density function) is just the conditional distribution sequence (or conditional density function). So, in such case the conditional expectation is well-defined. But for a general random vector (ξ,η), the above limit may not exist. To define the conditional expectation, it is needed to use measure theory.

If (ξ,η) is a continuous random vector, then following the lines of the proof (3.11) in Section 3.1 we can find that, for $m(x) = E(\eta \mid \xi = x)$ and any Borel set A,

$$\int_{\xi\in A} m(\xi)\,\mathrm{dP} = \int_{\xi\in A}\eta\,\mathrm{dP}. \tag{3.57}$$

Here, the integral on an event A as $\int_{\xi\in A}\eta\,\mathrm{dP}$ is defined to be the expectation $E[\eta I\{\xi\in A\}]$, where $I\{\xi\in A\}$ is the indicator of $\{\xi\in A\}$.

Conversely, if there is a function $m(\xi)$ of ξ such that the above equality holds for any Borel set A, then we will have $P(m(\xi)\neq E(\eta\mid\xi))=0$. In fact, choosing $A=\{x: m(x)\geqslant E(\eta\mid\xi=x)\}$, we obtain

$$\begin{aligned}\int (m(\xi)-E(\eta\mid\xi))^+\,dP &= \int_{\xi\in A} m(\xi)\,dP - \int_{\xi\in A} E(\eta\mid\xi)\,dP \\ &= \int_{\xi\in A}\eta\,dP - \int_{\xi\in A}\eta\,dP = 0.\end{aligned}$$

Similarly,

$$\int (m(\xi)-E(\eta\mid\xi))^-\,dP = 0.$$

Hence

$$\int |m(\xi) - E(\eta \mid \xi)| \, \mathrm{d}P = 0.$$

Therefore, $P(m(\xi) \neq E(\eta \mid \xi)) = 0$.

So, if disregarding the values on a null event, the conditional expectation is a unique function $E(\eta \mid \xi)$ of ξ such that (3. 57) holds for any Borel set A. This conclusion remains true for any a general random variable. We state the conclusion as the following theorem whose proof is beyond the scope of this text book.

Theorem 7 *Let $\eta, \xi_1, \cdots\cdots, \xi_d$ be random variables with $E|\eta| < \infty$. Denote $\boldsymbol{\xi} = (\xi_1, \cdots, \xi_d)$. Then there is a Borel function $m(\boldsymbol{x})$ such that for any d-dimensional Borel set A,*

$$\int_{\boldsymbol{\xi} \in A} m(\boldsymbol{\xi}) \mathrm{d}P = \int_{\boldsymbol{\xi} \in A} \eta \mathrm{d}P. \tag{3.58}$$

Further, if disregarding the values on a null event, such a function $m(\boldsymbol{\xi})$ is unique.

Denition 11 *The random variable $m(\boldsymbol{\xi})$ as in (3. 58) is said to be the conditional expectation of η given $\boldsymbol{\xi}$ or the conditional expectation of η with respect to $\boldsymbol{\xi}$, written as $E(\eta \mid \boldsymbol{\xi})$. And $m(\boldsymbol{x})$ is said to be the conditional expectation of η given $\boldsymbol{\xi} = \boldsymbol{x}$, written as $E(\eta \mid \boldsymbol{\xi} = \boldsymbol{x})$.*

Example 52 Suppose that ξ and η are independent and $E|\eta| < \infty$. Prove $E(\eta|\xi) = E\eta E\xi$.

Proof By the definition, it is sufficient to show that for any Borel set A we have

$$\int_{\xi \in A} E\eta \mathrm{d}P = \int_{\xi \in A} \eta \mathrm{d}P.$$

While, the independence of ξ and η obviously implies that

$$\int_{\xi \in A} = E[\eta I\{\xi \in A\}] = E\eta P(\xi \in A) = \int_{\xi \in A} E\eta \mathrm{d}P.$$

Example 53 Suppose that $E|\eta| < \infty$ and τ is function of ξ. Prove

$$E(\eta \mid \tau) = E[E(\eta \mid \xi)\tau].$$

Proof By the definition of $E(\eta \mid \tau)$, it is sucient to show that

$$\int_{\tau \in A} \eta \mathrm{d}P = \int_{\tau \in A} E(\eta \mid \xi) \mathrm{d}P.$$

Denote $\tau = g(\xi)$. Note that $\tau \in A$ if and only if $\xi \in g^{-1}(A)$. Hence, by the definition of $E(\eta \mid \xi)$,

$$\begin{aligned}\int_{\tau \in A} E(\eta \mid \xi) \mathrm{d}P &= \int_{\xi \in g^{-1}(A)} E(\eta \mid \xi) \mathrm{d}P = \int_{\xi \in g^{-1}(A)} \eta \mathrm{d}P \\ &= \int_{\tau \in A} \eta \mathrm{d}P.\end{aligned}$$

The result follows.

8. Martingale

The martingale is an important concept in probability theory and a useful tool to study various random phenomena. We give the definition of martingale as follows.

Definition 12 *Let $\{S_n; n \geqslant 1\}$ be a sequence of random variables with finite expectations. If*

$$E(S_n \mid S_1, S_2, \cdots, S_{n-1}) = S_{n-1}, n \geqslant 2,$$

then $\{S_n; n \geqslant 1\}$ is called a martingale. If denote $S_0 = 0, X_n = S_n - S_{n-1}$, then

$$E(X_n \mid X_1, X_2, \cdots, X_{n-1}) = 0, n \geqslant 2.$$

A sequence of random variables $\{Xn;\ n \geqslant 1\}$ having the above property is said to be martingale difference sequence.

By the definition, if $\{S_n; n \geqslant 1\}$ is a martingale, then for any $1 \leqslant k \leqslant n-1$,

$$\begin{aligned} E(S_n \mid S_1, \cdots, S_k) &= E[E(S_n \mid S_1, S_2, \cdots, S_{n-1}) \mid S_1, \cdots, S_k] \\ &= E(S_{n-1} \mid S_1, \cdots, S_k) = \cdots = S_k, \end{aligned}$$

where the first equality is due to the conclusion in Example 53 in which $\tau = (S_1, \cdots, S_k)$ is a function of $\xi = (S_1, \cdots, S_{n-1})$.

Example 54 Let $\{X_n; n \geqslant 1\}$ be a sequence of independent random variables with means 0. Then $\{X_n; n \geqslant 1\}$ is a martingale difference.

In fact, by independence it is obvious that

$$E(X_n \mid X_1, \cdots, X_{n-1}) = EX_n = 0.$$

Example 55 Let $\{X_n; n \geqslant 1\}$ be a sequence of random variables with common distribution sequence as

$$P(X_n = 1) = p, P(X_n = -1) = 1 - p, n \geqslant 1,$$

where $0 < p < 1$. Denote $S_n = X_1 + \cdots + X_n$ and $Y_n = \frac{1-p}{p}^{S_n}$. Prove that $\{Y_n; n \geqslant 1\}$ is a martingale.

Proof Since $Y_n = Y_{n-1} \frac{1-p}{p}^{X_n}$, we have

$$E(Y_n \mid Y_1, \cdots, Y_{n-1}) = Y_{n-1} E\left(\left(\frac{1-p}{p}\right)^{X_n} \mid Y_1, \cdots, Y_{n-1}\right).$$

Noting independence between X_n and $Y_1, \cdots, Y_{n-1}$, we obtain

$$\begin{aligned} E\left(\left(\frac{1-p}{p}\right)^{X_n} \mid Y_1, \cdots, Y_{n-1}\right) &= E\left(\frac{1-p}{p}\right)^{X_n} \\ &= \left(\frac{1-p}{p}\right) P(X_n = 1) + \left(\frac{1-p}{p}\right)^{-1} P(X_n = -1) \\ &= \left(\frac{1-p}{p}\right) \cdot p + \frac{p}{1-p} \cdot (1-p) = 1. \end{aligned}$$

Hence, $E(Y_n \mid Y_1, \cdots, Y_{n-1}) = Y_{n-1}$. Therefore, $\{Y_n; n \geqslant 1\}$ is a martingale.

Example 56 (Likelihood ratio) Let $\{X_n; n \geqslant 1\}$ be a sequence of independent and identically distributed random variables with a common density function being one of f and g. Suppose that f and g are two density functions and have the same support, i. e., $\{x: f(x) > 0\} = \{x: g(x) > 0\}$. The following likelihood ratio is a useful test statistic to judge which is the true density function according to the observed values of $X_1, \cdots, X_n$.

$$Y_n = \frac{\prod_{k=1}^{n} g(X_k)}{\prod_{k=1}^{n} f(X_k)}.$$

Prove that $\{Y_n; n \geqslant 1\}$ is a martingale when f is the true density function.

Proof For convenience, we suppose that f is positive everywhere. Then

$$\begin{aligned} E(Y_n \mid Y_1, \cdots, Y_{n-1}) &= E\left(Y_{n-1} \frac{g(X_n)}{f(X_n)} \mid Y_1, \cdots, Y_{n-1}\right) \\ &= Y_{n-1} E\left(\frac{g(X_n)}{f(X_n)} \mid Y_1, \cdots, Y_{n-1}\right) \\ &= Y_{n-1} E \frac{g(X_n)}{f(X_n)} \\ &= Y_{n-1} \int_{-\infty}^{\infty} \frac{g(x)}{f(x)} f(x) \mathrm{d}x \\ &= Y_{n-1} \int_{-\infty}^{\infty} g(x) \mathrm{d}x = Y_{n-1}. \end{aligned}$$

The conclusion follows.

Example 57 (Polya's urn) Suppose that an urn has a white balls and b black balls initially. Repeat the procedure as that, a ball is drawn randomly from the urn and replaced to the urn with c balls of color the same as the drawn ball. Let Y_n be the proportion of black balls after the n-th draw, and $Y_0 = \frac{b}{a+b}$. Prove that $\{Y_n; n \geqslant 1\}$ is a martingale.

Proof Let α be number of black balls after the n-th draw, and β be total number of balls. Then, no matter what value of (α, β) is, under the conditon $Y_n = \frac{\alpha}{\beta}$, we have

$$E\left(Y_{n+1} \mid Y_n = \frac{\alpha}{\beta}\right) = \frac{\alpha + c}{\beta + c} \cdot \frac{\alpha}{\beta} + \frac{\alpha}{\beta + c} \cdot \left(1 - \frac{\alpha}{\beta}\right) = \frac{\alpha}{\beta} = Y_n.$$

So, $E(Y_{n+1} \mid Y_1, \cdots, Y_n) = Y_n$. Therefore, $\{Y_n; n \geqslant 1\}$ is a martingale.

9. Branching process

We consider the model in Example 42 of Section 1.4, Chapter 1 again. In a population consisting of individuals able to produce offspring of the same kind, suppose that each individual will, by the end of his lifetime, have produced k new offspring with probability p_k, $k = 0, 1, 2\cdots$, independently of the number produced by any other individual. Assume that the number of individuals which initially presents (the 0-th generation) is 1. Denote ξ_n to be the number of individuals of the n-th generation, and η_{ni} to be the number of offspring of the i-th individual of the n-th generation. Then $\xi_0 = 1$ and

$$\xi_{n+1} = \sum_{i=1}^{\xi_n} \eta_{ni}, n \geqslant 0.$$

Here, $\sum_{i=1}^{0}(\cdot)$ is defined to be 0. As assumed, each individual produces offspring independently in a same manner. That is, $\{\eta_{ni};\ n \geqslant 0,\ i \geqslant 1\}$ is an array of independent and identically distributed random variables taking non-negative integer values with a common distribution sequence as

$$p_k = P(\eta_{ni} = k); k = 0, 1, \cdots.$$

The above random sequence $\{\xi_n; n \geqslant 0\}$ is called G-W branching process named after English mathematicians Galton (Francis Galton, 1822 — 1911) and Watson (Henry William Watson, 1827 — 1903) who introduced it in 1874, and is called branching process for abbreviation. Important questions for this random sequence are the distribution of ξ_n, the growth trend of ξ_n as n increases to innity, and the probability that the population will eventually die out (i. e., $\xi_n = 0$ for some n).

(1) Expectation and variance.

We first consider the expectation and variance of ξ_n. We will find that $E\xi_n = m^n$ if $m = E\xi_1 = \sum_{k=1}^{\infty} kp_k$ exists, and

$$\mathrm{Var}(\xi_n) = \begin{cases} n\sigma^2, & \text{when } m = 1; \\ \dfrac{\sigma^2 m^{n-1}(m^n - 1)}{m - 1}, & \text{when } m \neq 1 \end{cases}$$

if Var $(\xi_1) = \sigma^2$ exists.

For showing the results, by taking the conditional expectation we obtain

$$E(\xi_n \mid \xi_{n-1} = k) = E(\sum_{i=1}^{k})\eta_{n-1,i} \mid \xi_{n-1} = k) = \sum_{i=1}^{k} E\eta_{n-1,i} = km.$$

So

$$E(\xi_n \mid \xi_{n-1}) = m\xi_{n-1}. \tag{3.59}$$

Hence we obtain the expectation as

$$E\xi_n = mE\xi_{n-1} = \cdots = m^n.$$

For deriving the variance, write ξ_n as

$$\xi_n = \sum_{i=1}^{\xi_{n-1}} (\eta_{n-1,i} - m) + m\xi_{n-1}.$$

Then

$$\begin{aligned} E(\xi_n^2 \mid \xi_{n-1} = k) &= E\Big[\sum_{i=1}^{k} (\eta_{n-1,i} - m)\Big]^2 + (mk)^2 + 2mkE\Big[\sum_{i=1}^{k} (\eta_{n-1,i} - m)\Big] \\ &= k\sigma^2 + m^2 k^2. \end{aligned}$$

That is

$$E(\xi_n^2 \mid \xi_{n-1}) = \sigma^2 \xi_{n-1} + m^2 \xi_{n-1}^2.$$

Therefore,

$$E\xi_n^2 = \sigma^2 E\xi_{n-1} + m^2 E\xi_{n-1}^2 = \sigma^2 m^{n-1} + m^2 E\xi_{n-1}^2,$$

$$E\left(\frac{\xi_n}{m^n}\right)^2 = E\,\frac{\xi_{n-1}}{m^{n-1}}^2 + \frac{\sigma^2}{m^{n+1}}.$$

Since $E\dfrac{\xi_n}{m^n}=1$, we conclude that

$$\begin{aligned}\operatorname{Var}\left(\frac{\xi_n}{m^n}\right)^2&=\frac{\sigma^2}{m^{n+1}}+\operatorname{Var}\left(\frac{\xi_{n-1}}{m^{n-1}}\right)\\&=\frac{\sigma^2}{m^{n+1}}+\frac{\sigma^2}{m^n}+\cdots+\frac{\sigma^2}{m^{1+1}}+\operatorname{Var}\xi_0\\&=\begin{cases}n\sigma^2, & \text{when } m=1;\\ \dfrac{m^{-1}-m^{-n-1}}{m-1}\sigma^2, & \text{when } m\neq 1.\end{cases}\end{aligned}$$

The results follow.

Remark By (3.59), $\left\{\dfrac{\xi_n}{m^n};n\geqslant 1\right\}$ is a martingale.

(2) Generating function.

Let $g(s)=p_0+\sum_{k=1}^{\infty}p_k s^k$ be the generating function of η_{ni}, and $g_n(s)=Es^{\xi_n}$ be the generating function of $\xi_n(n\geqslant 1)$. The equation $g_1(s)=g(s)$ is obvious and we will have following iterates

$$g_n(s)=g(g_{n-1}(s)=g_{n-1}(g(s)),\ |s|\leqslant 1. \tag{3.60}$$

In fact, taking the conditional expectation yields

$$E(s^{\xi_n}\mid\xi_{n-1}=k)=E\Big(\prod_{i=1}^{k}s^{\eta_{n-1,i}}\mid\xi_{n-1}=k\Big)=\prod_{i=1}^{k}E(s^{\eta_{n-1,i}})=(g(s))^k.$$

That is $E[s^{\xi_n}\mid\xi_{n-1}]=(g(s))^{\xi_{n-1}}$. Therefore,

$$g_n(s)=E[E(s^{\xi_n}\mid\xi_{n-1})]=E[(g(s))^{\xi_{n-1}}]=g_{n-1}(g(s)),$$

which is the second equality in (3.60). The first equality of (3.60) can be shown by the induction and the above equation. In fact, when $n=2$, since $g_1\equiv g$, we have $g_2(s)=g_1(g(s))=g(g_1(s))$ as desired. Suppose that the conclusion is true for $n-1$, then

$$g_n(s)=g_{n-1}(g(s))=g[g_{n-2}(g(s))]=g[g_{n-1}(s)].$$

(3.60) is proved.

The expectation and variance of ξ_n can be derived from its generating function. The details are omitted here.

(3) Extinction probability.

The extinction probability is the probability that the population will eventually die out, and so is

$$q=P(\lim_{n\to 0}\xi_n=0).$$

Note $\{\lim_{n\to\infty}\xi_n=0\}=\bigcup_{n=1}^{\infty}\{\xi_n=0\}$, $\{\xi_n=0\}\subset\{\xi_{n+1}=0\}$. By the continuity of the probability,

$$q=\lim_{n\to\infty}P(\xi_n=0)=\lim_{n\to\infty}g_n(0). \tag{3.61}$$

Denote $q_n=\mathrm{P}(\xi_n=0)=g_n(0)$. Then q_n is a non-increasing sequence. By (3.60), we have $q_n=g(q_{n-1})$. Letting $n\to\infty$ and applying the continuity of the function $g(s)$, we conclude that the extinction probability will satisfy the following equation

$$q = g(q).$$

This coincides with the conclusion in Chapter 1. If $p_1 = 1$, then $\xi_n \equiv 1$ and so $q = 0$. When $p_1 < 1$ and $m = E\xi_1 \leqslant 1$, in Example 42 of Section 1.4, Chapter 1, it is shown that the above equation has a unique solution 1. So, the extinction probability $q = 1$.

Now, assume $m > 1$ (in this case, $p_0 + p_1 < 1$). We will find that the extinction probability q is then the unique solution of (3.62) on $[0,1)$. Hence, $0<q<1$ if $p_0 > 0$, and $q = 0$ if $p_0 = 0$.

In fact, since $(g(s) - s)'' = g''(s) > 0$ for $s \in (0,1)$, $g'(1) - 1 = m - 1 > 0$ and $g'(0) - 0 = p_1 < 1$, there is a unique point $s_0 \in (0,1)$ such that $g'(s_0) - 1 = 0$. It follows that the function $g(s) - s$ is strictly decreasing on $[0, s_0]$, and strictly increasing on $[s_0, 1]$. And so, it has at least one zero point on each of the two intervals. Obviously, $s = 1$ is a zero point on $[s_0, 1]$, and so it must be the unique zero point on this interval. Further, $g(s_0) - s_0 < g(1) - 1 = 0$ and $g(0) - 0 = p_0 \geqslant 0$. So, $g(s) - s$ has one and only one zero point on $[0, s_0)$. We conclude that, on the interval $[0,1)$ the equation (3.61) has one unique solution π.

For verifying that the extinction probability q is just the unique solution π of the equation, it is sufficient to show that $q_n \leqslant \pi$. Since $g(s)$ is a non-decreasing function of $s \in [0,1]$, we have $q_1 = g(0) \leqslant g(\pi) = \pi$. By the induction, $q_n = g(q_{n-1}) \leqslant g(\pi) = \pi$. Therefore, we must have $q = \pi < 1$.

If $p_0 = 0$, then $s = 0$ is a solution of $g(s) - s = 0$, and so $q = 0$. If $p_0 > 0$, then $g(0) - 0 = p_0 \neq 0$, and so $0 < q < 1$.

Summing up the above arguments, we have the following theorem.

Theorem 8 *The extinction probability q is the smallest solution of $g(s) = s$ on $[0,1]$. Furthermore,* (i)$q = 1$ *if* $m < 1$; (ii) $q = 0$ *if* $m = 1$ *and* $p_1 = 1$, *and* $q = 1$ *if* $m = 1$ *and* $p_1 < 1$; (iii) $q = 0$ *if* $m > 1$ *and* $p_0 = 0$, *and* $0 < q < 1$ *if* $m > 1$ *and* $p_0 > 0$.

(4) Limit of the generating function.

Next, we consider the limit of the generating function $g_n(s)$. If $0 \leqslant s \leqslant q$, then $g_1(s) = g(s) \leqslant g(q) = q$ by the monotonicity of g. By the induction, $g_n(s) = g(g_{n-1}(s)) \leqslant q$. On the other hand, $g_n(s) \geqslant g_n(0) = q_n$. So

$$\lim_{n\to\infty} g_n(s) = q, 0 \leqslant s \leqslant q.$$

If $m \leqslant 1$ and $p_1 < 1$, then $q = 1$, and so the above equation is just

$$\lim_{n\to\infty} g_n(s) = q, 0 \leqslant s < 1.$$

In the case of $m > 1$, since q and 1 are unique two solutions of the equation $g(s) = s$, and $g(s) < s$ in a neighborhood of 1, we must have $q = g(q) \leqslant g(s) < s < 1$ if $q < s < 1$. Note $g_n(s) = g(g_{n-1}(s))$. By the induction,

$$q \leqslant g_n(s) < g_{n-1}(s) < \cdots < 1, q < s < 1,$$

which means that $g_n(s)$ is a non-increasing sequence with values between q and 1. So,

$$q \leqslant \lim_{n\to\infty} g_n(s) =: \alpha < 1.$$

On the other hand, since $g_n(s) = g(g_{n-1}(s))$, α must be a solution of $g(s) = s$ on $[0; 1)$. It follows $\alpha = q$.

Summing up the above arguments, we conclude that: if $p_1 < 1$, then

$$\lim_{n\to\infty} g_n(s) = q, 0 \leqslant s < 1.$$

When $p_1 = 1$, the above equation fails because $q = 0$ and $g_n(s) \equiv s$.

Note

$$g_n(s) = P(\xi_n = 0) + \sum_{k=1}^{\infty} P(\xi_n = k)s^k.$$

When $p_1 < 1$, its coeffcients will satisfy

$$P(\xi_n = 0) \to q,$$

$$P(\xi_n = k) \to 0, k = 1, 2, \cdots.$$

The latter shows that the probability that the n-th generation has a positive number of individuals goes to 0.

Exercise 3

1. Assume that the random variable ξ has the following distribution, find $E\xi$.

(1) $P(\xi = k) = 1/5$, $k = 1,2,3,4,5$;

(2) $P(\xi = k) = a^k/(1+a)^{k+1}$, where $a > 0$ is a constant, $k = 0,1,2,\cdots$.

2. There are k balls labelled by k in a bag, $k = 1,2,\cdots,n$. Now draw at random a ball out of the bag, find the expected value of the label of the drawn ball.

3. Somebody has n keys but only one can open the door of his house.

Now he chooses at random a key, find the expected value of number of keys used until he opens the door when either the key used is not placed back or the key used is placed back.

4. Assume that ξ is a nonnegative integer-valued random variable and the expected value exists. Show $E\xi = \sum_{k=1}^{\infty} P(\xi \geqslant k)$.

5. There are N buses labelled from 1 to N in a city. If one writes down at random the labels of r buses and denotes by ξ the biggest number, find $E\xi$ (Hint: Applying Exercise 4).

6. Assume that the random variable ξ has the following density function, find $E\xi$.

(1) $$p(x) = \begin{cases} \dfrac{2}{\pi}\cos^2 x, & -\dfrac{\pi}{2} \leqslant x \leqslant \dfrac{\pi}{2}; \\ 0, & \text{otherwise}; \end{cases}$$

(2) $$p(x) = \begin{cases} x, & 0 \leqslant x < 1; \\ 2 - x, & 1 \leqslant x < 2; \\ 0, & \text{otherwise}; \end{cases}$$

(3) $p(x)=\frac{1}{2\lambda}e^{-\frac{|x-\mu|}{\lambda}}$, $-\infty<x<\infty$,

where $\lambda>0$, λ, μ are constants.

7. Assume that ξ is uniformly distributed over $[-1/2,1/2]$, find the expectation of $\eta=\sin(\pi\xi)$.

8. Assume that the density of the speed of molecule motion is given by the Maxwell law

$$p(x)=\begin{cases}\frac{4x^2}{a^3\sqrt{\pi}}e^{-\frac{x^2}{a^2}}, & x>0;\\ 0, & x\leqslant 0.\end{cases}$$

Now let the mass of molecule be m, and find the average speed and the average kinetic energy.

9. Suppose you toss a fair coin repeatedly until a tail turns up. You will win 2^n dollars if you end with n heads. Find the expectation of dollars you win.

10. Assume that ξ_1 and ξ_2 are independent with a common normal distribution $N(a,\sigma^2)$. Show

$$E\max(\xi_1,\xi_2)=a+\frac{\sigma}{\sqrt{\pi}}.$$

11. Assume that the event A occurs in the i-th trial with probability p_i, and μ is the number of A occurring in n independent trials. Find $E\mu$.

12. There are n cards labelled by $1,2,\cdots,n$ in a box. Now draw randomly k cards with replacement, find the expected value of the sum μ of numbers of the drawn cards.

13. Each product made in a streamline is bad with probability p. The machine must be repaired whenever there are k bad products. Find the expected value of the number of products made during two repairs.

14. There are n freshmen in mathematics department; the teacher takes their identification cards from the office room and gives at random each student a card. How many students from the average have their own identification cards?

15. Choose randomly two points M_1 and M_2 in a line segment of length a, and find the expected value of the length of M_1M_2.

16. There are N balls in a pocket, where the number of white balls is random variable with the expectation a. Find the probability a white ball comes up in a ramdom draw.

17. Assume that $(\xi,\eta)\sim N(0,0,1,1,r)$, show

$$E\max(\xi,\eta)=\sqrt{\frac{1-r}{\pi}}.$$

18. Find the variance of each random variable in Exercise 1.

19. Find the variance of the random variable in Exercise 2.

20. Find the variance of the number of keys used in Exercise 3.

21. Find the variance of each random variable in Exercise 6.

22. Find the variance of η in Exercise 7.

23. Find Varμ in Exercise 11.

24. Find Varμ in Exercise 12.

25. Suppose that the random variable ξ is such that $E|\xi|^r < \infty$ for $r > 0$. Show for an arbitrary $\varepsilon > 0$ it follows that

$$P(|\xi| > \varepsilon) \leqslant \frac{E|\xi|^r}{\varepsilon^r}.$$

26. Suppose that $f(x)(x \geqslant 0)$ is nondecreasing and positive. If the random variable ξ is such that $Ef(|\xi|) < \infty$, show for an arbitrary $x > 0$

$$P(|\xi| \geqslant x) \leqslant \frac{Ef(|\xi|)}{f(x)}.$$

27. Assume that ξ takes values only on the interval $[a, b]$, show Var$\xi \leqslant (b-a)^2/4$.

28. Assume that $\xi_1, \cdots, \xi_n$ are independent and Var$\xi_i = \sigma_i^2$. Under what condition on the weights $a_1, \cdots, a_n$ with $\sum_{i=1}^{n} a_i = 1$, the variance of $\sum_{i=1}^{n} a_i\xi_i$ attains its minimum.

29. In automobile insurance, the claim amount B depends on loss. Now assume that B is a continuous random variable over the interval $0 \leqslant x < L$ and has the density function

$$p(x) = \begin{cases} \lambda e^{-\lambda x}, & 0 \leqslant x < L; \\ 0, & x < 0 \end{cases}$$

while $P(B = L) = e^{-\lambda L}$ at point $x = L$, and the maximum call amount is not greater than L. In addition, the probability that an automobile occurs loss is 0.10. Find the expectation and the variance of the claim amount B.

30. A fortune insurance company undertakes to provide 160 fire insurances for certain buildings; the corresponding maximum claim amount and the number of policies are as follows(see Table 3-2):

Table 3-2

Type k	Maximum claims	Number of policies
1	10 000	80
2	20 000	35
3	30 000	25
4	50 000	15
5	100 000	5

Assume that each building occurs fire independently with probability 0.04, and the claim amount of the k-th type of fire insurance is uniformly distributed over $(0, L_k)$.

Denote by S the total claim amount, and calculate the expectation and variance of S.

31. Find the expectations and variances of the following random variables:

(1) ξ is distributed as $\chi^2(n)$;

(2) ξ is distributed as $t(n)$;

(3) ξ is distributed as $F(m,n)$.

32. Assume that the 2-dimensional random vector ξ has the following density

(1) $p(x,y) = \begin{cases} 2-x-y, & 0<x<1,\ 0<y<1; \\ 0, & \text{otherwise}; \end{cases}$

(2) $p(x,y) = \begin{cases} 6xy^2, & 0<x<1, 0<y<1; \\ 0, & \text{otherwise}. \end{cases}$

Find its covariance matrix.

33. Recall Example 21 in 1. 3. 2. Suppose that a person types n letters, types the corresponding addresses on n envelopes, and then places the n letters in the n envelopes randomly. Let X be the number of letters which are placed in their correct envelopes. Show that $EX=1$ and $\mathrm{Var}(X)=1$.

34. Let $X_1, X_2, \cdots, X_n$ be identically distributed random variables. Show that

$$\mathrm{Var}\left(\frac{X_1+\cdots+X_n}{n}\right) \leqslant \mathrm{Var}(X_1).$$

35. Let $f(x)$ and $g(x)$ be two monotonic non-decreasing (or non-increasing) functions and X be a random variable. Suppose that the means and variances of $f(X)$ and $g(X)$ are finite. Show that

$$\mathrm{Cov}(f(X), g(X)) \geqslant 0.$$

Hint: Let Y be an independent copy of X. Then $E[f(X)g(X)] = E[f(Y)g(Y)]$, $E[f(X)g(Y)] = E[f(Y)g(X)] = E[f(X)] \cdot E[g(X)]$.

36. Assume that $U = a\xi + b$, $V = c\eta + d$, where a, b, c, d are constants and $ac \geqslant 0$. Show that U and V have the same correlation coeffcient as ξ and η.

37. Assume that ξ and η are independent identically distributed with a common normal distribution $N(a, \sigma^2)$, find the correlation coefficient of $p\xi + q\eta$ and $u\xi + v\eta$.

38. Assume that $\xi_1, \cdots, \xi_{m+n}$ $(n > m)$ are independent with a common distribution and the variances exist. Find the correlation coefficient of $S = \xi_1 + \cdots + \xi_n$ and $T = \xi_{m+1} + \cdots + \xi_{m+n}$.

39. Assume that $\xi_1, \cdots, \xi_{2n}$ have mean 0 and variance 1, and the pairwise correlation coefficient is ρ. Find the correlation coefficient of $\eta = \xi_1 + \cdots + \xi_n$ and $\zeta = \xi_{n+1} + \cdots + \xi_{2n}$.

40. Assume that (ξ, η) is uniformly distributed over the unit disc $\{(x,y): x^2+y^2 \leqslant 1\}$. Show that ξ and η are uncorrelated but not independent.

41. Assume that ξ has an even density function and $E\xi^2 < \infty$, show that $|\xi|$ and ξ are uncorrelated but not independent.

42. Assume that both ξ and η take only two values, show that if they are

uncorrelated then they must be independent.

43. Assume that ξ and η are independent and their variances exist. Show that $\mathrm{Var}(\xi\eta) = \mathrm{Var}\xi\mathrm{Var}\eta + (E\xi)^2\mathrm{Var}\eta + \mathrm{Var}\xi(E\eta)^2$.

44. Assume that any pair of $\xi_1, \cdots, \xi_n$ has correlation coefficient ρ. Show that $\rho \geqslant -1/(n-1)$.

45. Find the following characteristic functions:

(1) $P(\xi = k) = pq^{k-1}$, $k = 1, 2, \cdots$, $q = 1 - p$;

(2) ξ is uniform over $[-a, a]$;

(3) ξ is exponentially distributed with parameter λ;

(4) Γ is distributed as $G(\lambda, r)$, (Hint: For any complex number $z = b + ic$, $\int_0^{+\infty} x^{r-1}e^{-zx}dx = \Gamma(r)/z^r$ whenever $r > 0$);

(5) ξ has density

$$p(x) = \begin{cases} \dfrac{2+x}{4}, & -2 \leqslant x < 0; \\ \dfrac{2-x}{4}, & 0 \leqslant x < 2; \\ 0, & \text{otherwise}; \end{cases}$$

(6) $\eta = a\xi + b$, where ξ is uniform over $[0, 1]$;

(7) $\eta = \ln \xi$, where ξ is uniform over $[0, 1]$.

46. If a distribution function F is such that $F(x) = 1 - F(-x-0)$, then we call it symmetric. Show that a distribution function is symmetric if and only if its characteristic function is a real and even function.

47. Assume that $\varphi(t)$ is a characteristic function, show that so are the following functions.

(1) $[\varphi(t)]^n$, where n is a positive integer;

(2) $[\varphi(t) \sin at]/at$, where $a > 0$.

48. Show that the following functions are characteristic functions and find out the corresponding distribution functions.

(1) $\cos^2 t$;

(2) $(1 + it)^{-1}$;

(3) $(\sin t/t)^2$;

(4) $(2e^{-it} - 1)^{-1}$;

(5) $(1 + t^2)^{-1}$.

49. Try to give a $\varphi(t)$ such that $\varphi(-t) = \overline{\varphi(t)}$ and $|\varphi(t)| \leqslant \varphi(0) = 1$ but $\varphi(t)$ is not a characteristic function.

50. Is $\varphi(t) = (1 - i|t|)^{-1}$ a characteristic function? Why?

51. Show that

$$\varphi(t)=\begin{cases}1-\dfrac{|t|}{a}, & |t|<a;\\ 0, & |t|\geqslant a,\end{cases}\qquad a>0,$$

is a characteristic function, and find the corresponding distribution function.

52. Show that if $\varphi(t)$ is continuous and satisfies the following conditions, then $\varphi(t)$ must be a characteristic function:

(1) $\varphi(t)=\varphi(-t)$;

(2) $\varphi(t+2a)=\varphi(t)$;

(3) $\varphi(t)=(a-t)/a\ (0\leqslant t\leqslant a)$.

53. Assume that ξ is a random variable taking integer values with the distribution $P(\xi=k)=p_k$, $k=0,\pm1,\pm2,\cdots$, and the characteristic function $f(t)$. Show that

$$p_k=\frac{1}{2\pi}\int_{-\pi}^{\pi}\mathrm{e}^{-ikt}f(t)\,\mathrm{d}t.$$

54. Let X and Y be two independent random variables. Suppose that X and $X+Y$ have the same distribution. Prove that $P(Y=0)=1$.

55. Assume that $(\xi,\eta)\sim N(a,b,\sigma_1^2,\sigma_2^2,r)$, find the distribution function of $\zeta=\xi+\eta$.

56. Assume that $\xi_1,\cdots,\xi_n$ are independent with a common normal distribution $N(a,\sigma^2)$.

(1) find the distribution, expectation and covariance matrix of $\boldsymbol{\xi}=(\xi_1,\cdots,\xi_n)'$;

(2) find the distribution of $\bar{\xi}=\dfrac{1}{n}\sum_{i=1}^{n}\xi_i$.

57. Suppose that components of a multivariate normal distribution are independent and have the same variance. Show that so do components of the multivariate normal distribution through an orthogonal transform.

58. Suppose that $\boldsymbol{\xi}=(\xi_1,\xi_2,\xi_3)'\sim N(\boldsymbol{a},\boldsymbol{B})$, where $\boldsymbol{a}=(a_1,a_2,a_3)^{\mathrm{T}}$, $\boldsymbol{B}=(b_{ij})_{3\times3}$. Let

$$\begin{cases}\eta_1=\dfrac{\xi_1}{2}-\xi_2+\dfrac{\xi_3}{2},\\[2mm] \eta_2=-\dfrac{\xi_1}{2}-\dfrac{\xi_3}{2}.\end{cases}$$

Find the distribution function of $\boldsymbol{\eta}=(\eta_1,\eta_2)'$.

59. Suppose that $\boldsymbol{\xi}=(\xi_1,\xi_2)'\sim N(\boldsymbol{a},\boldsymbol{B})$ is a $2n$-dimensional normal variable, where $\boldsymbol{\xi}_1$ and $\boldsymbol{\xi}_2$ are n-dimensional vectors,

$$\boldsymbol{B}=\begin{pmatrix}\boldsymbol{B}_{11} & \boldsymbol{B}_{12}\\ \boldsymbol{B}_{21} & \boldsymbol{B}_{22}\end{pmatrix},$$

where $\boldsymbol{B}_{22}=\boldsymbol{B}_{11}$, $\boldsymbol{B}_{21}=\boldsymbol{B}_{12}$. Show $\boldsymbol{\xi}_1+\boldsymbol{\xi}_2$ and $\boldsymbol{\xi}_1-\boldsymbol{\xi}_2$ are independent.

60. Suppose that $\boldsymbol{\xi}=(\xi_1,\xi_2)'\sim N(0,\boldsymbol{I})$, where $\boldsymbol{I}$ is a 2×2 unit matrix. Find the conditional distribution of ξ_1 given $\xi_1+\xi_2=x_1+x_2$. (Hint: Find the joint distribution of $\eta_1=\xi_1+\xi_2$ and $\eta_2=\xi_1$.)

61. Suppose that $\boldsymbol{\xi} = (\xi_1, \xi_2)' \sim N(0, \boldsymbol{B})$, where

$$\boldsymbol{B} = \begin{pmatrix} 4 & 2 \\ 2 & 1 \end{pmatrix}.$$

Find an orthogonal transform $\boldsymbol{U}$ and two numbers d_1, d_2 such that $\boldsymbol{\eta} = \boldsymbol{U}\boldsymbol{\xi} \sim N(\boldsymbol{a}_1, \boldsymbol{B}_1)$, where

$$\boldsymbol{B}_1 = \begin{pmatrix} d_1 & 0 \\ 0 & d_2 \end{pmatrix}.$$

62. Suppose that the random variables X and Y have the joint density

$$g(x^2 + y^2), \ -\infty < x, y < \infty,$$

where $g(t)$ is a nonnegative function on $[0, \infty)$ with $\int_0^\infty g(t)\mathrm{d}t = 1/\pi$. Let $f(t_1, t_2)$ be the joint characteristic function of X and Y, and $f_X(t)$ be the characteristic function of X.

(1) Show that $f(t_1, t_2) = f_X\left(\sqrt{t_1^2 + t_2^2}\right)$.

(2) Suppose that the random variables U and V have the joint density

$$\frac{1}{\sigma_1\sigma_2\sqrt{1-r^2}} g\left(\frac{1}{1-r^2}\frac{(u-\mu_1)^2}{\sigma_1^2} + \frac{2r(u-\mu_1)(v-\mu_2)}{\sigma_1\sigma_2} + \frac{(v-\mu_2)^2}{\sigma_2^2}\right),$$

$$-\infty < u, v < \infty,$$

where $-\infty < \mu_1, \mu_2 < \infty$, $0 < \sigma_1, \sigma_2 < \infty$ and $|r| < 1$. Show that the joint characteristic function of U and V is

$$\mathrm{e}^{\mathrm{i}(t_1\mu_1 + t_2\mu_2)} f_X\left(\sqrt{t_1^2\sigma_1^2 + 2rt_1t_2\sigma_1\sigma_2 + t_2^2\sigma_2^2}\right).$$

(3) Show that, if U and V are independent, then U and V are normal random variables and $r = 0$.

Hint: If $(\xi, \eta) = (X, Y)Q$ where Q is a 2×2 orthogonal matrix, then (ξ, η) and (X, Y) have the same distribution.

63. Let X and Y be two independent, nonnegative integer-valued, random variables with the property

$$\mathrm{P}(X = x \mid X + Y = x + y) = \frac{\binom{m}{x}\binom{n}{y}}{\binom{m+n}{x+y}}$$

for all $x = 0, 1, \cdots, m$ and $y = 0, 1, \cdots, n$, where m and n are given positive integers. Assume that $\mathrm{P}(X = 0)$ *and* $\mathrm{P}(Y = 0)$ are strictly positive. Show that both X and Y have binomial distribution with the same parameter p, the other parameters being m and n respectively.

64. Let X and Y be two independent, nonnegative integer-valued, random variables such that

$$\mathrm{P}(X = i) = p_X(i), \mathrm{P}(Y = i) = p_Y(i),$$

where

$$p_X(i) > 0, p_Y(i) > 0, i = 0,1,2\cdots.$$

And

$$\sum_{i=1}^{n} p_X(i) = \sum_{i=1}^{n} p_Y(i) = 1.$$

Assume

$$P(X = x \mid X+Y = l) = \begin{cases} \binom{l}{k} p^k (1-p)^{l-k}, & k = 0,1,\cdots,l, \\ 0, & k > l. \end{cases}$$

Prove that there is a constant $\lambda > 0$ such that $X \sim P(\lambda\alpha)$ and $Y \sim P(\lambda)$, where $\alpha = p/(1-p)$.

Hint: Let $g_X(s) = \sum_{i=0}^{\infty} p_X(s) s^i$, $g_Y(s) = \sum_{i=0}^{\infty} p_Y(i) s^i$. Establish first the relation

$$g_X(t) g_Y(s) = g_X(pt + (1-p)s) g_Y(pt + (1-p)s).$$

65. Let X and Y be two independent random variables with finite means μ_1 and μ_2 respectively. Suppose that for given $X + Y = z$, the conditional distribution of X is $N(\mu + \beta z, \sigma_2)$, where $-\infty < \mu < \infty$, $0 < \beta < 1$ and $\sigma > 0$. Prove that $X \sim N(\mu_1, \sigma_1^2)$ and $Y \sim N(\mu_2, \sigma_2^2)$ with

$$\beta = \frac{\sigma_1^2}{\sigma_1^2 + \sigma_2^2}, \sigma^2 = \frac{\sigma_1^2 \sigma_2^2}{\sigma_1^2 + \sigma_2^2} \text{ and } \mu = (1-\beta)\mu_1 - \beta\mu_2.$$

Chapter 4

Probability Limit Theorems

In the early days, the aim of probability theory is to reveal the inherent rules of random phenomena caused by a large number of random factors. Bernoulli first realized the importance of studying an infinite sequence of random trials, and established the first limit theorem in probability theory—the law of large numbers. de Moivre and Laplace presented that the measurement error can be regarded as the summation of a large number of independent and slight errors, and proved that the distribution of the measurement error is approximated by a normal distribution—the central limit theorem. Later on there have appeared various kinds of limit theorems. These results and research methods have an important influence upon probability and mathematical statistics and find a lot of applications. In this chapter, the attention will be paid to the laws of large numbers, the central limit theorems and their related problems.

4.1 Convergence in distribution and central limit theorems

Let ξ be a random variable on a probability space $(\Omega, \mathscr{F}, P)$. As we know, the ξ can be characterized by its distribution function $F(x) = P(\xi \leqslant x)$. Hence, it is desirable to study the sequence of distribution functions before we move to the sequence of random variables.

4.1.1 Weak convergence of distribution functions

Definition 1 *Let F be a distribution function, $\{F_n, n \geqslant 1\}$ a sequence of distribution functions. We say that F_n converges weakly to F, denoted by $F_n \xrightarrow{w} F$, if $F_n(x) \to F(x)$ holds at every continuity point x of F as $n \to \infty$.*

Let ξ be a random variable, $\{\xi_n, n \geqslant 1\}$ a sequence of random variables, and we say ξ_n converges in distribution to ξ, denoted by $\xi_n \xrightarrow{d} \xi$, if the distribution functions of the ξ_n converge weakly to the distribution function of ξ.

Remark 1 The limit function of a sequence of distribution functions is not necessarily a distribution function. For example, let

$$F_n(x) = \begin{cases} 0, & x < n; \\ 1, & x \geqslant n. \end{cases}$$

This distribution function converges pointwise to 0, but $F(x) \equiv 0$ is not a distribution function.

Remark 2 In the definition of weak covergence, the condition at every continuity point x of F is a weak restriction. For example, let

$$F_n(x) = \begin{cases} 0, & x < \dfrac{1}{n}; \\ 1, & x \geqslant \dfrac{1}{n}, \end{cases} \qquad F(x) = \begin{cases} 0, & x < 0; \\ 1, & x \geqslant 0. \end{cases}$$

Then $F_n(x)$ converges pointwise to $F(x)$ except at the point $x = 0$, while $x = 0$ is a unique discontinuous point of $F(x)$. Thus it follows from Definition 1 that $F_n \xrightarrow{w} F$.

Remark 3 Since the set of discontinuous points of a distribution function F is at most countable, $F_n \xrightarrow{w} F$ means that F_n converges everywhere to F in a dense subset of **R** (a subset D is dense in **R** if for every $x_0 \in \mathbf{R}$ and an arbitrary small neighborhood of x_0, there must exist a point $x \in D$).

Next we shall give the Helly theorem, which is important to the study of weak convergence of distribution functions.

Theorem 1 (*Helly's first theorem*) *Let* $\{F_n, n \geqslant 1\}$ *be a sequence of distribution functions. Then there exists a non-decreasing right-continuous function* F *(not necessarily a distribution function) with* $0 \leqslant F(x) \leqslant 1$, $x \in \mathbf{R}$, *and a subsequence* $\{F_{n_k}\}$, *such that* $F_{n_k}(x) \to F(x)$ *for every continuity point* x *of* F *as* $k \to +\infty$.

The proof is given in Supplements and Remarks.

Theorem 2 (*Helly's second theorem*) *Let* F *be a distribution function,* $\{F_n, n \geqslant 1\}$ *a sequence of distribution functions such that* $F_n \xrightarrow{w} F$. *If* $g(x)$ *is a bounded continuous function in* **R**, *then*

$$\int_{-\infty}^{+\infty} g(x) \mathrm{d}F_n(x) \to \int_{-\infty}^{+\infty} g(x) \mathrm{d}F(x). \tag{4.1}$$

Let F *and* F_n *be non-decreasing right-continuous functions (not necessary to be distribution functions). Suppose that* $F_n(x) \to F(x)$ *as* $n \to \infty$ *for any continuous point* x *of* F. *If* $a < b$ *are continuous points of* F *and* $g(x)$ *is a continuous function on* $[a; b]$, *then*

$$\int_a^b g(x) \mathrm{d}F_n(x) \to \int_a^b g(x) \mathrm{d}F(x).$$

Proof We only give the proof of the first part of the conclusion. For the second part, we can apply the first part of conclusion after redefining F, F_n and g as

$$F^*(x) = \begin{cases} 1, & x \geqslant b, \\ \dfrac{F(x) - F(a)}{F(b) - F(a)}, & a \leqslant x \leqslant b, \\ 0, & x \leqslant a; \end{cases} \qquad F_n^*(x) = \begin{cases} 1, & x \geqslant b, \\ \dfrac{F_n(x) - F_n(a)}{F_n(b) - F_n(a)}, & a \leqslant x \leqslant b, \\ 0, & x \leqslant a; \end{cases}$$

$g^*(x) = g(b)$ if $x \geqslant b$, $g^*(x) = g(x)$ if $a \leqslant x \leqslant b$, and $g^*(x) = g(a)$ if $x \leqslant a$.

Since g is a bounded function, there must exist a constant $c > 0$ such that $|g(x)| < c$, $x \in \mathbf{R}$. Since all the continuity points of F is a dense set in $\mathbf{R}$, and $F(-\infty) = 0$, $F(+\infty) = 1$, then we can select $a > 0$ for every given $\varepsilon > 0$, such that $\pm a$ are continuity points of F, and

$$F(-a) < \frac{\varepsilon}{12c}, \qquad 1 - F(a) < \frac{\varepsilon}{12c}. \tag{4.2}$$

Since $F_n \xrightarrow{w} F$, there exists an $N_1(\varepsilon)$ such that

$$\begin{aligned} &| F_n(-a) - F(-a) | < \frac{\varepsilon}{12c}, \\ &| 1 - F_n(a) - (1 - F(a)) | < \frac{\varepsilon}{12c}, \end{aligned} \tag{4.3}$$

when $n \geqslant N_1(\varepsilon)$. Then we have

$$\begin{aligned} &\left| \int_{-\infty}^{-a} g(x)\mathrm{d}F_n(x) - \int_{-\infty}^{-a} g(x)\mathrm{d}F(x) \right. \\ &\left. + \int_a^{+\infty} g(x)\mathrm{d}F_n(x) - \int_a^{+\infty} g(x)\mathrm{d}F(x) \right| \\ &\leqslant c[F_n(-a) + F(-a) + 1 - F_n(a) + 1 - F(a)] \\ &\leqslant c[| F_n(-a) - F(-a) | + 2F(-a) \\ &\quad + | 1 - F_n(a) - (1 - F(a)) | + 2(1 - F(a))] < \frac{\varepsilon}{2}. \end{aligned} \tag{4.4}$$

Next, we turn to consider

$$\left| \int_{-a}^{a} g(x)\mathrm{d}F_n(x) - \int_{-a}^{a} g(x)\mathrm{d}F(x) \right|.$$

Since $g(x)$ is uniformly continuous on the close interval $[-a, a]$, we can choose a partition of $[-a, a]$: $-a = x_0 < x_1 < \cdots < x_m = a$ such that all the points x_i's are continuity points of F, and $\max\limits_{x_{i-1} < x \leqslant x_i} | g(x) - g(x_i) | < \varepsilon/8$. We have

$$\begin{aligned} &\left| \int_{-a}^{a} g(x)\mathrm{d}F_n(x) - \int_{-a}^{a} g(x)\mathrm{d}F(x) \right| \\ &\leqslant \sum_{i=1}^{m} \left| \int_{x_{i-1}}^{x_i} g(x)\mathrm{d}F_n(x) - \int_{x_{i-1}}^{x_i} g(x)\mathrm{d}F(x) \right| \\ &\leqslant \sum_{i=1}^{m} \int_{x_{i-1}}^{x_i} | g(x) - g(x_i) | \,\mathrm{d}F_n(x) \\ &\quad + \sum_{i=1}^{m} \int_{x_{i-1}}^{x_i} | g(x) - g(x_i) | \,\mathrm{d}F(x) \\ &\quad + \sum_{i=1}^{m} | g(x_i) | \cdot \left| \int_{x_{i-1}}^{x_i} \mathrm{d}F_n(x) - \int_{x_{i-1}}^{x_i} \mathrm{d}F(x) \right| \\ &\leqslant \frac{\varepsilon}{8} \sum_{i=1}^{m} \{F_n(x_i) - F_n(x_{i-1}) + F(x_i) - F(x_{i-1})\} \end{aligned}$$

$$
\begin{aligned}
&+2c\sum_{i=0}^{m} | F_n(x_i) - F(x_i) | \\
&= \frac{\varepsilon}{8}(F_n(a) - F_n(-a) + F(a) - F(-a)) \qquad (4.5)\\
&+2c\sum_{i=0}^{m} | F_n(x_i) - F(x_i) |.
\end{aligned}
$$

Since $F_n(a) - F_n(-a) \leqslant 1$, $F(a) - F(-a) \leqslant 1$, we can select an $N_2(\varepsilon)$ such that when $n \geqslant N_2(\varepsilon)$,

$$| F_n(x_i) - F(x_i) | < \frac{\varepsilon}{8mc}, \quad i = 0, 1, \cdots, m. \qquad (4.6)$$

So, the right hand side of (4.5) is less than $\varepsilon/2$. Hence

$$\left| \int_{-\infty}^{\infty} g(x)\mathrm{d}F_n(x) - \int_{-\infty}^{\infty} g(x)\mathrm{d}F(x) \right| < \varepsilon \qquad (4.7)$$

when $n \geqslant \max(N_1(\varepsilon), N_2(\varepsilon))$. The proof is complete.

Theorem 3 *(Lévy's continuity theorem) Let F be a distribution function, $\{F_n, n \geqslant 1\}$ a sequence of distribution functions. If $F_n \xrightarrow{w} F$, then the corresponding sequence of characteristic functions $f_n(t)$ converges to the characteristic function $f(t)$ of F uniformly in t on any finite interval.*

For any $b > 0$, just consider $| t | \leqslant b$. Let $g_t(x) = \mathrm{e}^{itx}$, $x \in \mathbf{R}$. Note that

$$| g_t(x) | = 1, \quad \sup_{|t| \leqslant b} | g_t(x) - g_t(y) | \leqslant | b | \cdot | x - y |.$$

Then the proof of the theorem is similar to that of Theorem 2, omitted here.

In the previous chapter, it is known that a characteristic function and a distribution function can be uniquely decided by each other. Similarly, the converse theorem of Lévy's continuity theorem also holds true.

Theorem 4 *(The converse limit theorem) Let $f_n(t)$ be characteristic function of distribution function $F_n(x)$. If $f_n(t) \to f(t)$ for every t, and $f(t)$ is continuous on $t = 0$, then $f(t)$ must be the characteristic function of some distribution function F, and $F_n \xrightarrow{w} F$.*

Since the proof of the theorem is rather intricate, we will state and postpone it to the Supplements and Remarks. But the theorem itself is significant in that it makes characteristic function an important tool in the study of probability limit theorems. We first illustrate how to use this theorem.

Example 1 Prove the Poisson approximation of binomial distributions by the method of characteristic function.

Proof Let ξ_n be a random variable with binomial distribution $B(n, p_n)$, and $\lim_{n \to +\infty} np_n = \lambda$. Then its characteristic function is $f_n(t) = (p_n \mathrm{e}^{it} + q_n)^n$, where $q_n = 1 - p_n$. So

$$\lim_{n\to+\infty} f_n(t) = \lim_{n\to+\infty}\left(1+\frac{np_n(e^{it}-1)}{n}\right)^n = e^{\lambda(e^{it}-1)}.$$

This is just the characteristic function of the Poisson distribution. It follows from the converse limit theorem that the binomial distribution $B(n, p_n)$ converges in distribution to the Poisson distribution $P(\lambda)$.

By Theorems 3 and 4, we have the following corollary on the convergence in distribution.

Corollary 1 *Let ξ be a random variables and $\{\xi_n\}$ be a sequence of random variables. The following statements are equivalent.*

(1) *$\xi_n \xrightarrow{d} \xi$, i.e., $F_n(x) \to F(x)$ for any continuous point x of F, where F and F_n are the distribution functions of ξ and ξ_n respectively;*

(2) *For any bounded continuous function $g(x)$, $\mathrm{E}g(\xi_n) \to \mathrm{E}g(\xi)$;*

(3) *For any bounded uniformly continuous function $g(x)$, $\mathrm{E}g(\xi_n) \to \mathrm{E}g(\xi)$;*

(4) *For any bounded continuous function $g(x)$ with bounded derivatives of each order, $\mathrm{E}g(\xi_n) \to \mathrm{E}g(\xi)$;*

(5) *For any real t, $f_n(t) \to f(t)$, where $f(t)$ and $f_n(t)$ are the characteristic functions of ξ and ξ_n respectively.*

Each of the statements can be used to verify the convergence in distribution and among them, the characteristic function is the most popular tool. But the characteristic function is a complex valued function. Sometimes, the convergence in distribution can be verified by the convergence of the density functions or probability mass functions.

Corollary 2 (*i*) *Suppose that the random variables ξ_n and ξ have the density functions $p_n(x)$ and $p(x)$ respectively. If $p_n(x) \to p(x)$ for any x, then $\xi_n \xrightarrow{d} \xi$;*

(*ii*) *Suppose that the distribution sequences of the discrete random variables ξ_n and ξ are $p_n(x_j) = P(\xi_n = x_j)$ and $p(x_j) = P(\xi_n = x_j)$, $j = 1, 2, \cdots$, respectively. If $p_n(x_j) \to p(x_j)$ for any j, then $\xi_n \xrightarrow{d} \xi$.*

Proof We only give the proof of (i) as that of (ii) is similar. Note $\int_{-\infty}^{\infty} p_n(x)\mathrm{d}x - \int_{-\infty}^{\infty} p(x)\mathrm{d}x = 0$. We have

$$\begin{aligned} |F(x) - F_n(x)| &= \left|\int_{-\infty}^{x} [p(y) - p_n(y)]\mathrm{d}y\right| \leqslant \int_{-\infty}^{\infty} |p(y) - p_n(y)|\,\mathrm{d}y \\ &= \int_{-\infty}^{\infty} [|p(y) - p_n(y)| + p(y) - p_n(y)]\mathrm{d}y \\ &= 2\int_{-\infty}^{\infty} (p(y) - p_n(y))^+\,\mathrm{d}y. \end{aligned}$$

Since $(p(y)-p_n(y))^+ \leqslant p(y)$, $(p(y)-p_n(y))^+ \to 0$, by the dominated convergence theorem we have

$$\sup_{x} \mid F(x) - F_n(x) \mid \leqslant 2\int_{-\infty}^{\infty} (p(y) - p_n(y))^+ \,\mathrm{d}y \to 0.$$

The proof is completed.

For example, in Chapter 2 we have found that the density function $p_n(x)$ of a t-distribution with freedom n satisfies

$$p_n(x) \to \frac{1}{\sqrt{2\pi}} \mathrm{e}^{-\frac{x^2}{2}}.$$

So, its distribution function $F_n(x) \xrightarrow{d} \Phi(x)$.

Besides the continuity theorem, the weak convergence of distribution functions possesses the following properties.

Property 1 Let $\{F_n, n \geqslant 1\}$ be a sequence of distribution functions. If $F_n \xrightarrow{w} F$, and F is a continuous distribution function, then $F_n(x)$ must converge uniformly to $F(x)$.

The proof is left to the reader.

Property 2 Let ξ be a random variable, $\{\xi_n, n \geqslant 1\}$ a sequence of random variables, $g(x)$ a continuous function on $\mathbf{R}$. If $\xi_n \xrightarrow{d} \xi$, then $g(\xi_n) \xrightarrow{d} g(\xi)$.

Proof Assume the distribution functions of ξ and ξ_n are F and F_n respectively. If $\xi_n \xrightarrow{d} \xi$, i. e., $F_n \xrightarrow{w} F$, then it follows from Theorem 2 that the characteristic function $\int_{-\infty}^{+\infty} \mathrm{e}^{itg(x)} \mathrm{d}F_n(x)$ converges to $\int_{-\infty}^{+\infty} \mathrm{e}^{itg(x)} \mathrm{d}F(x)$, which is just the characteristic function of $g(\xi)$. Similarly to Theorem 4, the distribution function of $g(\xi_n)$ converges weakly to the distribution function of $g(\xi)$, i. e., $g(\xi_n) \xrightarrow{d} g(\xi)$.

Property 3 Let $\{a_n, n \geqslant 1\}$ and $\{b_n, n \geqslant 1\}$ be two sequences of constants, F a distribution function, $\{F_n, n \geqslant 1\}$ a sequence of distribution functions. If $a_n \to a$, $b_n \to b$, $F_n \xrightarrow{w} F$, then

$$F_n(a_n x + b_n) \to F(ax + b),$$

where x is such that $ax + b$ is a continuity point of F.

Proof Let x be such that $ax + b$ is a continuity point of F, and let $\varepsilon > 0$ be such that F is continuous at the point $ax + b \pm \varepsilon$ (it is possible, for the continuity points of F are dense on $\mathbf{R}$). Obviously, $a_n x + b_n \to ax + b$. So, for n large enough,

$$ax + b - \varepsilon \leqslant a_n x + b_n \leqslant ax + b + \varepsilon, \tag{4.8}$$

which implies

$$F_n(ax + b - \varepsilon) \leqslant F_n(a_n x + b_n) \leqslant F_n(ax + b + \varepsilon).$$

Since $F_n \xrightarrow{w} F$, it is easy to see

$$\begin{aligned} F(ax + b - \varepsilon) &\leqslant \liminf_{n\to\infty} F_n(a_n x + b_n) \\ &\leqslant \limsup_{n\to\infty} F_n(a_n x + b_n) \leqslant F(ax + b + \varepsilon). \end{aligned}$$

Letting $\varepsilon \to 0$ and noting that F is continuous at $ax+b$, we conclude the proof.

Corollary 3 If $\xi_n \xrightarrow{d} \xi$, then $a_n\xi_n + b_n \xrightarrow{d} a\xi + b$ (a_n, $a \neq 0$).

Proof Assume $a>0$. $a_n \to a$ implies $a_n>0$ for sufficiently large n. Thus for any real number x,

$$P(a_n\xi_n + b_n \leqslant x) = F_n\left(\frac{x-b_n}{a_n}\right).$$

Applying Property 3 yields the desired result. The case a<0 is similar.

The above definitions, theorems and properties on the convergence in distribution for random variables and distributions can be easily extended to those for random vectors and multivariate distribution functions. For example, let $f_n(t)$ and $f(t)$ be the characteristic functions of n-dimensional random vectors ξ_n and ξ respectively. Then $\xi_n \xrightarrow{d} \xi$ if and only if $f_n(t) \to f(t)$ for any $t \in \mathbf{R}^n$.

4.1.2 Central limit theorems

Assume that the probability of success in a Bernoulli trial is p $(0<p<1)$. Let S_n denote the number of successes in n Bernoulli trials, and then $P(S_n=k)=b(k;n,p)$. In practice, people are usually interested in the probability that the number of successes is between two integers α and β $(\alpha<\beta)$. That is to calculate

$$P(\alpha < S_n \leqslant \beta) = \sum_{\alpha<k\leqslant\beta} b(k;n,p). \tag{4.9}$$

The computation of the right hand side of the equality is generally very complex. However, as mentioned in Section 2.1 of Chapter 2, it is found by de Moivre and Laplace that the binomial distribution can be well approximated by normal distribution when $n \to \infty$ (c.f. the Supplements and Remarks of Chapter 2).

Theorem 5 (*de Moivre-Laplace*) *Let $\Phi(x)$ be the standard normal distribution function. We have*

$$\lim_{n\to+\infty} P\left(\frac{S_n - np}{\sqrt{npq}} \leqslant x\right) = \Phi(x), \tag{4.10}$$

for $-\infty < x < +\infty$, *where* $q = 1-p$.

Note that $ES_n = np$, $\mathrm{Var}S_n = npq$. The left hand side of equation (4.10) is the limit of distribution functions of normalized sums. So, this theorem shows that binomial distributions converge in distribution to the normal distribution.

In history, people proved the above theorem by accurately estimating the value of a binomial distribution. But from the viewpoint of modern probability theory, this result is just a special case of more general central limit theorems which are introduced below.

A direct application of the theorem is as follows. When n is big enough and p is moderate, equation (4.9) can be approximately calculated by the normal distribution.

$$
\begin{aligned}
& P(\alpha < S_n \leqslant \beta) \\
&= P\left(\frac{\alpha - np}{\sqrt{npq}} < \frac{S_n - np}{\sqrt{npq}} \leqslant \frac{\beta - np}{\sqrt{npq}}\right) \\
&\approx \Phi\left(\frac{\beta - np}{\sqrt{npq}}\right) - \Phi\left(\frac{\alpha - np}{\sqrt{npq}}\right).
\end{aligned} \tag{4.11}
$$

Figure 4-1 describes its implication (In order to show it intuitively, the random variable in the figure is not standardized.): draw some small rectangles, the center of whose bottom is k ($\alpha \leqslant k \leqslant \beta$), the length is 1, and the height is $b(k; n, p)$. The sum of areas of these small rectangles is just $P(\alpha < S_n \leqslant \beta)$. Then draw the density curve of $N(np, npq)$, and the area covered by the curve in $[\alpha, \beta]$ is the right hand side of equality (4.11).

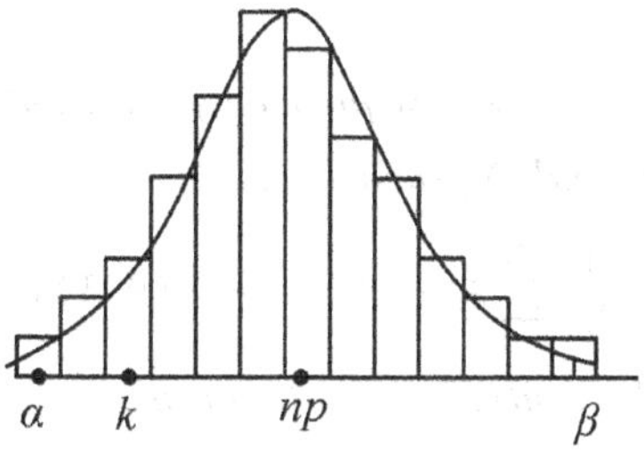

Figure 4-1

Remark In Chapter 2, it is proved that the Poisson distribution can approximate the binomial distribution. We point out that the Poisson approximation theorem does not contradict Theorem 5. It is required that $\lim_{n \to +\infty} np_n = \lambda$ is a constant in the Poisson theorem, while in Theorem 5, p is fixed. In practical applications, for n big enough

(1) if p is moderate, and we use the normal distribution $\Phi(x)$ to approximate the probability in left hand side of equation (4.10), the precision reaches $O(n^{-1/2})$;

(2) if p is close to 0 (or 1), and np is not big (or not small), then the declination of the curve of the binomial distribution is rather big. It's not good enough to approximate by the normal distribution. In this case, the precision will be higher when we use the Poisson distribution instead.

Example 2 Let $n = 10^4$, $p = 5 \times 10^{-3}$, and find $P(S_n \leqslant 70)$.

Solution Although p is small, np is very big. In this case, it is not suitable to approximate the binomial distribution by the Poisson distribution. Instead, Theorem 5 tells us

$$P(S_n \leqslant 70) = P\left(\frac{S_n - 50}{\sqrt{50}} \leqslant \frac{20}{\sqrt{50}}\right) \approx \Phi\left(\frac{20}{\sqrt{50}}\right) \approx 0.997.$$

Example 3 Rolling a fair coin, how many trials are needed to ensure the probability that the proportion of heads is between 0.4 and 0.6 is not smaller than 90%.

Solution Let n be the number of times of rolling the coin, S_n the number of times of appearing head, and then $S_n \sim B(n, 1/2)$. Note that n is to satisfy

$$P(0.4 < \frac{S_n}{n} \leqslant 0.6) \geqslant 0.9.$$

From Theorem 5, the left side of the above inequality equals

$$\begin{aligned} P(\frac{0.4n - n/2}{\sqrt{n/4}} < \frac{S_n - n/2}{\sqrt{n/4}} \leqslant \frac{0.6n - n/2}{\sqrt{n/4}}) \\ \approx \Phi(0.2\sqrt{n}) - \Phi(-0.2\sqrt{n}) \\ = 2\Phi(0.2\sqrt{n}) - 1. \end{aligned}$$

Take $n \geqslant 69$ such that the above $\geqslant 0.9$.

The significance of the de Moivre-Laplace theorem is far beyond the above numerical calculations. In fact, this theorem and its generalization are the central problems in the early study of probability theory.

Definition 2 *Let $\{\xi_n, n \geqslant 1\}$ be a sequence of random variables. If there exist two sequences of constants $B_n > 0$ and A_n such that*

$$\frac{1}{B_n}\sum_{k=1}^{n}\xi_k - A_n \xrightarrow{\mathrm{d}} N(0,1),$$

then we say that $\{\xi_n\}$ obeys the central limit theorem.

Theorem 6 *(Lindeberg-Lévy) Let $\{\xi_n, n \geqslant 1\}$ be a sequence of independent and identically distributed random variables. Let $S_n = \sum_{k=1}^{n}\xi_k$, $E\xi_1 = a$, $\mathrm{Var}\xi_1 = \sigma^2$. Then the central limit theorem holds true, i. e.,*

$$\frac{S_n - na}{\sqrt{n}\sigma} \xrightarrow{\mathrm{d}} N(0,1) \quad \text{as} \quad n \to \infty. \tag{4.12}$$

Proof We shall use the method of characteristic function. Let $f(t)$ and $f_n(t)$ be characteristic functions of $\xi_1 - a$ and $S_n - na/\sqrt{n}\sigma$ respectively. Since $\xi_1, \xi_2, \cdots, \xi_n$ are independent and identically distributed, we have $f_n(t) = (f(t/\sqrt{n}\sigma))^n$. And note that $E\xi_1 = a$, $\mathrm{Var}\xi_1 = \sigma^2$, so the characteristic function $f(t)$ has continuous derivative of 2-order. Using Taylor's expansion for f, we have

$$f(t) = f(0) + f'(0)t + \frac{1}{2}f''(0)t^2 + o(t^2) \quad \text{as } t \to 0.$$

For given $t \in \mathbf{R}$,

$$f\left(\frac{t}{\sqrt{n}\sigma}\right) = 1 - \frac{t^2}{2n} + o(\frac{1}{n}) \quad \text{as } n \to \infty.$$

Therefore $f_n(t) \to \mathrm{e}^{-\frac{t^2}{2}}$. The latter is the characteristic function of the standard normal distribution. Then Theorem 6 follows from Theorem 4.

The central limit theorem has a wide range of applications. In practice, if n is big enough, the standardized sum of independent and identically distributed random variables can be regarded as a normal variable.

Example 4 There are several kinds of methods to produce normal random numbers. Besides the method introduced in Section 2.5, the following method is often used: suppose $\{\xi_k\}$ is independent identically distributed and obeys the uniform distribution on $[0, 1]$, then $E\xi_k = 0.5$, $\sigma = \sqrt{\mathrm{Var}\xi_k} = 1/\sqrt{12}$. By the central limit theorem,

$$\eta = \frac{\sum_{k=1}^{n} \xi_k - n/2}{\sqrt{n}/\sqrt{12}}$$

approximately obeys the standard normal distribution when n is very big. In fact, it is usually sufficient to take $n = 12$. In other words if we take 12 uniform random numbers in $[0, 1]$, $\eta = \sum_{k=1}^{n} \xi_k - 6$ is approximately a standard normal random number.

In Theorem 6, it is required that all ξ_k are identically distributed. This condition is a little strict sometimes. More generally, we have the following theorem. Let $B_n^2 = \sum_{k=1}^{n} \mathrm{Var}\xi_k$.

Theorem 7 (*Lindeberg-Feller*) *Suppose that* $\{\xi_k, k \geqslant 1\}$ *is a sequence of independent random variables, then*

$$\lim_{n\to\infty} \max_{1\leqslant k\leqslant n} \frac{\mathrm{Var}\xi_k}{B_n^2} = 0 \quad (\textit{Feller's condition})$$

and

$$\frac{\sum_{k=1}^{n} (\xi_k - E\xi_k)}{B_n} \xrightarrow{d} N(0,1)$$

hold if and only if the Lindeberg condition is satisfied: for any $\varepsilon > 0$

$$\frac{1}{B_n^2} \sum_{k=1}^{n} \int_{|x-E\xi_k|\geqslant \varepsilon B_n} (x - E\xi_k)^2 \mathrm{d}F_k(x) \to 0.$$

Note that for any $\varepsilon > 0$

$$\begin{aligned} P\left(\max_{1\leqslant k\leqslant n} \frac{|\xi_k - E\xi_k|}{B_n} \geqslant \varepsilon\right) &= P\left(\bigcup_{1\leqslant k\leqslant n} \left\{\frac{|\xi_k - E\xi_k|}{B_n} \geqslant \varepsilon\right\}\right) \\ &\leqslant \sum_{k=1}^{n} P(|\xi_k - E\xi_k| \geqslant \varepsilon B_n) \\ &= \sum_{k=1}^{n} \int_{|x-E\xi_k|\geqslant \varepsilon B_n} \mathrm{d}F_k(x) \\ &\leqslant \frac{1}{\varepsilon^2 B_n^2} \sum_{k=1}^{n} \int_{|x-E\xi_k|\geqslant \varepsilon B_n} (x - E\xi_k)^2 \mathrm{d}F_k(x) \to 0. \end{aligned}$$

Hence the Lindeberg condition implies that the absolute value of each summand in the sum $\sum_{k=1}^{n} (\xi_k - E\xi_k)/B_n$ converges to zero in probability uniformly in $1 \leqslant k \leqslant n$. Moreover we have also the converse fact: if, roughly speeking, none of summands plays a significant role, then the central limit theorem holds.

As a consequence of Lindeberg-Feller's theorem, we have

Theorem 8 (*Lyapunov*) *Suppose that* $\{\xi_k, k \geqslant 1\}$ *is a sequence of independent random variables, and suppose further.*

$$\frac{1}{\left(\sum_{k=1}^{n}\operatorname{Var}\xi_k\right)^{1+\delta/2}}\sum_{k=1}^{n}E\mid\xi_k - E\xi_k\mid^{2+\delta}\to 0 \qquad as\ n\to\infty,$$

where $\delta>0$, *then the central limit theorem holds true.*

These results explain the reason of universal existence of normal random variables in nature and human society.

Example 5 Suppose that $\{\xi_k\}$ is a sequence of independent random variables, and the distribution law of ξ_k is $\begin{pmatrix} -k & k \\ 0.5 & 0.5\end{pmatrix}$. It is easy to know that $E\xi_k = 0$, $\operatorname{Var}\xi_k = k^2$, $E\mid\xi_k\mid^3 = k^3$. Hence,

$$\frac{\sum_{k=1}^{n}E\mid\xi_k\mid^3}{\left(\sum_{k=1}^{n}\operatorname{Var}\xi_k\right)^{3/2}} = \frac{\sum_{k=1}^{n}k^3}{\left(\sum_{k=1}^{n}k^2\right)^{3/2}}\to 0 \quad \text{as } n\to\infty.$$

This means that the Lyapunov condition is satisfied. So, $\{\xi_k, k\geqslant 1\}$ obeys the central limit theorem.

In many branches of mathematical statistics, such as parameter (interval) estimation, hypothesis test, sample survey etc., the central limit theorem plays an important role. In fact, it is also one of theoretical foundations of some subjects, such as insurance actuarial, etc. Assume that an insurance company offers insurance operation for some kind of insurance. Now there are n customers sharing the insurance, and let X_i be the claim amount of the i-th policy which suffers risk. As for this insurance company, the random claim amount is the sum of all the policy's claim amounts, denoted by S_n. That is

$$S_n = \sum_{i=1}^{n}X_i.$$

Making clear the distribution of S_n is very important for the insurance company to premium pricing. In practical problems, it is usually assumed that all the policy claims are independent. Then, when the number n of policies is large enough, we do not ask for calculation of the accurate distribution of S_n (in the general case, it is difficult to calculate, even impossible). At this time, we can use the central limit theorem to approximate S_n by a normal variable in distribution: $(S_n - ES_n)/\sqrt{\operatorname{Var}S_n}$ approximately has the normal distribution $N(0, 1)$. With the help of this result, we can estimate some insurance parameters.

Example 6 An insurance company issues two kinds of one-year-term life insurance with random claim amounts 1 000 yuan and 2 000 yuan respectively. The claim probability q_k and the number of insurant n_k are denoted in Table 4-1 below.

Table 4-1

Type k	q_k	b_k	n_k
1	0.02	1	500
2	0.02	2	500
3	0.10	1	300
4	0.10	2	500

The insurance company hopes that the probability that the sum of claims exceeds the total premium is less than 0.05. Now the premium is priced according to the expectation value principle. That is, the premium of policy i is $\pi(X_i)=(1+\theta)EX_i$, and it is required to estimate the value of θ.

Solution Calculate the expectation and variance of $S=\sum_{i=1}^{1\,800}X_i$.

$$\begin{aligned} ES &= \sum_{i=1}^{1\,800}EX_i=\sum_{k=1}^{4}n_kb_kq_k \\ &= 500\times1\times0.02+500\times2\times0.02+300\times1\times0.10 \\ &\quad +500\times2\times0.10 \\ &= 160, \end{aligned}$$

$$\begin{aligned} \mathrm{Var}S &= \sum_{i=1}^{1\,800}\mathrm{Var}X_i=\sum_{k=1}^{4}n_kb_k^2q_k(1-q_k) \\ &= 500\times1^2\times0.02\times0.98+500\times2^2\times0.02\times0.98 \\ &\quad +300\times1^2\times0.10\times0.9+500\times2^2\times0.10\times0.9 \\ &= 256. \end{aligned}$$

Consequently we obtain the sum of premium

$$\pi(S)=(1+\theta)ES=160(1+\theta).$$

According to the request, we have $P(S\leqslant(1+\theta)ES)=0.95$, that is

$$P\left(\frac{S-ES}{\sqrt{\mathrm{Var}S}}\leqslant\frac{\theta ES}{\sqrt{\mathrm{Var}S}}\right)=P\left(\frac{S-ES}{\sqrt{\mathrm{Var}S}}\leqslant10\theta\right)=0.95.$$

One can approximately regard $(S-ES)/\sqrt{\mathrm{Var}S}$ as a standard normal variable. From the table of standard normal distribution function, we have $10\theta=1.645$, namely $\theta=0.1645$.

4.2 Convergence in probability and weak law of large numbers

4.2.1 Convergence in probability

Although distribution function completely characterizes the distribution law of the values of a random variable, it is possible that two different random variables have the same distribution function. For example, throwing a dot in [0, 1] randomly, the dot is located at any point in [0,1] with the same possibility. Let ω denote the location of the dot and define

$$\xi(\omega)=\begin{cases}1, & \omega\in[0,0.5];\\ 0, & \omega\in(0.5,1],\end{cases} \tag{4.13}$$

$$\eta(\omega)=\begin{cases}0, & \omega\in[0,0.5];\\ 1, & \omega\in(0.5,1].\end{cases}$$

Then ξ and η have the same distribution function

$$F(x)=\begin{cases}0, & x<0;\\ \dfrac{1}{2}, & 0\leqslant x<1;\\ 1, & x\geqslant 1.\end{cases} \tag{4.14}$$

If we define $\xi_n=\xi$ for $n\geqslant 1$, then $\xi_n\xrightarrow{d}\eta$, but $|\xi_n-\eta|\equiv 1$. It shows that the convergence of distribution function cannot reflect the degree of approximation between the values of random variables. So it is necessary to introduce other types of convergence.

Definition 3 *Suppose that ξ and $\{\xi_n, n\geqslant 1\}$ are defined on the same probability space $(\Omega,\mathscr{F},P)$. If for any $\varepsilon>0$*

$$\lim_{n\to+\infty}P(|\xi_n-\xi|\geqslant\varepsilon)=0 \tag{4.15}$$

or equivalently

$$\lim_{n\to+\infty}P(|\xi_n-\xi|<\varepsilon)=1, \tag{4.16}$$

then we say that ξ_n converges to ξ in probability, denoted by $\xi_n\xrightarrow{P}\xi$.

It is required in Definition 3 that all ξ and ξ_n have the same definition domain. $\xi_n\xrightarrow{P}\xi$ can be interpreted as follows: except for a small possibility, the values of ξ_n and ξ can approximate to any extent if n is big enough.

We can see from the above example that $\xi_n\xrightarrow{d}\xi$ cannot imply $\xi_n\xrightarrow{P}\xi$. As for the relation between these two types of convergence, we have the following results.

Property 1 Suppose that ξ and $\{\xi_n, n\geqslant 1\}$ are random variables defined on the probability space $(\Omega,\mathscr{F},P)$.

(1) If $\xi_n\xrightarrow{P}\xi$, then $\xi_n\xrightarrow{d}\xi$.

(2) If $\xi_n\xrightarrow{d}c$, where c is a constant, then $\xi_n\xrightarrow{P}c$.

Proof (1) Let F and F_n be the distribution functions of ξ and ξ_n respectively, and let x be a continuity point of F. For any $\varepsilon>0$,

$$\begin{aligned}(\xi\leqslant x-\varepsilon)&=(\xi\leqslant x-\varepsilon,\xi_n\leqslant x)\cup(\xi\leqslant x-\varepsilon,\xi_n>x)\\ &\subset(\xi_n\leqslant x)\cup(\xi_n-\xi\geqslant\varepsilon).\end{aligned}$$

Hence

$$F(x-\varepsilon) \leqslant F_n(x) + P(\xi_n - \xi \geqslant \varepsilon).$$

Since $\xi_n \xrightarrow{P} \xi$ as $n \to \infty$, we obtain

$$P(\xi_n - \xi \geqslant \varepsilon) \leqslant P(|\xi_n - \xi| \geqslant \varepsilon) \to 0 \quad \text{as } n \to \infty.$$

Thus

$$F(x-\varepsilon) \leqslant \liminf_{n\to\infty} F_n(x). \tag{4.17}$$

Similarly

$$\begin{aligned}(\xi_n \leqslant x) &= (\xi_n \leqslant x, \xi \leqslant x+\varepsilon) \cup (\xi_n \leqslant x, \xi > x+\varepsilon) \\ &\subset (\xi \leqslant x+\varepsilon) \cup (\xi - \xi_n \geqslant \varepsilon).\end{aligned}$$

Thus

$$F_n(x) \leqslant F(x+\varepsilon) + P(\xi - \xi_n \geqslant \varepsilon).$$

Then

$$\limsup_{n\to\infty} F_n(x) \leqslant F(x+\varepsilon). \tag{4.18}$$

Combining (4.17) and (4.18) implies

$$F(x-\varepsilon) \leqslant \liminf_{n\to\infty} F_n(x) \leqslant \limsup_{n\to\infty} F_n(x) \leqslant F(x+\varepsilon).$$

Letting $\varepsilon \to 0$, by continuity of F at x, we obtain

$$\lim_{n\to\infty} F_n(x) = F(x).$$

That is

$$\xi_n \xrightarrow{d} \xi.$$

(2) If $\xi_n \xrightarrow{d} c$, then

$$\lim_{n\to\infty} F_n(x) = \begin{cases} 0, & x < c; \\ 1, & x > c. \end{cases}$$

Hence for any $\varepsilon > 0$,

$$\begin{aligned}P(|\xi_n - c| \geqslant \varepsilon) &= P(\xi_n \geqslant c+\varepsilon) + P(\xi_n \leqslant c-\varepsilon) \\ &= 1 - P(\xi_n < c+\varepsilon) + P(\xi_n \leqslant c-\varepsilon) \\ &= 1 - F_n(c+\varepsilon-0) + F_n(c-\varepsilon) \\ &\to 0 \quad (n \to \infty).\end{aligned}$$

The proof is complete.

Example 7 Let $\{\xi_n\}$ be a sequence of independent and identically distributed random variables with a common uniform distribution in $[0, a]$. Let $\eta_n = \max\{\xi_1, \xi_2, \cdots, \xi_n\}$. Prove that $\eta_n \xrightarrow{P} a$.

Proof Let $G_n(x)$ be the distribution function of ξ_n, $D(x-a)$ the degenerate distribution function at a. From Property 1, we only need to prove $G_n(x) \xrightarrow{w} D(x-a)$ as $n \to \infty$.

In Chapter 2, we have verified that if $F(x)$ is the distribution function of ξ_k, then the distribution function of η_n is $G_n(x) = (F(x))^n$. Now the distribution function of ξ_k is

$$F(x) = \begin{cases} 0, & x < 0; \\ x/a, & 0 \leqslant x < a; \\ 1, & x \geqslant a. \end{cases} \tag{4.19}$$

Hence

$$G_n(x) = \begin{cases} 0, & x < 0; \\ (x/a)^n, & 0 \leqslant x < a; \\ 1, & x \geqslant a \end{cases}$$

$$\to D(x-a) = \begin{cases} 0, & x < a; \\ 1, & x \geqslant a \end{cases} \quad \text{as } n \to \infty.$$

The proof is complete.

Convergence in probability has a lot of properties similar to that of limit of a sequence of numbers in calculus. Now we give two examples to illustrate the proof method of such problems. The proofs of most of the properties are left to the readers as exercises.

Property 2 Let $\{\xi, \xi_n, n \geqslant 1\}$ be a sequence of random variables defined on the probability space $(\Omega, \mathscr{F}, P)$. Prove that

(1) if $\xi_n \xrightarrow{P} \xi$, $\xi_n \xrightarrow{P} \eta$, then $P(\xi = \eta) = 1$.

(2) if $\xi_n \xrightarrow{P} \xi$, f is a continuous function on $(-\infty, +\infty)$, then $f(\xi_n) \xrightarrow{P} f(\xi)$.

Proof (1) For any $\varepsilon > 0$, we have

$$(|\xi - \eta| \geqslant \varepsilon) \subseteq \left(|\xi_n - \xi| \geqslant \frac{\varepsilon}{2}\right) \cup \left(|\xi_n - \eta| \geqslant \frac{\varepsilon}{2}\right).$$

Thus

$$P(|\xi - \eta| \geqslant \varepsilon) \leqslant P\left(|\xi_n - \xi| \geqslant \frac{\varepsilon}{2}\right) + P\left(|\xi_n - \eta| \geqslant \frac{\varepsilon}{2}\right).$$

Note that the left side of the above inequality is independent of n and $\xi_n \xrightarrow{P} \xi$, $\xi_n \xrightarrow{P} \eta$ as $n \to \infty$. Therefore $P(|\xi - \eta| \geqslant \varepsilon) = 0$. Furthermore,

$$\begin{aligned} P(|\xi - \eta| > 0) &= P\left(\bigcup_{n=1}^{\infty} \left(|\xi - \eta| \geqslant \frac{1}{n}\right)\right) \\ &\leqslant \sum_{n=1}^{\infty} P\left(|\xi - \eta| \geqslant \frac{1}{n}\right) = 0, \end{aligned} \tag{4.20}$$

i.e., $P(\xi = \eta) = 1$.

(2) For any given $\varepsilon, \varepsilon' > 0$, there exists an $M > 0$ satisfying

$$P(|\xi| \geqslant M) \leqslant P\left(|\xi| \geqslant \frac{M}{2}\right) < \frac{\varepsilon'}{4}. \tag{4.21}$$

Since $\xi_n \xrightarrow{P} \xi$, when $n \geqslant N_1$ for some $N_1 \geqslant 1$,

$$P\left(|\xi_n - \xi| \geqslant \frac{M}{2}\right) < \frac{\varepsilon'}{4}.$$

Hence

$$P(|\xi_n| \geqslant M) \leqslant P\left(|\xi_n - \xi| \geqslant \frac{M}{2}\right) + P\left(|\xi| \geqslant \frac{M}{2}\right) < \frac{\varepsilon'}{4} + \frac{\varepsilon'}{4} = \frac{\varepsilon'}{2}. \tag{4.22}$$

And since $f(x)$ is a continuous function on $(-\infty, +\infty)$, $f(x)$ is uniformly continuous in $[-M, M]$. For given $\varepsilon > 0$, there exists $\delta > 0$ such that $|f(x) - f(y)| < \varepsilon$ for any $|x - y| < \delta$.

Thus

$$P(|f(\xi_n) - f(\xi)| \geqslant \varepsilon) \leqslant P(|\xi_n - \xi| \geqslant \delta) + P(|\xi_n| \geqslant M) + P(|\xi| \geqslant M). \tag{4.23}$$

For the above δ, when $n \geqslant N_2$ for some $N_2 \geqslant 1$,

$$P(|\xi_n - \xi| \geqslant \delta) < \frac{\varepsilon'}{4}. \tag{4.24}$$

Combining (4.21), (4.22), (4.23) and (4.24), we obtain

$$P(|f(\xi_n) - f(\xi)| \geqslant \varepsilon) < \frac{\varepsilon'}{4} + \frac{\varepsilon'}{2} + \frac{\varepsilon'}{4} = \varepsilon'$$

provided $n \geqslant \max\{N_1, N_2\}$. Thus $f(\xi_n) \xrightarrow{P} f(\xi)$.

In order to make a further discussion on the conditions of convergence in probability, we give the following generalization of Chebyshev's inequality (see Section 3.2). The proof is left to the reader.

Markov's inequality Let ξ be a random variable defined on the probability space $(\Omega, \mathscr{F}, P)$, $f(x)$ be a nonnegative monotone non-decreasing function on $[0, +\infty)$, and then for any $x > 0$,

$$P(|\xi| \geqslant x) \leqslant \frac{Ef(|\xi|)}{f(x)}. \tag{4.25}$$

Theorem 9 $\xi_n \xrightarrow{P} \xi$ *if and only if*

$$E\frac{|\xi_n - \xi|}{1 + |\xi_n - \xi|} \to 0.$$

Proof Sufficiency: Noting that $f(x) = |x|/(1 + |x|)$ is non-negative and non-decreasing on $[0, +\infty)$, according to Markov's inequality, we have

$$P(|\xi_n - \xi| \geqslant \varepsilon) \leqslant \frac{1+\varepsilon}{\varepsilon} E\frac{|\xi_n - \xi|}{1 + |\xi_n - \xi|} \to 0 \quad \text{as } n \to \infty.$$

That is $\xi_n \xrightarrow{P} \xi$.

Necessity: Let $F_n(x)$ denote the distribution function of $\xi_n - \xi$. For any $\varepsilon > 0$,

$$E\frac{|\xi_n - \xi|}{1 + |\xi_n - \xi|} = \int_{-\infty}^{+\infty} \frac{|x|}{1 + |x|} \mathrm{d}F_n(x)$$
$$= \int_{|x| < \varepsilon} \frac{|x|}{1 + |x|} \mathrm{d}F_n(x) + \int_{|x| \geqslant \varepsilon} \frac{|x|}{1 + |x|} \mathrm{d}F_n(x)$$

$$\leqslant \frac{\varepsilon}{1+\varepsilon} + \int_{|x| \geqslant \varepsilon} \mathrm{d}F_n(x)$$

$$= \frac{\varepsilon}{1+\varepsilon} + P(|\xi_n - \xi| \geqslant \varepsilon). \tag{4.26}$$

Since $\xi_n \xrightarrow{P} \xi$, first letting $n \to \infty$ at the both sides of (4.26) and then letting $\varepsilon \to 0$ yield

$$E\frac{|\xi_n - \xi|}{1 + |\xi_n - \xi|} \to 0.$$

4.2.2 Weak law of large numbers

Consider an event A in a random trial E. Suppose the probability of A occurring is p $(0 < p < 1)$. Now we conduct n independent trials—n-fold Bernoulli trial. Let

$$\xi_i = \begin{cases} 1, & A \text{ occurs at the } i\text{-th trial;} \\ 0, & A \text{ does not occur at the } i\text{-th trial,} \end{cases} \tag{4.27}$$

$1 \leqslant i \leqslant n$. Then $P(\xi_i = 1) = p$, $P(\xi_i = 0) = 1 - p$.

$S_n = \sum_{i=1}^{n} \xi_i$ denotes the number of A occurring in the n Bernoulli trials. S_n may take any one of $0, 1, 2, \cdots, n$ as its value, which depends on the outcome of the experiment. It is well known that $ES_n/n = p$. In Section 1.1 we pointed out that when $n \to \infty$, the frequency S_n/n stabilize at (in some sense of convergence) probability p. We want to know how much the difference between S_n/n and p is at all.

First we should know that for any $0 < \varepsilon < 1$ we cannot expect that $|S_n/n - p| \leqslant \varepsilon$ holds for all the trials, even if n is big enough. In fact for $0 < p < 1$,

$$P\left(\frac{S_n}{n} = 1\right) = P(\xi_1 = 1, \cdots, \xi_n = 1) = p^n,$$

$$P\left(\frac{S_n}{n} = 0\right) = P(\xi_1 = 0, \cdots, \xi_n = 0) = (1-p)^n.$$

Neither of them is zero. In the first case, taking $\varepsilon < 1 - p$, we have $|S_n/n - p| = 1 - p > \varepsilon$. In the second case, taking $\varepsilon < p$, we have $|S_n/n - p| = p > \varepsilon$.

However, if n is large enough, the probability of occurance of the events $\{S_n/n = 1\}$ and $\{S_n/n = 0\}$ is very small. Generally speaking, we hope that the probability of $\{|S_n/n - p| \geqslant \varepsilon\}$ can be as small as possible when n is large enough. This fact was first discovered by J. Bernoulli.

Theorem 10 *(Bernoulli) Let $\{\xi_n, n \geqslant 1\}$ be a sequence of independent and identically distributed random variables with $P(\xi_n = 1) = p$, $P(\xi_n = 0) = 1 - p$, $0 < p < 1$. Put $S_n = \sum_{i=1}^{n} \xi_i$. Then we have*

$$\frac{S_n}{n} \xrightarrow{P} p.$$

Following Bernoulli, people have attempted to establish similar results for more general random variables.

Definition 4 *Suppose that* $\{\xi_n, n \geqslant 1\}$ *is a sequence of random variables defined on the probability space* $(\Omega, \mathscr{F}, P)$. *If there exist constant sequences* $\{a_n, n \geqslant 1\}$ *and* $\{b_n, n \geqslant 1\}$ *such that*

$$\frac{1}{a_n}\sum_{k=1}^{n}\xi_k - b_n \xrightarrow{P} 0 \quad as \quad n \to \infty, \tag{4.28}$$

then $\{\xi_n\}$ *is said to obey the weak law of large numbers. In short* $\{\xi_n, n \geqslant 1\}$ *obeys the law of large numbers.*

Theorem 11 (*Chebyshev*) *Let* $\{\xi_n, n \geqslant 1\}$ *be a sequence of independent random variables defined on the probability space* $(\Omega, \mathscr{F}, P)$ *with* $E\xi_n = \mu_n$ *and* $\mathrm{Var}\xi_n = \sigma_n^2$. *If* $\sum_{k=1}^{n}\sigma_k^2/n^2 \to 0$, *then* $\{\xi_n, n \geqslant 1\}$ *obeys the weak law of large numbers, i.e.,*

$$\frac{1}{n}\sum_{k=1}^{n}\xi_k - \frac{1}{n}\sum_{k=1}^{n}\mu_k \xrightarrow{P} 0.$$

Proof By the independence of $\xi_1, \cdots, \xi_n$, we have

$$E\left(\frac{1}{n}\sum_{k=1}^{n}\xi_k\right) = \frac{1}{n}\sum_{k=1}^{n}\mu_k, \quad \mathrm{Var}\left(\frac{1}{n}\sum_{k=1}^{n}\xi_k\right) = \frac{1}{n^2}\sum_{k=1}^{n}\sigma_k^2.$$

Using the Chebyshev inequality in Section 3.2, we have

$$P\left(\left|\frac{1}{n}\sum_{k=1}^{n}(\xi_k - \mu_k)\right| \geqslant \varepsilon\right) \leqslant \frac{1}{\varepsilon^2}\mathrm{Var}\left(\frac{1}{n}\sum_{k=1}^{n}\xi_k\right)$$

$$= \frac{1}{\varepsilon^2 n^2}\sum_{k=1}^{n}\sigma_k^2 \to 0 \quad \text{as } n \to \infty.$$

The proof is complete.

Remark 1 The Bernoulli law of large numbers is the special case of the Chebyshev law of large numbers.

Remark 2 If the condition "$\{\xi_n, n \geqslant 1\}$ is independent" is replaced by "$\{\xi_n, n \geqslant 1\}$ is pairwise uncorrelated", then Theorem 11 still holds. More generally, it follows from the proof of this theorem that if the independence condition is not assumed, and the condition "$\sum_{k=1}^{n}\sigma_k^2/n^2 \to 0$" is replaced by "$\mathrm{Var}(\sum_{k=1}^{n}\xi_k)/n^2 \to 0$", the conclusion of Theorem 11 still holds. This conclusion is called "the Markov law of large numbers".

If $\{\xi_n\}$ is not only independent, but also identically distributed, then Theorem 11 can be improved as follows.

Theorem 12 (*Khinchine*) *Let* $\{\xi_n, n \geqslant 1\}$ *be a sequence of independent and identically distributed random variables defined on the probability space* $(\Omega, \mathscr{F}, P)$ *with* $E|\xi_1| < \infty$. *Let* $E\xi_1 = \mu$, $S_n = \sum_{k=1}^{n}\xi_k$. *Then* $\{\xi_n, n \geqslant 1\}$ *obeys the weak law of large numbers, i.e.,*

$$\frac{S_n}{n} \xrightarrow{P} \mu \quad as\ n \to \infty.$$

Proof Let $f(t)$ and $f_n(t)$ be the characteristic functions of ξ_1 and S_n/n respectively. Since $\{\xi_n, n \geqslant 1\}$ is independent and identically distributed, we have $f_n(t) = (f(t/n))^n$. Moreover, from the Taylor expansion formula, we have

$$f(t) = 1 + \mathrm{i}\mu t + o(t) \quad \text{as } t \to 0 \tag{4.29}$$

since $E\xi_1 = \mu$. Then for every $t \in \mathbf{R}$,

$$f(t/n) = 1 + \mathrm{i}\frac{\mu t}{n} + o\left(\frac{1}{n}\right) \quad \text{as } n \to \infty,$$

$$f_n(t) = \left(1 + \mathrm{i}\frac{\mu t}{n} + o\left(\frac{1}{n}\right)\right)^n \to \mathrm{e}^{\mathrm{i}\mu t} \quad \text{as } n \to \infty \tag{4.30}$$

Note that $\mathrm{e}^{\mathrm{i}\mu t}$ is just the characteristic function of the degenerate distribution function concentrated at the point μ. By the inverse limit theorem in Section 4.1 we know that $S_n/n \xrightarrow{d} \mu$, and by Property 1, we have $S_n/n \xrightarrow{P} \mu$. The proof is complete.

Example 8 Suppose that ξ_k has the distribution law $\begin{pmatrix} k^s & -k^s \\ 0.5 & 0.5 \end{pmatrix}$, where $s < 1/2$ is a constant, and $\{\xi_k, k \geqslant 1\}$ is independent. Prove that $\{\xi_k, k \geqslant 1\}$ obeys the weak law of large numbers.

Proof By the distribution law of ξ_k, we have $E\xi_k = 0$, $\mathrm{Var}\xi_k = k^{2s}$. When $s < 1/2$,

$$\frac{1}{n^2}\sum_{k=1}^{n}\mathrm{Var}\xi_k = \frac{1}{n^2}\sum_{k=1}^{n}k^{2s} < \frac{1}{n^2}\sum_{k=1}^{n}n^{2s} = n^{2s-1} \to 0 \quad \text{as } n \to \infty.$$

In addition, $\{\xi_k, k \geqslant 1\}$ is also independent, so $\{\xi_k, k \geqslant 1\}$ obeys the Chebyshev law of large numbers, i.e.,

$$\frac{1}{n}\sum_{k=1}^{n}\xi_k \xrightarrow{P} 0.$$

Example 9 Suppose that $\{\xi_k, k \geqslant 1\}$ is independent with a common density function

$$p(x) = \begin{cases} \dfrac{2}{x^3}, & x \geqslant 1; \\ 0, & x < 1. \end{cases} \tag{4.31}$$

Prove that $\{\xi_k, k \geqslant 1\}$ obeys the law of large numbers.

Proof Since $\{\xi_k, k \geqslant 1\}$ is independent identically distributed and $E\xi_k = \int_{-\infty}^{\infty} xp(x)\mathrm{d}x = 2$, $\{\xi_k, k \geqslant 1\}$ obeys the Khinchine law of large numbers.

Example 10 Let $\{\xi_k, k \geqslant 1\}$ be a sequence of independent identically distributed random variables with $E\xi_k = \mu$ and $\mathrm{Var}\xi_k = \sigma^2$. Denote

$$\bar{\xi}_n = \frac{1}{n}\sum_{k=1}^{n}\xi_k, \qquad S_n^2 = \frac{1}{n}\sum_{k=1}^{n}(\xi_k - \bar{\xi}_n)^2.$$

Prove that $S_n^2 \xrightarrow{P} \sigma^2$.

Proof

$$\begin{aligned} S_n^2 &= \frac{1}{n}\sum_{k=1}^{n}(\xi_k - \bar{\xi}_n)^2 \\ &= \frac{1}{n}\sum_{k=1}^{n}((\xi_k - \mu) - (\bar{\xi}_n - \mu))^2 \\ &= \frac{1}{n}\sum_{k=1}^{n}(\xi_k - \mu)^2 - (\bar{\xi}_n - \mu)^2. \end{aligned} \tag{4.32}$$

By the Khinchine law of large numbers, we have $\bar{\xi}_n \xrightarrow{P} \mu$. Thus $(\bar{\xi}_n - \mu)^2 \xrightarrow{P} 0$.

Moreover, since $\{(\xi_k - \mu)^2, k \geqslant 1\}$ is independent identically distributed and $E(\xi_k - \mu)^2 = \mathrm{Var}\xi_k = \sigma^2$, $\{(\xi_k - \mu)^2, k \geqslant 1\}$ also obeys the Khinchine weak law of large numbers, i. e. $\sum_{k=1}^{n}(\xi_k - \mu)^2/n \xrightarrow{P} \sigma^2$. From (4. 32) and the property of convergence in probability (c. f. Exercise 19), we know that $S_n^2 \xrightarrow{P} \sigma^2$.

Remark In mathematical statistics, $\bar{\xi}_n$ is called the sample mean and $nS_n^2/(n-1)$ is called the sample variance. The Khinchine law of large numbers tells us that the sample mean converges in probability to the population mean. The above example shows that the sample variance converges in probability to the population variance.

At last we give the definition of another convergence for random variables.

Definition 5 *Let ξ and $\{\xi_n, n \geqslant 1\}$ be random variables defined on a common probability space $(\Omega, \mathscr{F}, P)$ with $E|\xi|^r < \infty$ and $E|\xi_n|^r < \infty$ for some $r > 0$. If*

$$E|\xi_n - \xi|^r \to 0, \tag{4.33}$$

then we say that $\{\xi_n, n \geqslant 1\}$ converges in mean of order r to ξ, denoted by $\xi_n \xrightarrow{L_r} \xi$.

Suppose that there exists $0 < r < \infty$ such that $\xi_n \xrightarrow{L_r} \xi$. Letting $f(x) = |x|^r$ and applying Markov's inequality to $\xi_n - \xi$, we can get $\xi_n \xrightarrow{P} \xi$. But the following example shows that the converse does not hold.

Example 11 Define ξ_n by $P(\xi_n = n) = 1/\log(n+3)$, $P(\xi_n = 0) = 1 - 1/\log(n+3)$, $n = 1, 2, \cdots$. It is easy to know $\xi_n \xrightarrow{P} 0$, but for any $0 < r < +\infty$,

$$E|\xi_n|^r = \frac{n^r}{\log(n+3)} \to \infty.$$

That is, $\xi_n \xrightarrow{L_r} 0$ does not hold true.

4.3 Almost sure convergence and strong laws of large numbers

4.3.1 Almost sure convergence

As we know, a random variable is a function with real values defined on a sample space. So we can discuss convergence of a sequence of random variables at every sample point just like pointwise convergence of a sequence of functions in calculus. However, because of randomness, we hardly expect that a sequence of random variables always have a limit at every sample point. Thus, a question is whether the limit exists on a point set with probability one.

Definition 6 *Suppose that ξ and $\{\xi_n, n \geqslant 1\}$ are random variables defined on a common probability space $(\Omega, \mathscr{F}, P)$.*

(1) *If there exists a $\Omega_0 \in \mathscr{F}$ such that $P(\Omega_0) = 0$ and for any $\omega \in \Omega \backslash \Omega_0$, $\xi_n(\omega) \to \xi(\omega)$ $(n \to \infty)$, then we say that ξ_n converges with probability one or almost surely to ξ, denoted by* $\xi_n \to \xi$ a. s.

(2) *If there exists a $\Omega_0 \in \mathscr{F}$ such that $P(\Omega_0) = 0$ and for any $\omega \in \Omega \backslash \Omega_0$, the sequence of numbers $\{\xi_n(\omega)\}$ is a Cauchy fundamental sequence, i.e., $\xi_n(\omega) - \xi_m(\omega) \to 0$ $(n > m \to \infty)$, then we say that $\{\xi_n\}$ is a Cauchy fundamental sequence with probability one.*

Remark $\xi_n \to \xi$ a. s. means that ξ_n pointwise converges to ξ at most except a set of probability zero. Since every Cauchy fundamental sequence must have a limit, ξ_n converges almost surely if and only if $\{\xi_n\}$ is a Cauchy fundamental sequence with probability one.

Now we give a criterion for almost sure convergence.

Theorem 13 *Suppose that ξ and $\{\xi_n, n \geqslant 1\}$ are random variables defined on a probability space $(\Omega, \mathscr{F}, P)$.*

(1) $\xi_n(\omega) \to \xi(\omega)$ a. s. *if and only if exits for any $\varepsilon > 0$,*

$$\lim_{n \to \infty} P\left(\sup_{k \geqslant n} |\xi_k - \xi| \geqslant \varepsilon\right) = 0$$

or equivalently

$$\lim_{n \to \infty} P\left(\bigcup_{k \geqslant n} (|\xi_k - \xi| \geqslant \varepsilon)\right) = 0.$$

(2) *$\{\xi_n, n \geqslant 1\}$ is a Cauchy fundamental sequence with probability one if and only if for any $\varepsilon > 0$,*

$$\lim_{n \to \infty} P\left(\sup_{k \geqslant 0} |\xi_{k+n} - \xi_k| \geqslant \varepsilon\right) = 0$$

or equivalently

$$\lim_{n\to\infty} P\Big(\bigcup_{k\geqslant 0} (|\xi_{k+n} - \xi_k| \geqslant \varepsilon)\Big) = 0.$$

Proof (1) For any $\varepsilon > 0$, let $A_n^\varepsilon = \{|\xi_n - \xi| \geqslant \varepsilon\}$ and $A^\varepsilon = \bigcap_{n=1}^{\infty}\bigcup_{k\geqslant n} A_k^\varepsilon$. Then

$$\{\xi_n \not\to \xi\} = \bigcup_{m=1}^{\infty} A^{\frac{1}{m}}.$$

By the continuity theorem, we have

$$P(A^\varepsilon) = P\Big(\bigcap_{n=1}^{\infty}\bigcup_{k\geqslant n} A_k^\varepsilon\Big) = \lim_{n\to\infty} P\Big(\bigcup_{k\geqslant n} A_k^\varepsilon\Big),$$

which implies that the following relations hold:

$$0 = P(\xi_n \not\to \xi) \Leftrightarrow P\Big(\bigcup_{m=1}^{\infty} A^{\frac{1}{m}}\Big) = 0$$

$$\Leftrightarrow P\Big(A^{\frac{1}{m}}\Big) = 0 \text{ for each } m \geqslant 1$$

$$\Leftrightarrow P\Big(\bigcup_{k\geqslant n} A_k^{1/m}\Big) \to 0 \text{ for each } m \geqslant 1$$

$$\Leftrightarrow P\Big(\bigcup_{k\geqslant n}\Big(|\xi_k - \xi| \geqslant \frac{1}{m}\Big)\Big) \to 0 \text{ for each } m \geqslant 1$$

$$\Leftrightarrow P\Big(\bigcup_{k\geqslant n}(|\xi_k - \xi| \geqslant \varepsilon)\Big) \to 0 \text{ for any } \varepsilon \geqslant 0.$$

(2) For any $\varepsilon > 0$, let $B_{n,k}^\varepsilon = \{|\xi_{k+n} - \xi_k| \geqslant \varepsilon\}$ and $B^\varepsilon = \bigcap_{m=1}^{\infty}\bigcup_{n\geqslant m}\bigcup_{k\geqslant 1} B_{n,k}^\varepsilon$. Then $\{\xi_n$ is not a Cauchy fundamental sequence $\} = \bigcup_{\varepsilon>0} B^\varepsilon$.

The remainder of proof is similar to that of (1).

Corollary If for any $\varepsilon > 0$, $\sum_{n=1}^{\infty} P(|\xi_n - \xi| \geqslant \varepsilon) < \infty$, then

$$\xi_n \to \xi \quad \text{a. s.}$$

Proof Note that

$$P\Big(\bigcup_{k\geqslant n}(|\xi_k - \xi| \geqslant \varepsilon)\Big) \leqslant \sum_{k=n}^{\infty} P(|\xi_k - \xi| \geqslant \varepsilon) \to 0.$$

The proof is complete.

Remark Theorem 13 shows that $\xi_n \to \xi$ a. s. implies $\xi_n \xrightarrow{P} \xi$. However, there are examples to show that, $\xi_n \xrightarrow{P} \xi$ does not imply $\xi_n \to \xi$ a. s. (see Supplements and Remarks).

4.3.2 Strong laws of large numbers

What is closely related to almost sure convergence is the so-called strong law of large numbers.

Definition 7 *Suppose that* $\{\xi_n, n \geqslant 1\}$ *is a sequence of random variables defined on a common probability space* $(\Omega, \mathscr{F}, P)$. *If there exist constant sequences* $\{a_n, n \geqslant 1\}$ *and* $\{b_n, n \geqslant 1\}$ *such that*

$$\frac{1}{a_n}\sum_{k=1}^{n}\xi_k - b_n \to 0 \qquad \text{a. s.}$$

we say that $\{\xi_n\}$ *obeys the strong law of large numbers.*

Since almost sure convergence is stronger than convergence in probability, the strong law of large numbers is deeper than the weak law of large numbers. In Section 4. 2 we know that Bernoulli obtained the Bernoulli law of large numbers via some accurate calculation of the binomial distribution. That is, the frequency in Bernoulli random trials converges in probability to probability of the event. Not until 1909 did Borel prove the following stronger result.

Theorem 14 *(Borel) Suppose that* $\{\xi_n\}$ *is a sequence of independent identically distributed random variables defined on a common probability space* $(\Omega, \mathscr{F}, P)$ *with* $P(\xi_n=1)=p$, $P(\xi_n=0)=1-p$, $0<p<1$. *Let* $S_n=\sum_{k=1}^{n}\xi_k$, *and then*

$$\frac{S_n}{n}-p \to 0 \qquad \text{a. s.} \tag{4. 34}$$

Proof According to Theorem 13, we shall prove for $\varepsilon>0$,

$$\lim_{n\to\infty}P\left(\sup_{k\geqslant n}\left|\frac{S_k}{k}-p\right|>\varepsilon\right)=0.$$

Since

$$P\left(\sup_{k\geqslant n}\left|\frac{S_k}{k}-p\right|>\varepsilon\right)\leqslant\sum_{k=n}^{\infty}P\left(\left|\frac{S_k}{k}-p\right|>\varepsilon\right),$$

let $\eta_k=\xi_k-p$. Then $E\eta_k=0$. By the Markov inequality and independence of $\eta_1, \eta_2, \cdots$, we have

$$\begin{aligned}
P\left(\left|\frac{S_k}{k}-p\right|>\varepsilon\right)&\leqslant\frac{1}{\varepsilon^4}E\left(\frac{S_k}{k}-p\right)^4\\
&=\frac{1}{\varepsilon^4k^4}\sum_{i,j,l,m=1}^{k}E[\eta_i\eta_j\eta_l\eta_m]\\
&=\frac{1}{\varepsilon^4k^4}(kE\eta_1^4+k(k-1)(E\eta_1^2)^2)\\
&=\frac{1}{k^3\varepsilon^4}(p(1-p)(p^3+(1-p)^3)+(k-1)p^2(1-p)^2),
\end{aligned}$$

which is $O\left(\frac{1}{k^2}\right)$ and so the corresponding series is convergent. (4. 34) is now proved.

Theorem 14 further clarifies the meaning of frequency stabilizing toward probability. Kolmogorov generalized the above result to general random variables in 1930.

Theorem 15 *(Kolmogorov) Suppose that* $\{\xi_n, n\geqslant 1\}$ *is a sequence of independent identically distributed random variables defined on a probability space* $(\Omega, \mathscr{F}, P)$ *with* $E|\xi_1|<\infty$, $E\xi_1=\mu$. *Let* $S_n=\sum_{k=1}^{n}\xi_k$, *and then*

$$\frac{S_n}{n} - \mu \to 0 \quad \text{a. s.} \tag{4.35}$$

In fact, the converse of Theorem 15 also holds: if there exists a constant μ such that (4.35) holds, then the expectation of ξ_1 exists and equals to μ.

The proofs of Theorem 15 and its converse can be found in 12 at Supplements and Remarks.

Example 13 (The Monte Carlo method) Let $f(x)$ be a continuous function defined on $[0, 1]$ with values in $[0, 1]$, and let $\xi_1, \eta_1, \xi_2, \eta_2, \cdots$ be a sequence of independent random variables with a common uniform distribution in $[0, 1]$. Define

$$\rho_i = \begin{cases} 1, & \text{if } f(\xi_i) \geqslant \eta_i; \\ 0, & \text{if } f(\xi_i) < \eta_i. \end{cases}$$

Then $\{\rho_i, i \geqslant 1\}$ are also independent identically distributed random variables. Furthermore,

$$\begin{aligned} E\rho_1 &= P(f(\xi_1) \geqslant \eta_1) = \iint\limits_{y \leqslant f(x)} \mathrm{d}x\mathrm{d}y \\ &= \int_0^1 \left(\int_0^{f(x)} \mathrm{d}y \right) \mathrm{d}x = \int_0^1 f(x)\mathrm{d}x. \end{aligned}$$

By Theorem 15, we have

$$\frac{1}{n}\sum_{k=1}^{n} \rho_k \to \int_0^1 f(x)\mathrm{d}x \quad \text{a. s.} \tag{4.36}$$

Hence we can calculate the value of $\int_0^1 f(x)\mathrm{d}x$ by simulation. The idea is to throw a dot randomly in the square $\{0 \leqslant x \leqslant 1, 0 \leqslant y \leqslant 1\}$ in the plane xOy and to count the number of points falling in the range $\{0 \leqslant x \leqslant 1, 0 \leqslant y \leqslant f(x)\}$, (i. e., the left hand side of(4.36)). The frequency will approximate well enough to the integral.

So far we have studied some classical limit theorems in probability theory.

Supplements and Remarks

1. Limit theorems

Limit theorems had been the central subject of the study of probability theory in 18th and 19th centuries. The Bernoulli law of large numbers, which appeared in the famous book *The Art of Conjecturing* due to Bernoulli in 1713, is the first probability theorem proved rigorously. But the terminology of the law of large numbers was attributed to Poisson in 1837, and the terminology of the central limit theorem was given by Pólya in 1920. It is a sort of theorems which state that the distributions of the partial sums of a sequence of random variables approximate to the normal distribution. It is the most important sort of theorems in probability theory, and still has extensive practical background. Early central limit theorems are related to the n-fold Bernoulli trial. In 1716, de Moivre studied the case of $p = 1/2$, and later Laplace generalized it to the case of $0 < p < 1$. From the middle of the 19th century to the beginning of the 20th century, a number of famous Russian mathematicians made great contribution to the development of probability theory. They generalized the Bernoulli law of large numbers, the de Moivre-Laplace central limit theorem to the sums of general random variables by using rigorous and powerful mathematical tools, such as Fourier transform.

2. The proof of Helly's first theorem

Let $r_1, r_2, \cdots$, denote the set of rational numbers. That $0 \leqslant F(x) \leqslant 1$ means that $\{F_n(r_1)\}$ is a bounded sequence. So, there exists a convergent subsequence $\{F_{1n}(r_1)\}$. Denote the limit by $G(r_1) = \lim_{n\to\infty} F_{1n}(r_1)$.

Then, consider the bounded sequence $\{F_{1n}(r_2)\}$. There exists a further convergent subsequence $\{F_{2n}(r_2)\}$. Denote the limit by $G(r_2) = \lim\limits_{n\to\infty} F_{2n}(r_2)$.

Repeating this procedure, we obtain

$$\{F_{kn}\} \subset \{F_{k-1,n}\},\ G(r_k) = \lim_{n\to\infty} F_{kn}(r_k),\ k \geqslant 2.$$

Now, consider the diagonal sequence $\{F_{nn}\}$. Obviously, $\lim\limits_{n\to\infty} F_{nn}(r_k) = G(r_k)$ holds true for all nonnegative integers k. In addition, since F_n is non-decreasing, then $G(r_i) \leqslant G(r_j)$ if $r_i < r_j$. So, $G(r)$ is a bounded non-decreasing function defined on the set of rational numbers. Let

$$F(x) = \lim_{r_j \leqslant x,\ r_j \to x} G(r_j),\ x \in \mathbf{R}. \tag{4.37}$$

This function is identical to $G(x)$ on the set of rational numbers, and is also bounded and non-decreasing.

In the following, it will be proved that

$$\lim_{n\to\infty} F_{nn}(x) = F(x) \tag{4.38}$$

for every continuity point x of F.

For every $\varepsilon > 0$ and each continuity point x of F, select $h > 0$ such that

$$F(x+h) - F(x-h) < \frac{\varepsilon}{2}.$$

By the density of rational numbers, there exist rational numbers r_i and r_j, satisfying

$$x - h < r_i < x < r_j < x + h,$$

which in turn implies

$$F(x-h) \leqslant F(r_i) \leqslant F(x) \leqslant F(r_j) \leqslant F(x+h). \tag{4.39}$$

Moreover, there exists an $N(\varepsilon)$ such that

$$| F_{nn}(r_j) - F(r_j) | < \frac{\varepsilon}{2},\ | F_{nn}(r_i) - F(r_i) | < \frac{\varepsilon}{2}, \tag{4.40}$$

whenever $n \geqslant N(\varepsilon)$.

Furthermore, in view of monotonicity of F_n and F, we have when $n \geqslant N(\varepsilon)$,

$$F_{nn}(x) \leqslant F_{nn}(r_j) \leqslant F(r_j) + \frac{\varepsilon}{2}$$

$$\leqslant F(x+h)+\frac{\varepsilon}{2}\leqslant F(x)+\varepsilon,$$

$$F_{mi}(x)\geqslant F_{mi}(r_i)\geqslant F(r_i)-\frac{\varepsilon}{2}$$

$$\geqslant F(x-h)-\frac{\varepsilon}{2}\geqslant F(x)-\varepsilon$$

whenever $n\geqslant N(\varepsilon)$.

Combining these inequalities together, we obtain

$$|F_{mi}(x)-F(x)|<\varepsilon. \tag{4.41}$$

So equality (4.38) holds true. F is by definition right continuous on the set of discontinuous points. The proof of Theorem 1 is complete.

3. The proof of the converse limit theorem

To prove Theorem 4 in Section 4.1, we need a property of the characteristic function as follows.

Lemma 1 *Let $f(t)$ be the characteristic function of the distribution function $F(x)$. Then for any $\lambda>0$,*

$$\int_{|x|\geqslant 2\lambda}\mathrm{d}F(x)\leqslant 2\lambda\int_0^{1/\lambda}[1-\mathrm{Re}(f(t))]\mathrm{d}t. \tag{4.42}$$

Proof By the definition of the characteristic function $f(t)=\int_{-\infty}^{\infty}\mathrm{e}^{itx}\mathrm{d}F(x)$, we have

$$\lambda\int_0^{1/\lambda}[1-\mathrm{Re}(f(t))]\mathrm{d}t=\lambda\int_0^{1/\lambda}\int_{-\infty}^{\infty}[1-\cos tx]\mathrm{d}F(x)\mathrm{d}t$$

$$=\int_{-\infty}^{\infty}\lambda\int_0^{1/\lambda}[1-\cos tx]\mathrm{d}t\mathrm{d}F(x)=\int_{-\infty}^{\infty}\left[1-\frac{\sin(x/\lambda)}{x/\lambda}\right]\mathrm{d}F(x)$$

$$\geqslant\int_{|x/\lambda|\geqslant 2}\left[1-\frac{\sin(x/\lambda)}{x/\lambda}\right]\mathrm{d}F(x)\geqslant\frac{1}{2}\int_{|x|\geqslant 2\lambda}\mathrm{d}F(x).$$

The lemma is proved.

Now, we begin the proof of the converse limit theorem. For any subsequence $\{F_{n'}\}$ of $\{F_n\}$, by Helly's first theorem, there is a further subsequence $\{F_{n''}\}\subset\{F_{n'}\}$ and a non-decreasing right-continuous function $F(x)$ such that

$$F_{n''}{}^{(x)}\to F(x)\text{ for any continuous point }x\text{ of }F. \tag{4.43}$$

Next, it is sufficient to show that F is a distribution function. In fact, if having shown that F is a distribution function, then the above convergence is just $F_{n''}\xrightarrow{w}F$. And then, $f_{n''}(t)\to f_F(t)$ by Levy's continuity theorem, where $f_F(t)$ is the characteristic function of the distribution function F. We must have $f_F(t)=f(t)$ since $f_{n''}(t)\to f(t)$ by the given condition. Hence, $f(t)$ is a characteristic function which uniquely determines the limit distribution function F. Therefore, for any subsequence $\{F_{n'}\}$ of $\{F_n\}$, there is a further subsequence which converges to a same distribution function F determined by $f(t)$. So, $F_n\xrightarrow{w}F$.

From (4.43), it is trivial that $0 \leqslant F(x) \leqslant 1$. To prove that $F(x)$ is a distribution function, it is sufficient to show that $\lim_{a\to\infty}(F(a)-F(-a))=1$. By (4.43), if $\pm a$ are continuous points of F, then

$$\int_{|x|\leqslant a} dF_{n''}(x) \to F(a) - F(-a) n \to \infty.$$

So, it is sufficient to show that $\lim_{a\to\infty}\lim_{n''\to\infty}\int_{|x|\geqslant a} dF_{n''}(x) = 0$, which is implied by

$$\lim_{\lambda\to\infty}\limsup_{n\to\infty}\int_{|x|\geqslant\lambda} dF_n(x) = 0. \tag{4.44}$$

$\{F_n\}$ is called uniformly tight if the above equation holds. Now, by (4.42), $f(0)=1$ and the continuity of $f(t)$ at $t=0$. We conclude that

$$\lim_{\lambda\to\infty}\limsup_{n\to\infty}\int_{|x|\geqslant 2\lambda} dF_n(x) \leqslant \lim_{\lambda\to\infty}\limsup_{n\to\infty} 2\lambda\int_0^{1/\lambda}[1-\mathrm{Re}(f(t))]dt$$

$$= \lim_{\lambda\to\infty} 2\lambda\int_0^{1/\lambda}[1-\mathrm{Re}(f(t))]dt \leqslant 2\lim_{\lambda\to\infty}\sup_{0\leqslant t\leqslant 1/\lambda}[1-\mathrm{Re}(f(t))] = 0.$$

The proof of the theorem is completed.

4. Taylor's expansion for a characteristic function

As a corollary to the results of Chapter 3, when the distribution function $F(x)$ has finite moment of order r, its characteristic function $f(t)$ has continuous derivative of order r (some positive integer). Hence we can expand $f(t)$ around $t=0$ using Taylor's expansion.

Theorem 16 *Suppose that the random variable ξ has finite moment of order r. Let α_1, α_2, $\cdots$, α_r be these moments. Then its characteristic function $f(t)$ can be expanded around $t=0$, with either of the two forms of its remainder*:

$$f(t) = \begin{cases} 1 + \sum_{k=1}^{r}\alpha_k\frac{(it)^k}{k!} + o(|t|)^r, & r\geqslant 1; \\ 1 + \sum_{k=1}^{r-1}\alpha_k\frac{(it)^k}{k!} + \beta_r\theta_r\frac{|t|^r}{r!}, & r>1, \end{cases} \tag{4.45}$$

where $\beta_r = E|\xi|^r$, $|\theta_r|\leqslant 1$.

5. The Monte Carlo method

Credit for inventing the Monte Carlo method often goes to Ulam, a Polish born mathematician who worked for von Neumann in the United States—Manhattan Project during World War II. Ulam is primarily known for designing the hydrogen bomb with E. Teller in 1951. He invented the Monte Carlo method in 1946 while pondering the probabilities of winning a card game of solitaire.

The Monte Carlo method, as it is understood today, encompasses any technique of statistical sampling employed to approximate solutions to quantitative problems. Ulam did not invent statistical sampling. This had been employed to solve quantitative problems before, with physical processes such as dice tosses or card draws being used to generate samples. Ulam's contribution was to recognize the potential for the newly invented electronic computer to automate such sampling. Working with von Neumann and Metropolis, he developed algorithms for computer implementations, as well as explored means of transforming non-random problems into random forms that would facilitate their solution via statistical sampling. This work transformed statistical sampling from a mathematical curiosity to a formal methodology applicable to a wide variety of problems. It was Metropolis who named the new methodology after the casinos of Monte Carlo.

To understand the Monte Carlo method theoretically, it is useful to think of it as a general technique of numerical integration. It can be shown, at least in a trivial sense, that every application of the Monte Carlo method can be represented as a definite integral. As one example, consider our needle dropping experiment. It is nothing more than an elaborate means of estimating the integral.

Suppose we need to evaluate a multi-dimensional definite integral of the form:

$$J=\int_0^1\cdots\int_0^1 f(u_1,\cdots,u_n)\mathrm{d}u_1\cdots\mathrm{d}u_n=\int_{(0,1)^n} f(\boldsymbol{u})\mathrm{d}\boldsymbol{u}. \tag{4.46}$$

Most integrals can be converted to this form with a suitable change of variables, so we can consider this to be a general application suitable for the Monte Carlo method.

The integral represents a non-random problem, but the Monte Carlo method approximates a solution by introducing a random vector $\boldsymbol{U}$ that is uniformly distributed on the region of integration. Applying the function f to $\boldsymbol{U}$, we obtain a random variable $f(\boldsymbol{U})$. This has expectation:

$$Ef(\boldsymbol{U})=\int_{(0,1)^n} f(\boldsymbol{u})p(\boldsymbol{u})\mathrm{d}\boldsymbol{u}, \tag{4.47}$$

where p is the probability density function of $\boldsymbol{U}$. Because p equals 1 on the region of integration, (4.47) becomes

$$Ef(\boldsymbol{U})=\int_{(0,1)^n} f(\boldsymbol{u})\mathrm{d}\boldsymbol{u}. \tag{4.48}$$

Comparing (4.46) and (4.48), we obtain a probabilistic expression for the integral J:

$$J=Ef(\boldsymbol{U}). \tag{4.49}$$

So random variable $f(\boldsymbol{U})$ has mean J and a standard deviation σ.

To estimate J with a standard error lower than σ, let us produce m samples $\boldsymbol{U}_1,\cdots,\boldsymbol{U}_m$. Applying the function f to each of these yields m independent and identically distributed random variables $f(\boldsymbol{U}_1),\cdots,f(\boldsymbol{U}_m)$, each with expectation J and standard deviation σ. If we have a realization $\boldsymbol{u}_1,\cdots,\boldsymbol{u}_m$ for our sample, we may estimate J by

$$\tilde{J}=\frac{1}{m}\sum_{k=1}^{m} f(\boldsymbol{u}_k).$$

For instance, find $\int_{-\infty}^{\infty}\mathrm{e}^{-x^2/2}\mathrm{d}x=\sqrt{2\pi}$ again.

6. Convergence in probability cannot imply the convergence with probability one

Here is an example.

Example 14 Let $\Omega = [0, 1]$, $\mathscr{F}$ be the σ-field of all the Borel sets in $[0, 1]$, and P be Lebesgue measure on $[0, 1]$. Define

$$\eta_n^i = \begin{cases} 1, & \omega \in [\frac{i-1}{n}, \frac{i}{n}]; \\ 0, & \text{otherwise} \end{cases} \tag{4.50}$$

where $i = 1,2,\cdots,n$, $n = 1,2,\cdots$. Think about the sequence of random variables $\{\eta_1^1, \eta_2^1, \eta_2^2, \eta_3^1, \eta_3^2, \eta_3^3, \cdots\}$ and rewrite it as $\{\xi_n, n \geqslant 1\}$. Note that for any $\xi > 0$

$$\max_{1 \leqslant i \leqslant n} P(|\eta_n^i| \geqslant \xi) \leqslant \frac{1}{n} \to 0,$$

i. e., $\xi_n \xrightarrow{P} 0$.

On the other hand, for any $\omega \in \Omega$, there are infinitely many 1 and infinitely many 0 in $\xi_n(\omega)$, $n = 1, 2, \cdots$. So $\xi_n(\omega)$ does not have a limit.

7. Moments and moment generating functions

Suppose that F and F_n are distribution functions and $F_n \xrightarrow{w} F$. The following convergence may be not true

$$m_{n,k} := \int_{-\infty}^{\infty} x^k \mathrm{d}F_n(x) \to \int_{-\infty}^{\infty} x^k \mathrm{d}F(x) =: m_k, \tag{4.51}$$

even when the k-th moments $m_{n,k}$ of F_n and m_k of F are finite. However, if the following condition is satisfied as

$$\lim_{\lambda \to \infty} \limsup_{n \to \infty} \int_{|x| \geqslant \lambda} |x|^k \mathrm{d}F_n(x) = 0, \tag{4.52}$$

then (4.51) is true. The condition (4.52) is referred as the uniform integrability of x^k with respect to $\{F_n\}$. In fact, Let a and b be the continuous points of F such that $a < 0 < b$. Then by Helly's second theorem,

$$\int_a^b x^k \mathrm{d}F_n(x) \to \int_a^b x^k \mathrm{d}F(x).$$

Hence

$$\limsup_{n \to \infty} \left| \int_{-\infty}^{\infty} x^k \mathrm{d}F_n(x) - \int_a^b x^k \mathrm{d}F(x) \right|$$
$$\leqslant \limsup_{n \to \infty} \int_{|x| \geqslant \min(|a|, b)} |x|^k \mathrm{d}F_n(x).$$

Letting $\min\{|a|, b\} \to \infty$ yields(4.51).

Conversely, $F_n \xrightarrow{w} F$ may fail even though (4.51) holds for any k. However, if (4.51) holds for any k and all the moments in the right hand of (4.51) uniquely determines a distribution function F (for example, the condition of Theorem 6 in the Supplements and Remarks of Chapter 3 is satised. It can be verified that such a condition is satisfied for the normal distribution), then $F_n \xrightarrow{w} F$. This is the moment method for showing the convergence in distribution.

Theorem 17 (Frechet-Shohat) *Suppose*

$$m_{n,k} = \int_{-\infty}^{\infty} x^k \mathrm{d}F_n(x) \to m_k, \quad k = 1,2,\cdots.$$

If there is a unique distribution function F such that

$$\int_{-\infty}^{\infty} x^k \mathrm{d}F(x) = m_k, \quad k = 1,2,\cdots,$$

then $F_n \xrightarrow{w} F$.

Proof Firstly, by the Chebyshev inequality we have

$$\int_{|x|\geqslant\lambda} \mathrm{d}F_n(x) \leqslant \frac{m_{n,2}}{\lambda^2}, \quad \int_{|x|\geqslant\lambda} |x|^k \mathrm{d}F_n(x) \leqslant \frac{m_{n,2k}}{\lambda^k}.$$

So, (4.44) and (4.52) hold, i.e., $\{F_n\}$ is uniformly tight, and the function x^k is uniformly integrable with respect to $\{F_n\}$.

Now, by Helly's first theorem, for any subsequence $\{F_{n'}\}$ of $\{F_n\}$, there is a further subsequence $\{F_{n''}\} \subset \{F_{n'}\}$ and a non-decreasing right continuous function $H(x)$ for which

$$F_{n''}(x) \to H(x) \text{ for any continuous point } x \text{ of } H.$$

By the tightness of $\{F_n\}$, H must be a distribution function. And so $\{F_{n''}\} \xrightarrow{w} H$, which together with the uniform integrability of x^k with respect to $\{F_n\}$, implies

$$\int_{-\infty}^{\infty} x^k \mathrm{d}F_{n''}(x) \to \int_{-\infty}^{\infty} x^k \mathrm{d}H(x).$$

Hence

$$\int_{-\infty}^{\infty} x^k \mathrm{d}H(x) = m_k.$$

Note that $\{m_k\}$ uniquely determines the distribution function F. So, $H \equiv F$. The proof of the theorem is completed.

As mentioned in the Supplements and Remarks of Chapter 3, the moment generating function $M_\xi(t) = E\mathrm{e}^{t\xi}$ of a random variable ξ is also an important tool for studying the properties of the random variable and its distribution function. Many properties of moment generating functions are similar to those of characteristic functions. For example, the moment generating function uniquely determines its distribution function. Namely, if $M_\xi(t) = M_\eta(t)$ for all t in an interval $(-t_0, t_0)$, then ξ and η have the same distribution. Also, if ξ and η are independent, then $M_{\xi+\eta}(t) = M_\xi(t)M_\eta(t)$. The following theorem shows that the convergence of the moment generating functions will determine the convergence in distribution.

Theorem 18 *Suppose that the random variables* ξ_n *and* ξ *have finite moment generating functions* $M\xi_n(t) = \mathrm{E}\mathrm{e}^{t\xi_n}$ *and* $M_\xi(t) = \mathrm{E}\mathrm{e}^{t\xi}$ *on an interval* $(-t_0, t_0)$, *and on this interval* $(-t_0, t_0)$,

$$M_{\xi_n}(t) \to M_\xi(t).$$

Then $\xi_n \xrightarrow{d} \xi$.

Proof Note that the moment generating function $M_\xi(t)$ uniquely determines the distribution function F. Similar to the proof of the Frechet-Shohat theorem, it is sufficient to show the uniform integrability of $\{e^{|t\xi_n|}\}$, which can follow from the following inequality

$$\int_{|\xi_n|\geqslant\lambda} e^{|t\xi_n|}\,dP \leqslant e^{-|t|\varepsilon\lambda}[Ee^{(1+\varepsilon)|t|\xi_n} + Ee^{-(1+\varepsilon)|t|\xi_n}]$$

$$\leqslant e^{-|t|\varepsilon\lambda}[M_{\xi_n} - ((1+\varepsilon)\mid t\mid) + M_{\xi_n}(-(1+\varepsilon)\mid t\mid)].$$

The moment method and the moment generating function method can be used to show the central limit theorem. Unfortunately, the moment generating function or moments of high orders are not always finite for all random variables. The importance of the characteristic function is, besides its many good properties, that it exists for any random variables. The characteristic function (moment generating function) is exactly the Fourier transformation (Laplace transform) of the distribution function which will transform the convolution of distribution functions to the scalar product of related characteristic functions (moment generating functions) so that it provides a powerful technique for dealing with independent random variables. However, when the random variables are not independent, it is much harder to compute characteristic functions and moment generating functions than moments, so the moment method is still an important method to show the convergence in distribution even today.

8. The proof of the Lindeberg-Feller central limit theorem

Here we only give the proof of the sufficency part of Theorem 7 in Section 4.1, i.e. if $\xi_1, \xi_2, \cdots$, are independent random variables with finite means and variances, and the following Lindeberg condition is satisfied

$$\frac{1}{B_n^2}\sum_{k=1}^{n}\int_{|x-E\xi_k|\geqslant\varepsilon B_n}(x-E\xi_k)^2\,dF_k(x) \to 0 \text{ for any } \varepsilon > 0, \tag{4.53}$$

then the Feller condition

$$\frac{\max_{k\leqslant n}\mathrm{Var}(\xi_k)}{B_n^2} \to 0 \tag{4.54}$$

is satisfied and

$$P\left(\frac{S_n - ES_n}{B_n} \leqslant x\right) \to \Phi(x) \tag{4.55}$$

for any x. Here, $S_n = \sum_{k=1}^{n}\xi_k$, $B_n^2 = \mathrm{Var}(S_n) = \sum_{k=1}^{n}\mathrm{Var}(\xi_k)$.

Without loss of generality, we assume $E\xi_k = 0$. Denote $\sigma_k^2 = \mathrm{Var}(\xi_k)$. Then $B_n^2 = \sum_{k=1}^{n}\sigma_k^2$. Firstly, by noting

$$\frac{\sigma_k^2}{B_n^2} \leqslant \varepsilon^2 + \frac{1}{B_n^2}\int_{|x|\geqslant \varepsilon B_n} x^2 \mathrm{d}F_k(x) \leqslant \varepsilon^2 + \frac{1}{B_n^2}\sum_{k=1}^n \int_{|x|\geqslant \varepsilon B_n} x^2 \mathrm{d}F_k(x),$$

the Feller condition follows from the Lindeberg condition.

Next, we show the central limit theorem. We apply a replacement technique instead of the characteristic function. Define a function $\psi(t)$ such that $\psi(t) = 1$ on $(-\infty, 0]$, $\psi(t) = 0$ on $[1, \infty)$, and

$$\psi(t) = \alpha^{-1}\int_t^1 \exp\left\{-\frac{1}{s(1-s)}\right\}\mathrm{d}s,\ \alpha = \int_0^1 \exp\left\{-\frac{1}{s(1-s)}\right\}\mathrm{d}s$$

on $[0,1]$. Then $0 \leqslant \psi(t) \leqslant 1$ and ψ have bounded continuous derivatives of all orders. Let η be a standard normal random variable, and x be given. For any $\varepsilon > 0$, choosing $g_\varepsilon(t) = \psi\left(\frac{t-x}{\varepsilon}\right)$ yields

$$\begin{aligned} P\left(\frac{S_n}{B_n} \leqslant x\right) - \Phi(x) &\leqslant Eg_\varepsilon\left(\frac{S_n}{B_n}\right) - Eg_\varepsilon(\eta) + \int_x^{x+\varepsilon} g_\varepsilon(y)\mathrm{d}\Phi(y) \\ &\leqslant Eg_\varepsilon\left(\frac{S_n}{B_n}\right) - Eg_\varepsilon(\eta) + \frac{\varepsilon}{\sqrt{2\pi}}. \end{aligned}$$

Similarly, choosing $g_\varepsilon(y) = \psi\left(\frac{y(x-\varepsilon)}{\varepsilon}\right)$ yields

$$P\left(\frac{S_n}{B_n} \leqslant x\right) - \Phi(x) \geqslant Eg_\varepsilon\left(\frac{S_n}{B_n}\right) - Eg_\varepsilon(\eta) - \frac{\varepsilon}{\sqrt{2\pi}}.$$

Hence, it is sufficient to prove that, for any bounded continuous function $g(x)$ which has bounded continuous derivatives of all orders, we have

$$Eg\left(\frac{S_n}{B_n}\right) \to Eg(\eta). \tag{4.56}$$

If ξ_k's are normal random variables, the above convergence is obvious. We will replace ξ_k in the sum S_n by normal random variable η_k progressively. Let

$$h(t) = \sup_x \left| g(x+t) - g(x) - g'(x)t - \frac{1}{2}g''(x)t^2 \right|.$$

Since the first three derivatives of $g(t)$ are bounded, there is a constant $K > 0$ for which

$$h(t) \leqslant K \min\{t^2, |t|^3\}.$$

Then

$$\left| g(x+t_1) - g(x+t_2) - \left[g'(x)(t_1 - t_2) + \frac{1}{2}g''(x)(t_1^2 - t_2^2)\right] \right| \leqslant h(t_1) + h(t_2). \tag{4.57}$$

Let $\{\eta_k\}$ be a sequence of independent normal random variables with $\eta_k \sim N(0, \sigma_k^2)$, and independent of ξ_k. Denote

$$\zeta_{nk} = \sum_{1\leqslant i<k} \xi_i + \sum_{k<i\leqslant n} \eta_i,\ 1 \leqslant k \leqslant n.$$

Then $\zeta_{nn} + \xi_n = S_n$, $\zeta_{n1} + \eta_1$ and $B_n\eta$ are identically distributed, and

$$\left| Eg\left(\frac{S_n}{B_n}\right) - Eg(\eta) \right| \leqslant \sum_{k=1}^n \left| E\left[g\left(\frac{\zeta_{nk} + \xi_k}{B_n}\right) - g\left(\frac{\zeta_{nk} + \eta_k}{B_n}\right)\right] \right|.$$

Since for each k, ζ_{nk}, ξ_k and η_k are independent, we have

$$E\left\{g'\left(\frac{\zeta_{nk}}{B_n}\right)(\xi_k - \eta_k)\right\} = E\left\{g'\left(\frac{\zeta_{nk}}{B_n}\right)\right\}(E\xi_k - E\eta_k) = 0,$$

$$E\left\{g''\left(\frac{\zeta_n k}{B_n}\right)(\xi_k^2 - \eta_k^2)\right\} = E\left\{g''\left(\frac{\zeta_n k}{B_n}\right)\right\}(E\xi_k^2 - E\eta_k^2) = E\left\{g''\left(\frac{\zeta_n k}{B_n}\right)\right\}(\sigma_k^2 - \sigma_k^2) = 0.$$

By applying (4.57), we obtain

$$\left| Eg\left(\frac{S_n}{B_n}\right) - Eg(\eta) \right| \leqslant \sum_{k=1}^{n}\left\{Eh\left(\frac{\xi_k}{B_n}\right) + Eh\left(\frac{\eta_k}{B_n}\right)\right\}.$$

Next, it is sufficient to show

$$\sum_{k=1}^{n} Eh\left(\frac{\eta_k}{B_n}\right) \to 0, \tag{4.58}$$

$$\sum_{k=1}^{n} Eh\left(\frac{\xi_k}{B_n}\right) \to 0. \tag{4.59}$$

For (4.58), by (4.54) and the fact that $h(t) \leqslant K \mid t \mid^3$ we have

$$\text{left hand of (4.58)} \leqslant K \frac{1}{B_n^3}\sum_{k=1}^{n} E \mid \eta_k \mid^3 \leqslant K \frac{1}{B_n^3}\sum_{k=1}^{n} \sigma_k^3 E \mid \eta \mid^3$$

$$\leqslant KE \mid \eta \mid^3 \frac{\max_{k\leqslant n}\sigma_k}{B_n^3}\sum_{k=1}^{n}\sigma_k^2 = KE \mid \eta \mid^3 \frac{\max_{k\leqslant n}\sigma_k}{B_n} \to 0.$$

For (4.59), by splitting the expectation to two integrals on $\{\mid \xi_k \mid < \varepsilon B_n\}$ and $\{\mid \xi_k \mid \geqslant \varepsilon B_n\}$, and respectively applying the inequalities $g(t) \leqslant K \mid t \mid^3$ and $g(t) \leqslant K \mid t \mid^2$, we obtain

$$\text{left hand of (4.59)} \leqslant K\varepsilon \sum_{k=1}^{n}\frac{E\xi_k^2}{B_n^2} + K\frac{1}{B_n^2}\sum_{k=1}^{n}\int_{|x|\geqslant \varepsilon B_n} x^2 \mathrm{d}F_k(x)$$

$$= K\varepsilon + K\frac{1}{B_n^2}\sum_{k=1}^{n}\int_{|x|\geqslant \varepsilon B_n} x^2 \mathrm{d}F_k(x).$$

By the Lindeberg condition, (4.59) follows. And so (4.56) is proved.

9. The brief history of the central limit theorem

The history of probability theory is actually a history of limit theory. An elementary question is, what do you get if you toss a coin 10,000 times? If the coin is fair, can you come to the conclusion that the frequency that the outcome will be heads is about 1/2? The conclusion seems simple and obvious, but to give a rigorous theoretical proof in theory is not an easy work. Jacob Bernoulli (1654—1705) is the first one, taking over 20 of his years, to develop a sufficiently rigorous mathematical proof which was published in his *Ars Conjectandi* (*The Art of Conjecturing*) in 1713, 8 years after his death, by his nephew Nicholas Bernoulli. He named this his "Golden Theorem" but it became generally known as "Bernoulli's Theorem", the first version of the law of large numbers (c. f. Theorem 10 in Section 4.2).

To further calculate the probability related to the number of success in the n-fold Bernoulli experiment and the values of binomial distribution, mathematicians have cudgeled their brains, because it involves the calculation of the factorial $n!$. When n is large, it is unimaginable to calculate its value in that time when there was no computer. French mathematician de Moivre (Abraham de Moivre, 1667 — 1754) found an approximation of $n!$ as $n! = cn^{n+1/2}e^{-n}$ in 1733. This formula is called Stirling's approximation (or Stirling's formula) named after Scotland mathematician James Stirling who obtained the constant $c = \sqrt{2\pi}$, though it was first stated by de Moivre. In 1730s—1800s, de Moivre and another French mathematician Laplace (de Laplace, 1749—1827) proved the central limit theorem of the binomial distribution $B(n,p)$ for special $p = 1/2$ and arbitrary p, successively. This is so called the de Moivre-Laplace central limit theorem (c. f. Theorem 5 in Section 4.1).

The central limit theorem for general independent random variables was established by Russian mathematician Chebyshev (Pafnuty Lvovich Chebyshev, 1821—1894) who introduced the concept of general random variables in 1866 and later in 1887, discovered a "central limit theorem".

Chebyshev's central limit theorem *Suppose that* $\xi_1, \xi_2, \cdots,$ *are independent random variables with*

$$|\xi_k| \leqslant C, \ k = 1, 2, \cdots,$$

where C is a constant. Then (4.55) *holds.*

It should be mentioned that, this theorem is not actually correct. The conclusion is true if $B_n \to \infty$, while it fails if $B_n \not\to \infty$. In fact, Chebyshev only showed that the moments of $\dfrac{S_n - ES_n}{B_n}$ of each order converged to corresponding moments of the standard distribution $N(0,1)$, but the convergence of the distribution was not proved. Though his proof is not completely rigorous, Chebyshev is the first one to consider the central limit theorem of general random variables, and his method is the original idea of the moment method as introduced in Subsection 8. To make Chebyshev's proof rigorous, Markov (Andrey Markov, 1856 — 1922), one of Chebyshev's students, proved the following central limit theorem in 1898.

Markov's central limit theorem *Suppose that* $\xi_1, \xi_2, \cdots,$ *are independent random variables with finite means and variances, and for all* $r \geqslant 3$,

$$\frac{1}{B_n^r}\sum_{k=1}^{n} E|\xi_k|^r \to 0. \tag{4.60}$$

Then (4.55) *holds.*

In Markov's theorem, ξ_n is assumed to have finite moments of any order. Later, Lyapunov (Aleksandr Lyapunov, 1857—1918), another student of Chebyshev, relaxed the condition in Markov's theorem a lot in 1901 by showing that (4.55) remains true if (4.60) is satifised for some $r > 2$. This is so called Lyapunov's central limit theorem, seeing Theorem 8 in Section 4.1. Lyapunov applied the characteristic function to prove his theorem, which opened a door for studying limit theorems.

Finnish mathematician Lindeberg (Jarl Waldemar Lindeberg, 1876 — 1932) published his first paper on the central limit theorem in 1920 which was similar to that obtained earlier by Lyapunov whose work he did not then know. But his convolution argument was quite different from Lyapunov's characteristic function approach. Two years later, Lindeberg used his method to obtain a very general central limit theorem, that is so called Lindeberg's central limit theorem, namely, (4.55) holding under the so-called Lindeberg condition (4.53). The importance of Lindeberg's central limit theorem is that it reveals the reason for the ubiquity of normal random variables. The Lindeberg condition is quite a flexible condition. It can be easily verified that the condition (4.60) in Lyapunov's central limit theorem implies the Lindeberg condition (4.53). In 1935, Croatian-American mathematician Feller (William Feller, 1906—1970) proved that if his Feller condition (4.54) was satisfied, then the Lindeberg condition (4.53) was also a necessary condition for the central limit theorem (4.55) to hold.

10. The convergence rate of the central limit theorem

Let $\{\xi_n\}$ be a sequence of independent and identically distributed random variables with mean a and variance σ^2, $S_n = \xi_1 + \cdots + \xi_n$. By the Lindeberg-Feller central limit theorem,

$$P\left(\frac{S_n - na}{\sqrt{n}\sigma} \leqslant x\right) \to \Phi(x) \text{ as } n \to \infty,$$

for any x. Since the normal distribution function $\Phi(x)$ is a non-decreasing continuous function, the above convergence holds uniformly in x, i. e.,

$$\sup_{-\infty<x<\infty} \left| P\left(\frac{S_n - na}{\sqrt{n}\sigma} \leqslant x\right) - \Phi(x) \right| \to 0 \text{ as } n \to \infty.$$

A natural question is what the rate is at which this convergence takes place. The convergence rate is of special importance when we consider the approximation of a distribuiton of an estimator or test statistic in statistics. A classical result on this problem is as follows. If $E|X_1|^3 < \infty$, then

$$\sup_{-\infty<x<\infty} \left| P\left(\frac{S_n - na}{\sqrt{n}\sigma} \leqslant x\right) - \Phi(x) \right| \leqslant \frac{A}{\sqrt{n}} E|X_1|^3,$$

where A is a universal constant. This result was independently discovered by two mathematicians, Andrew C. Berry (in 1941) and Carl-Gustav Esseen (1942), and so was named the Berry-Esseen theorem or Berry-Esseen inequality. The convergence order $n^{-\frac{1}{2}}$ is general unimprovable. The determination of the best constant A is still an open problem. Esseen (1956) showed that $A \geqslant (\sqrt{10}+3)/6\sqrt{2\pi} > 0.4097$, and it is known that $A < 0.4748$ (Shevtsova, 2011).

11. Interpretation in number theory language

Let x be a real number in the interval $[0, 1]$, and let

$$x = 0.a_1a_2a_3\cdots \tag{4.61}$$

be its decimal expansion (so that each a_j stands for one of the digits $0, 1, \cdots, 9$). This expansion is unique except for numbers of the form $a/10$, which can be written either by means of a terminating expansion (containing infinitely many zeros) or by means of an expansion containing infinitely many nines. For clarity we will not use the latter form.

The decimal expansions are connected with Bernoulli trials with $p = 1/10$, the digit 0 representing success and all other digits failure. If we replace in (4.61) all zeros by the letters S and all other digits by F, then (4.61) represents a possible outcome of an infinite sequence of Bernoulli trials with $p = 1/10$. Conversely, an arbitrary sequence of letters S and F can be obtained in the described manner from the expansion of certain numbers x. In this way every event in the sample space of Bernoulli trials is represented by a certain aggregate of numbers x. Accordingly, our probabilities will always coincide with the measure of the corresponding aggregate of points on the x-axis. We have thus a tool of translating all limit theorems for Bernoulli trials with $p = 1/10$ into theorems concerning decimal expansions. The phrase with probability one is equivalent to for almost all x or almost everywhere.

We have considered the random variable S_n which gives the number of successes in n trials. Here it is more convenient to emphasize the fact that S_n is a function of the sample point, and we write $S_n(x)$ for the number of zeros among the first n decimals of x. The ratio $S_n(x)/n$ is called the frequency of zeros among the first n decimals of x.

In the language of ordinary measure theory the weak law of large numbers asserts that $S_n(x)/n \to 1/10$ in measure, while the strong law states that $S_n(x)/n \to 1/10$ almost everywhere. It gives an answer to a problem treated in a series of papers initiated by Hausdorff (1913) and Hardy and Littlewood (1914).

Instead of the digit zero we may consider any other digit and can formulate the strong law of large numbers to the effect that the frequency of each of the ten digits tends to 1/10 for almost all x. A similar theorem holds if the base 10 of the decimal system is replaced by any other base. This fact is discovered by Borel (1909) and is usually expressed by saying that almost all numbers are normal.

12. Proof of Theorem 15 and its converse

Borel-Cantelli Lemma. (1) Assume that $A_n, n\geqslant 1$, is a sequence of events. If $\sum\limits_{n=1}^{\infty} P(A_n) < \infty$, then

$$\lim_{n\to\infty} P\left(\bigcup_{k=n}^{\infty} A_k\right) = 0.$$

(2) Assume $A_n, n\geqslant 1$, is a sequence of independent events. If $\sum_{n=1}^{\infty} P(A_n) = \infty$, then

$$\lim_{n\to\infty} P\left(\bigcup_{k=n}^{\infty} A_k\right) = 1.$$

Proof (1) It follows from the subadditivity that

$$P\left(\bigcup_{k=n}^{\infty} A_k\right) \leqslant \sum_{k=n}^{\infty} P(A_k).$$

In addition, $\sum_{n=1}^{\infty} P(A_n) < \infty$ implies $\sum_{k=n}^{\infty} P(A_k) \to 0$. So

$$\lim_{n\to\infty} P\left(\bigcup_{k=n}^{\infty} A_k\right) = 0.$$

(2) Since A_n are mutaully independent, for any N,

$$P\left(\bigcup_{k=n}^{N} A_k\right) = 1 - P\left(\bigcap_{k=n}^{N} \overline{A}_k\right)$$

$= 1 - \prod_{k=n}^{N} P(\overline{A}_k) = 1 - \prod_{k=n}^{N} (1 - P(A_k))$. Applying the elementary inequality $1 - x \leqslant e^{-x}$ to get

$$\prod_{k=n}^{N} (1 - P(A_k)) \leqslant e^{-\sum_{k=n}^{N} P(A_k)}.$$

By virtue of the fact $\sum_{n=1}^{\infty} P(A_n) = \infty$, it follows for any given n

$$\lim_{N\to\infty} \sum_{k=n}^{N} P(A_k) = \infty,$$

which in turn implies $\lim_{N\to\infty} \prod_{k=n}^{N} (1 - P(A_k)) = 0$. Hence

$$\lim_{n\to\infty} P\left(\bigcup_{k=n}^{\infty} A_k\right) = \lim_{n\to\infty} \lim_{N\to\infty} (1 - \prod_{k=n}^{N} (1 - P(A_k)) = 1.$$

The lemma is now proven.

Lemma 2 Let ξ be a random variable, and then the three following statements are equivalent:

(1) $E|\xi| < \infty$;

(2) $\sum_{n=0}^{\infty} P(|\xi| \geqslant n) < \infty$;

(3) $\sum_{n=1}^{\infty} nP(n-1 \leqslant |\xi| < n) < \infty$.

Proof First, it is easy to see that (1) and (3) are equivalent. Indeed,

$$\begin{aligned}\sum_{n=0}^{\infty} P(|\xi| \geqslant n) &= \sum_{n=0}^{\infty} \sum_{k=n}^{\infty} P(k \leqslant |\xi| < k+1) \\ &= \sum_{k=0}^{\infty} \sum_{n=0}^{k} P(k \leqslant |\xi| < k+1) \\ &= \sum_{k=0}^{\infty} (k+1) P(k \leqslant |\xi| < k+1).\end{aligned}$$

Next we shall prove (1) and (2) are equivalent. Let

$$1_{\{n-1 \leqslant |\xi| < n\}} = \begin{cases} 1, & n-1 \leqslant |\xi| < n, \\ 0, & \text{otherwise.} \end{cases}$$

Then

$$E|\xi| = \sum_{n=1}^{\infty} E|\xi| 1_{\{n-1\leqslant|\xi|<n\}}$$
$$\leqslant \sum_{n=1}^{\infty} nE1_{\{n-1\leqslant|\xi|<n\}}$$
$$= \sum_{n=1}^{\infty} nP(n-1\leqslant|\xi|<n).$$

So, (3) implies (1). Conversely,

$$E|\xi| \geqslant \sum_{n=1}^{\infty}(n-1)E1_{\{n-1\leqslant|\xi|<n\}}$$
$$= \sum_{n=1}^{\infty} nP(n-1\leqslant|\xi|<n) - 1.$$

Hence (1) implies (3).

Proof of Theorem 15. Let us first prove (4.35). For any real number x, put $x^+ = \max(x,0)$, $x^- = \max(-x,0)$, then $x = x^+ - x^-$. Applying the above equation to each ξ_k to yield $\xi_k = \xi_k^+ - \xi_k^-$. Observe that $S_n = \sum_{k=1}^n \xi_k^+ - \sum_{k=1}^n \xi_k^-$ and $\mu = E\xi_k^+ - E\xi_k^-$. It is sufficient to show

$$\frac{1}{n}\sum_{k=1}^{n}\xi_k^+ \to E\xi_1^+,\ \frac{1}{n}\sum_{k=1}^{n}\xi_k^- \to E\xi_1^-,\quad \text{a. s.}$$

Without loss of generality, assume $\xi \geqslant 0$. Thus S_n is nondecreasing in n. For any $\theta>1$, define $k_m=[\theta^m]$, $m\geqslant 0$, where $[x]$ denotes the greatest integer which is not larger than x. When $k_m \leqslant n < k_{m+1}$,

$$\frac{S_{k_m}}{k_{m+1}} \leqslant \frac{S_n}{n} \leqslant \frac{S_{k_{m+1}}}{k_m}.$$

When $m\to\infty$, $k_m\to\infty$ and $k_{m+1}/k_m\to\theta$. If

$$\frac{S_{k_m}}{k_m}\to\mu \quad \text{a. s.} \tag{4.62}$$

it follows

$$\frac{\mu}{\theta} \leqslant \liminf_{n\to\infty}\frac{S_n}{n} \leqslant \limsup_{n\to\infty}\frac{S_n}{n} \leqslant \mu\theta \quad \text{a. s.}$$

Since θ is arbitrary, letting $\theta\to 1$ yields (4.35) of Theorem 15.

To prove (4.62), define for $m\geqslant 1$ and $1\leqslant k\leqslant k_m$,

$$\bar{\xi}_{k,m} = \begin{cases} \xi_k, & \xi_k \leqslant k_m, \\ 0, & \xi_k > k_m, \end{cases}$$

and set $\bar{S}_{k_m} = \sum_{k=1}^{k_m} \bar{\xi}_{k,m}$. $\bar{\xi}_{k,m}$ is often called the truncated random variable of ξ_k. Its moment of any order is finite. Now decomposing S_{k_m} is to get

$$\frac{S_{k_m}}{k_m} = \frac{\bar{S}_{k_m} - E\bar{S}_{k_m}}{k_m} + \frac{S_{k_m} - \bar{S}_{k_m}}{k_m} + \frac{E\bar{S}_{k_m}}{k_m}. \tag{4.63}$$

We shall prove each summand of the right hand side of (4.63) tends to 0 a. s. as $m\to\infty$. First, it is easy to see

$$\frac{E\overline{S}_{k_m}}{k_m}=E\overline{\xi}_{1,m}\to\mu. \tag{4.64}$$

Also, $S_{k_m}=\overline{S}_{k_m}$ unless $\xi_k>k_m$ for some $1\leqslant k\leqslant k_m$. Thus for $\varepsilon>0$,

$$P\left(\left|\frac{S_{k_m}-\overline{S}_{k_m}}{k_m}\right|>\varepsilon\right)\leqslant P\left(\bigcup_{k=1}^{k_m}\{\xi_k>k_m\}\right)$$

$$\leqslant\sum_{k=1}^{k_m}P(\xi_k>k_m)=k_mP(\xi_1>k_m).$$

Moreover,

$$\sum_{m=1}^{\infty}k_mP(\xi_1>k_m)=\sum_{m=1}^{\infty}k_m\sum_{l=m}^{\infty}P(k_l<\xi_1\leqslant k_{l+1})$$

$$=\sum_{l=1}^{\infty}\sum_{m=1}^{l}k_mP(k_l<\xi_1\leqslant k_{l+1}).$$

It follows easily from $k_l=[\theta^l]$ that $\sum_{m=0}^{l}k_mk_l\leqslant Ck_l$ where $C>0$ is a constant. Thus

$$\sum_{m=1}^{\infty}k_mP(\xi_1>k_m)\leqslant C\sum_{l=1}^{\infty}k_lP(k_l<\xi_1\leqslant k_{l+1})$$

$$\leqslant CE\xi_1<\infty,$$

where the last inequality follows from Lemma 2. Therefore

$$\sum_{m=1}^{\infty}P\left(\left|\frac{S_{k_m}-\overline{S}_{k_m}}{k_m}\right|>\varepsilon\right)<\infty.$$

We have by the Borel-Cantelli lemma and corollary of Theorem 13

$$\frac{S_{k_m}-\overline{S}_{k_m}}{k_m}\to 0\quad\text{a. s.} \tag{4.65}$$

Finally, for $\varepsilon>0$, it follows from Chebyshev's inequality

$$P\left(\left|\frac{\overline{S}_{k_m}-E\overline{S}_{k_m}}{k_m}\right|>\varepsilon\right)\leqslant\frac{\operatorname{Var}(\overline{\xi}_{1,m})}{k_m\varepsilon^2}$$

$$\leqslant\frac{E(\overline{\xi}_{1,m})^2}{k_m\varepsilon^2}.$$

By the definition of $\overline{\xi}_{1,m}$, we have

$$E(\overline{\xi}_{1,m})^2\leqslant\sum_{l=1}^{m}k_l^2P(k_{l-1}\leqslant\xi_1\leqslant k_l).$$

Exchanging the order of summation and using the definition of k_l, we have

$$\sum_{m=1}^{\infty}\frac{E(\overline{\xi}_{1,m})^2}{k_m}\leqslant\sum_{m=1}^{\infty}\frac{1}{k_m}\sum_{l=1}^{m}k_l^2P(k_{l-1}\leqslant\xi_1\leqslant k_l)$$

$$=\sum_{l=1}^{\infty}\left(\sum_{m=l}^{\infty}\frac{1}{k_m}\right)k_l^2P(k_{l-1}\leqslant\xi_1\leqslant k_l)$$

$$\leqslant C\sum_{l=1}^{\infty} k_{l-1} P(k_{l-1} \leqslant \xi_1 \leqslant k_l)$$
$$\leqslant CE\xi_1 < \infty.$$

Thus

$$\sum_{m=1}^{\infty} P\left(\left|\frac{\overline{S}_{k_m} - E\overline{S}_{k_m}}{k_m}\right| > \varepsilon\right) < \infty.$$

Again by the Borel-Cantelli lemma and corollary of Theorem 13,

$$\frac{\overline{S}_{k_m} - E\overline{S}_{k_m}}{k_m} \to 0 \quad \text{a. s.} \tag{4.66}$$

Combining (4.63), (4.64), (4.65) and (4.66) gives (4.62), and so does (4.35).

Next we shall prove the converse holds true. The key is to show $E|\xi_1| < \infty$. Since (4.35) is true,

$$\frac{\xi_n}{n} = \frac{S_n - S_{n-1}}{n} = \frac{S_n}{n} - \frac{S_{n-1}}{n-1} \cdot \frac{n-1}{n} \to 0 \quad \text{a. s.}$$

We have by (1) of Theorem 13,

$$\lim_{n\to\infty} P\left(\bigcup_{k=n}^{\infty}\left\{\frac{|\xi_k|}{k} > 1\right\}\right) = 0.$$

$\xi_n, n \geqslant 1$, are independent, so are the events $\left\{\frac{|\xi_k|}{k} > 1\right\}, k \geqslant 1$. It follows from the Borel-Cantelli lemma that

$$\sum_{n=1}^{\infty} P\left(\frac{|\xi_n|}{n} > 1\right) < \infty.$$

Moreover, since ξ_n are identically distributed random variables,

$$\sum_{n=1}^{\infty} P(|\xi_1| > n) < \infty.$$

Thus it follows from Lemma 2 that $E|\xi_1| < \infty$. At last we show $E\xi_1 = \mu$. By (4.35)

$$\frac{S_n}{n} \to E\xi_1 \quad \text{a. s.}$$

Therefore it follows from uniqueness that $E\xi_1 = \mu$.

13. Simulate the central limit theorem

Example 15 Suppose $X_1, \cdots, X_n$ are independent identically distributed random variables, and $X_i \sim U(0,1)$. Steps:

(1) Generate n random numbers $x_1, \cdots, x_n$ of $U(0,1)$ (see Supplements and Remarks 13 of Chapter 2).

(2) Compute $s_n = \sum_{i=1}^{n} x_i$ and $z_n = \frac{s_n - n/2}{\sqrt{n/12}}$.

(3) Repeat Steps 1 and 2 N times to obtain N observation values $z_n^1, \cdots, z_n^N$ of z_n.

(4) Draw a histogram for $z_n^1, \cdots, z_n^N$ and a fitting curve of standard normal $N(0,1)$ density.

These steps can be realized in Matlab as follows:

(1) Open the Edit window in Matlab and input the following programs. See Figure 4-2.

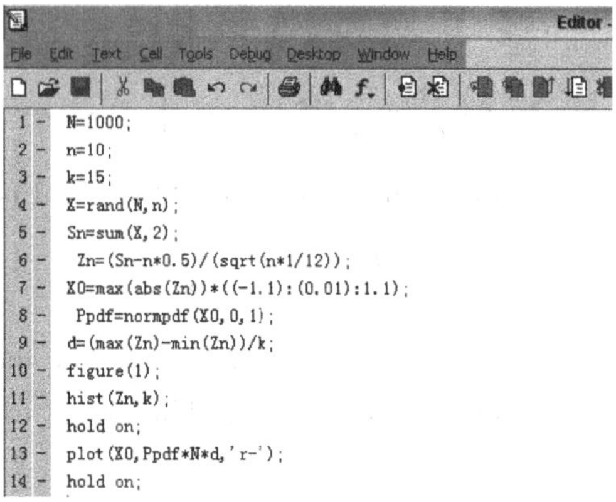


```
N=1000;
n=10;
k=15;
X=rand(N,n);
Sn=sum(X,2);
 Zn=(Sn-n*0.5)/(sqrt(n*1/12));
X0=max(abs(Zn))*((-1.1):(0.01):1.1);
 Ppdf=normpdf(X0,0,1);
d=(max(Zn)-min(Zn))/k;
figure(1);
hist(Zn,k);
hold on;
plot(X0,Ppdf*N*d,'r-');
hold on;
```

Figure 4-2

Note: k stands for the number of intervals in the histogram.

(2) Save the above program to "work" fold of the Matlab system, and name it as, say, b1.

(3) Run this program in Command window, and then output the graph as follows. See Figure 4-3.

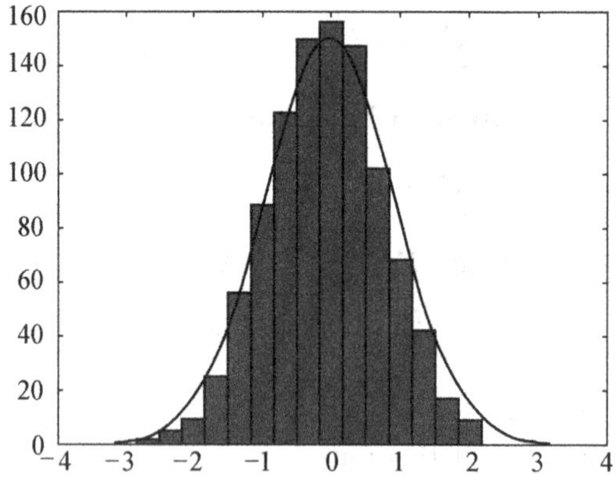

Figure 4-3

One can do simlarly for varying n, N and k. For instance,

$n=5$, $N=1000$, $k=15$ (see Figure 4-4);

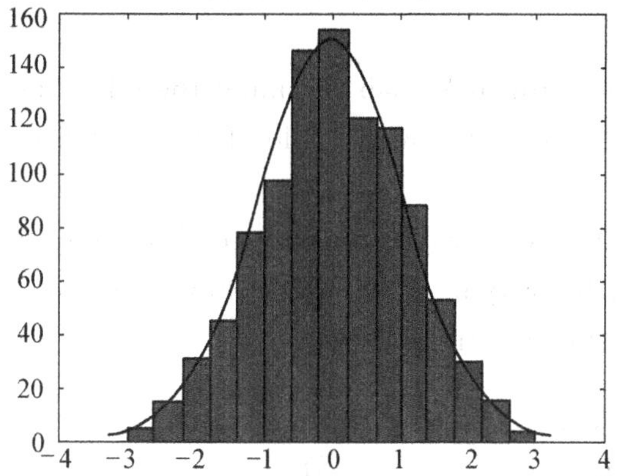

Figure 4-4

$n=25$, $N=1000$, $k=20$ (see Figure 4-5).

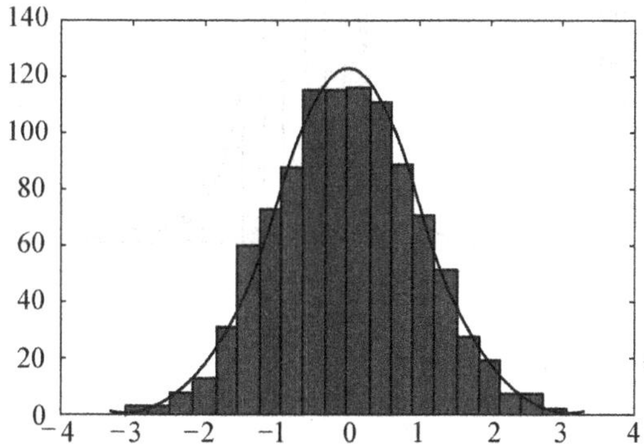

Figure 4-5

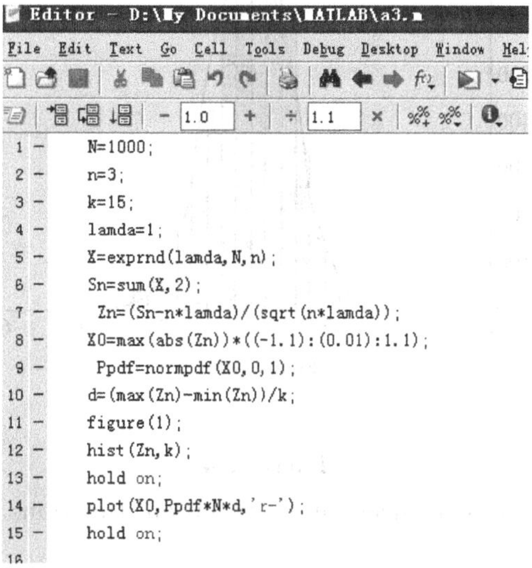

Figure 4-6

Example 16 Suppose $X_1, \cdots, X_n$ are independent identically distributed random variables, and $X_i \sim \exp(1)$.

(1) Open the Edit window in Matlab and input the following programs.

(2) Save the above program to "work" fold of the Matlab system, and name it as, say, b2.

(3) Run this progran in Command window, and then output the graph as follows.

One can do simlarly for varying n, N and k. For instance,

$n=8$, $N=5000$, $k=20$ (see Figure 4-7);

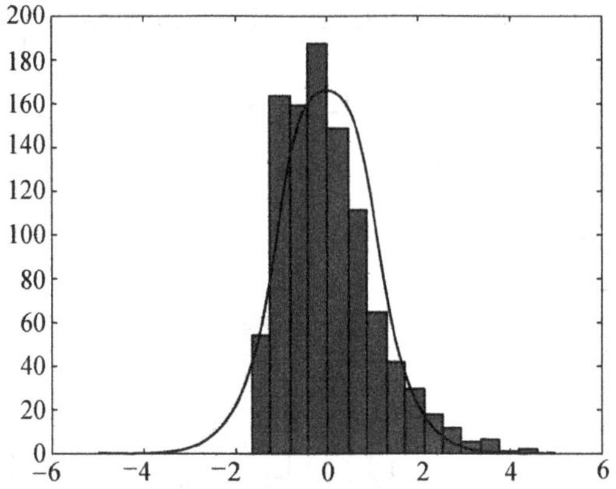

Figure 4-7

$n=15$, $N=5000$, $k=20$ (see Figure 4-8).

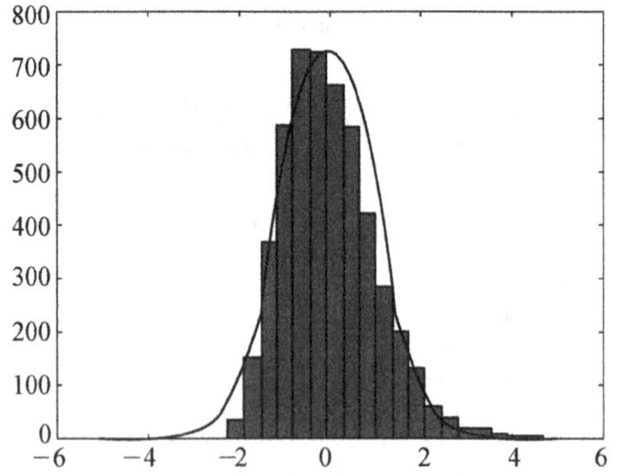

Figure 4-8

Exercise 4

1. Do the following sequences of distribution functions converge weakly to a distribution function?

(1) $F_n(x) = 0$ for $x < 1/n$; $F_n(x) = 1$ for $x \geqslant 1/n$.

(2)

$$F_n(x)=\begin{cases}0, & x<-n;\\ \dfrac{x+n}{2n}, & -n\leqslant x<n;\\ 1, & x\geqslant n.\end{cases}$$

2. Suppose that the distribution law of ξ_n is

$$P(\xi_n=0)=1-\frac{1}{n},\ P(\xi_n=n)=\frac{1}{n},\ n=1,2,\cdots.$$

Prove that the corresponding sequence of distribution functions converges to a distribution function, but $E\xi_n$ does not converge to the corresponding expectation.

3. Suppose that $\{\xi_n, n\geqslant 1\}$ is a sequence of independent and identically distributed random variables with a common distribution law

$$\begin{pmatrix}-1 & 1\\ 0.5 & 0.5\end{pmatrix}.$$

Let $\eta_n=\sum\limits_{k=1}^{n}\xi_k/2^k$. Prove that the distribution of η_n converges to the uniform distribution in $[-1, 1]$.

4. A computer system has 120 terminals.

(1) Every terminal is used for 5% of time, and all terminals are used independently. Find the probability that at least 10 terminals are used simultaneously.

(2) What about if every terminal is used for 20% of time?

5. Consider a certain amount of seeds. 1/6 of them are good. Take randomly 6 000 pieces out of them. How high is the probability that the difference between the proportion of good seeds in 6 000 pieces and 1/6 is smaller than 1%?

6. There are 200 lathes in a plant. Every lathe runs for 60% of work time, and consume 1kW per hour. Ask how much electric power should be provided to the plant so as to ensure normal operation with probability 0.999?

7. In an insurance company, 10 000 people buy an accident insurance. Everyone pays £ 12 as premium each year. The probability that this kind of accident happens in a year is 0.006. Each person claims £1 000 from company when the accident happen.

(1) For this kind of insurance, what is the probability that the insurance company is to be at a loss?

(2) For this kind of insurance, what is the probability that the insurance company earns profits no less than £6 000?

8. A fire insurance company undertakes to provide insurance for 160 houses, and the maximum payment is different from each other (see Table 4-2).

Table 4-2

Maximum payments (ten thousand yuan)	Numbers of houses insured
10	80
20	35
30	25
50	15
100	5

Suppose that

(1) the probability that every house requests claim in one year is 0. 04, but the probability that every house requests two claims or more is 0;

(2) every house fires independently;

(3) when the claim occurs, the claim amount obeys a uniform distribution between 0 to the maximum payment.

Let N be the number of times of claims in one year, and S the total amounts of claims.

(a) Find expectation and variance of N.

(b) Find expectation and variance of S.

(c) Determine the relative additive coeffcient θ, where θ=(the premium income of every policy − the average claim amount)/(the average claim amount), so that the probability that the premium income amount is larger than the total amount of claims is equal to 0. 99.

9. An insurance company offers five kinds of life insurances. The payment amount b_k of every kind of insurances (in case that the insured dies) and number n_k of the insured are shown as Table 4-3 below.

Table 4-3

Type k	Payments b_k (ten thousand yuan)	Numbers n_k of the insured
1	1	8 000
2	2	3 500
3	3	2 500
4	5	1 500
5	10	500

Assume that deaths of the insured are independent and the probability is 0. 02. The insurance company will look for reinsurance for every insurant for the sake of the company safety. The policy is as follows: Decide a retention, say, 20 thousand yuan; if someone's claim amount is below 20 thousand yuan, and then this insurance company will compensate it; if the claim amount exceeds 20 thousand yuan, then the extra part will be compensated by the reinsurance company. The reinsurance rate is 2.5% of total benefit. This insurance company (relative to the reinsurance company, it is also called ceding company) hopes its total expense (i. e. real claim amount S plus reinsurance premium) does not exceed 8 250 thousand yuan. What is the probability that real expenses exceeds the upper bound?

10. Let $\{\xi_n\}$ be independent and identically distributed, and the distribution is either of the following two:

(1) uniform distribution in $[-a, a]$;

(2) Poisson distribution.

Let $\eta_n = \sum_{k=1}^{n}(\xi_k - E\xi_k)/\sqrt{\sum_{k=1}^{n}\operatorname{Var}\xi_k}$. Find the characteristic function of η_n and its limit when $n \to +\infty$. Prove that the Lindeberg-Lévy theorem holds true in these cases.

11. Use the de Moivre-Laplace theorem to prove: if $0 < p < 1$, then no matter how large the constant k is, we have

$$P(|\mu_n - np| < k) \to 0, \quad n \to \infty.$$

12. Prove the standardized random variables with the Poisson distribution approximate to the standard normal random variable in distribution when parameter $\lambda \to \infty$.

13. Prove

$$e^{-n}\sum_{k=1}^{n}\frac{n^k}{k!} \to \frac{1}{2}, \qquad n \to \infty.$$

14. Assume that $\{\xi_n, n \geqslant 1\}$ and $\{\eta_n, n \geqslant 1\}$ are sequences of independent and identically distributed random variables with $E\xi_n = 0$, $\operatorname{Var}\xi_n = 1$ and $P(\eta_n = \pm 1) = 1/2$. Prove the distribution function of $S_n = \sum_{k=1}^{n}\xi_k\eta_k/\sqrt{n}$ converges weakly to $N(0, 1)$.

15. Let ξ and η be two independent and identically distributed random variables with finite variances σ^2. Suppose that $\alpha\xi + \beta\eta$ and ξ are identically distributed for any real numbers α and β with $\alpha^2 + \beta^2 = 1$. Prove that $\xi \sim N(0, \sigma^2)$.

16. Let $\{\xi_n, n \geqslant 1\}$ be a sequence of independent random variables with a common uniform distribution in $(0, \pi)$. Put $\eta_n = A_n \cos \xi_n$, where $A_n > 0$ and satisfies

$$\frac{\sum_{k=1}^{n} A_k^3}{\left(\sum_{k=1}^{n} A_k^2\right)^{3/2}} \to 0, \qquad n \to \infty.$$

Prove that $\{\eta_n, n \geqslant 1\}$ obeys the central limit theorem.

17. Assume that ξ_n obeys the Cauchy distribution with density function

$$p_n(x) = \frac{n}{\pi(1 + n^2x^2)}, \qquad n \geqslant 1.$$

Prove that $\xi_n \xrightarrow{P} 0$.

18. Assume that $\{\xi_n\}$ is a sequence of independent and identically distributed random variables with a common density function

$$p(x) = \begin{cases} e^{-(x-a)}, & x > a; \\ 0, & x \leqslant a. \end{cases}$$

Let $\eta_n = \min\{\xi_1, \xi_2, \cdots, \xi_n\}$. Prove that $\eta_n \xrightarrow{P} a$.

19. Prove that

(1) if $\xi_n \xrightarrow{P} \xi$, $\eta_n \xrightarrow{P} \eta$, then $\xi_n \pm \eta_n \xrightarrow{P} \xi \pm \eta$;

(2) if $\xi_n \xrightarrow{P} \xi$, $\eta_n \xrightarrow{P} \eta$, then $\xi_n \eta_n \xrightarrow{P} \xi\eta$;

(3) if $\xi_n \xrightarrow{P} \xi$, $\eta_n \xrightarrow{P} c$, where c is a constant, and neither η_n nor c is 0, and then $\xi_n/\eta_n \xrightarrow{P} \xi/c$;

(4) if $\xi_n \xrightarrow{d} \xi$, $\eta_n \xrightarrow{P} c$, where c is a constant, and then $\xi_n + \eta_n \xrightarrow{d} \xi + \eta$, $\xi_n/\eta_n \xrightarrow{d} \xi/c$ $(c \neq 0)$.

20. Prove that the following sequences of independent random variables obey the law of large numbers.

(1) $P(\xi_k = \sqrt{\ln k}) = P(\xi_k = -\sqrt{\ln k}) = \dfrac{1}{2}$;

(2) $P(\xi_k = 2^k) = P(\xi_k = -2^k) = 2^{-(2k+1)}$, $P(\xi_k = 0) = 1 - 2^{-2k}$;

(3) $P(\xi_k = 2^n/n^2) = 1/2^n$, $n = 1, 2, \cdots$;

(4) $P(\xi_k = n) = c/(n^2 \ln^2 n)$, $n = 2, 3, \cdots$, c is a constant.

21. Let ξ_1, ξ_2, $\cdots$, be independent and identically distributed random variables with finite variances, and $S_n = \xi_1 + \cdots + \xi_n$. Suppose that

$$\frac{S_n}{\sqrt{n}} \xrightarrow{P} \eta.$$

Prove that $P(\eta = 0) = 1$ and $P(\xi_i = 0) = 1$.

22. Suppose that $\{\xi_k\}$ obeys a common distribution with $\mathrm{Var}\xi_k < \infty$. Assume that ξ_k and ξ_{k+1} are correlated, $k = 1, 2, \cdots$, while ξ_k and ξ_l are independent as $|k - l| \geqslant 2$. Prove that the law of large numbers holds true.

23. (Bernstein's theorem) Assume that variances of $\{\xi_k\}$ are bounded: $\mathrm{Var}\xi_k \leqslant c$, $k = 1, 2, \cdots$, and $\mathrm{Cov}(\xi_i, \xi_j) \to 0$, when $|i - j| \to \infty$. Prove that $\{\xi_k\}$ obeys the law of large numbers.

24. In Bernoulli trials, the probability that the event A appears is p. Let

$$\xi_k = \begin{cases} 1, & \text{if } A \text{ appears in both } k\text{-th and } (k+1)\text{-th trial}; \\ 0, & \text{otherwise}. \end{cases}$$

Prove that $\{\xi_k \geqslant 1\}$ obeys the law of large numbers.

25. Assume that $\{\xi_k, k \geqslant 1\}$ is a sequence of independent and identically distributed random variables with a common uniform distribution in $[0, 1]$. Let $\eta_n = \left(\prod_{k=1}^{n} \xi_k\right)^{1/n}$,

prove that $\eta_n \xrightarrow{P} c$ (constant) and find c.

26. Assume that $\{\xi_k, k \geqslant 1\}$ is a sequence of independent and identically distributed random variables with $E\xi_k = a$, $\mathrm{Var}\xi_k < \infty$, prove that

$$\frac{2}{n(n+1)} \sum_{k=1}^{n} k\xi_k \xrightarrow{P} a.$$

27. Assume that $\{\xi_k, k \geqslant 1\}$ is a sequence of independent and identically distributed random variables with a common $N(0,1)$ distribution. Let $\eta_n = n\xi_{n+1} / \sum_{k=1}^{n} \xi_k^2$. Prove that the distribution function of $\eta_n \xrightarrow{w} N(0, 1)$.

28. Assume that $\{\xi_k, k \geqslant 1\}$ is a sequence of independent and identically distributed random variables with $\mathrm{Var}\xi_k < \infty$, and that $\sum_{n=1}^{\infty} a_n$ is an absolutely convergent series. Let $\eta_n = \sum_{k=1}^{n} \xi_k$. Then $\{a_n \eta_n\}$ obeys the law of large numbers.

29. Assume that $\{\xi_k, k \geqslant 1\}$ is a sequence of independent and identically distributed random variables with expectation 0 and variance 1 and $\{a_n\}$ is a sequence of constants with $a_n \to \infty$. Prove

$$\sum_{k=1}^{n} \frac{\xi_k}{\sqrt{n}\, a_n} \xrightarrow{P} 0.$$

30. Assume that $\{\xi_k, k \geqslant 1\}$, $\{\eta_k, k \geqslant 1\}$ are independent of each other, and both of them follow $N(0, 1)$ distribution. Let $\{a_n, n \geqslant 1\}$ be a sequence of numbers. Prove that

$$\frac{1}{n} \sum_{k=1}^{n} a_k \xi_k + \sum_{k=1}^{n} \eta_k \xrightarrow{P} 0 \text{ if and only if } \frac{1}{n^2} \sum_{k=1}^{n} a_k^2 \to 0.$$

31. Assume that $\{\xi_k, k \geqslant 1\}$ is a sequence of independent and identically distributed random variables with a common $N(0, 1)$ distribution. Prove that

$$\eta_n = \frac{\xi_1 + \xi_2 + \cdots + \xi_n}{\sqrt{\xi_1^2 + \xi_2^2 + \cdots + \xi_n^2}} \xrightarrow{d} N(0, 1).$$

32. Assume that $\{\xi_k\}$ is a sequence of independent identically distributed random variables with a common uniform distribution on $[-1, 1]$. Prove that

(1) $\{\xi_n^2\}$ obeys the law of large numbers;

(2) $U_n = \sum_{k=1}^{n} \xi_k \Big/ \sqrt{\sum_{k=1}^{n} \xi_k^2} \xrightarrow{d} N(0, 1)$.

33. Assume that $\{\xi_k, k \geqslant 1\}$ is a sequence of independent and identically distributed random variables and the central limit theorem holds. Then $\{\xi_k, k \geqslant 1\}$ obeys the law of large numbers if and only if $\mathrm{Var}(\sum_{k=1}^{n} \xi_k) = o(n^2)$.

34. Take $\Omega = [0, 1]$, and let $\mathscr{F}$ be the σ-field generated by all the Borel sets in $[0, 1]$. For any event $A = \{\omega \in (a, b)\}$ where $(a, b) \subset \Omega$, define $P(A) = b - a$. Now define

$\xi(\omega) \equiv 0$ and

$$\xi_n(\omega) = \begin{cases} n^{1/r}, & 0 < \omega \leqslant \dfrac{1}{n}; \\ 0, & \dfrac{1}{n} < \omega \leqslant 1. \end{cases}$$

Prove that $\xi_n \xrightarrow{P} \xi$, but $\xi_n \xrightarrow{L_r} \xi$ does not hold.

35. Let ξ_1, ξ_2, …, be a sequence of random variables. Suppose that

$$\xi_n - \xi_m \xrightarrow{P} 0 \text{ as } n, m \to \infty.$$

Prove that there exists a random variable ξ such that

$$\xi_n \xrightarrow{P} \xi \text{ as } n \to \infty.$$

36. Let ξ_1, ξ_2, …, be a sequence of random variables with $E|\xi_n|^r < \infty$, $n = 1, 2, \cdots$, where $r > 0$. Suppose that

$$\xi_n - \xi_m \xrightarrow{Lr} 0 \text{ as } n, m \to \infty.$$

Prove that there exists a random variable ξ with $E|\xi|^r < \infty$ such that

$$\xi_n \xrightarrow{Lr} \xi \text{ as } n \to \infty.$$

Bibliography

CHEN X R. Probability Theory and Mathematical Statistics [M]. Beijing: Science Press, 1996.

CHOW Y S, TEICHER H. Probability: Independence, Interchangeability, Martingales [M]. 2nd ed.

New York: Springer-Verlag, 1988.

CHUNG K L. Elementary Probability Theory with Stochastic Processes (Undergraduate Text in Mathematics) [M]. 3rd ed. New York: Springer-Verlag, 1979.

FELLER W. An Introduction to Probability Theory and Its Applications (Vol. 1) [M]. 3rd ed. New York: John Wiley and Sons, Inc. , 1968.

FELLER W. An Introduction to Probability Theory and Its Applications (Vol. 2) [M]. New York: John Wiley and Sons, Inc. , 1971.

GNEDENKO B V. The Theory of Probability[M]. Moscow: Mir Publishers, 1976.

GRIMMETT G R, STIRZAKER D R. Probability and Random Processes[M]. 3rd ed. Oxford: Oxford University Press, 2001.

LI X. Foundations of Probability Theory[M]. Beijing: Higher Education Press,1997.

LIN Z Y, LU C R, SU Z G. Foundations of Probability Limit Theory[M]. 2nd ed. Beijing: Higher Education Press, 2015.

LIN Z Y, SU Z G, ZHANG L X. Probability Theory[M]. 3rd ed. Hangzhou: Zhejiang University Press, 2016.

MAO S S, CHENG Y M, PU X L. A Course in Probability Theory and Mathematical Statistics[M]. Beijing: Higher Education Press, 2004.

QIAN M P, YE J. Random Mathematics[M]. Beijing: Higher Education Press, 2002.

ROSS S M. A First Course in Probability[M]. 7th ed. London: Pearson Education, Inc. ,2006.

SHIRYAYEV A N. Probability (Graduate Texts in Mathematics) [M]. New York: Springer-Verlag, 1984.

SU C. Probability Theory[M]. Beijing: Science Press, 2004.

WANG Z K. Probability Theory and Its Applications [M]. Beijing: Science Press, 1976.

Appendix A

Distribution of Typical Random Variables

In the following table, No. 1—13 are 1-dimensional continuous distributions, and No. 20—21 are k-dimensional continuous distributions, where the distribution density is with respect to Lebesgue measure. No. 14—19 are 1-dimensional discrete distributions, and No. 22—24 are k-dimensional discrete distributions, where the density function $P(x)$ means the probability mass at the point x.

The table contains the characteristic function, mean, and variance.

A. 1
Distribution of Typical Random Variables

No.	Name	Symbol	Density or Probability Function	Domains
1	Normal	$N(\mu,\sigma^2)$	$\frac{1}{(2\pi\sigma^2)^{1/2}}\exp\left[-\frac{(x-\mu)^2}{2\sigma^2}\right]$	$-\infty < x < \infty$
2	Logarithmic normal	—	$\frac{1}{(2\pi\sigma^2)^{1/2}}\frac{1}{y}\exp\left[-\frac{(\log y-\mu)^2}{2\sigma^2}\right]$	$0 < y < \infty$
3	Gamma	$\Gamma(p,\sigma)$	$[\Gamma(p)]^{-1}\sigma^{-p}x^{p-1}e^{-x/\sigma}$	$0 < x < \infty$
4	Exponential	$e(\mu,\sigma)$	$(1/\sigma)\exp(-(x-\mu)/\sigma)$	$\mu < x < \infty$
5	Two-sided exponential	—	$(1/2\sigma)e^{-\lvert x\rvert/\sigma}$	$-\infty < x < \infty$
6	Chi square	$\chi^2(n)$	$2^{-n/2}[\Gamma(n/2)]^{-1}x^{(n/2)-1}e^{-x/2}$	$0 < x < \infty$
7	Beta	$B(p,q)$	$[B(p,q)]^{-1}x^{p-1}(1-x)^{q-1}$	$0 < x < 1$
8	F	$F(m,n)$	$2K_F x^{(m/2)-1}[1+(mx/n)]^{-(m+n)/2}$ $K_F \equiv [B(m/2,n/2)]^{-1}(m/n)^{m/2}$	$0 < x < \infty$
9	z	$z(m,n)$	$K_F e^{mz}[1+(me^{2z}/n)]^{-(m+n)/2}$ $K_F \equiv [B(m/2,n/2)]^{-1}(m/n)^{m/2}$	$-\infty < z < \infty$
10	t	$t(n)$	$[\sqrt{n}B(n/2,1/2)]^{-1}[1+(t^2/n)]^{-(n+1)/2}$	$-\infty < t < \infty$
11	Cauchy	$C(\mu,\sigma)$	$(\pi\sigma)^{-1}\left[1+\frac{(x-\mu)^2}{\sigma^2}\right]^{-1}$	$-\infty < x < \infty$
12	One-side stable with exponent 1/2	—	$c(2\pi)^{-1/2}x^{-3/2}\exp(-c^2/2x)$	$0 < x < \infty$
13	Uniform	$U(\alpha,\beta)$	$1/(\beta-\alpha)$	$\alpha < x < \beta$
14	Binomial	$Bin(n,p)$	$\binom{n}{x}p^x q^{n-x}$	$x = 0,1,2,\cdots,n$
15	Poisson	$P(\lambda)$	$e^{-\lambda}\lambda^x/x!$	$x = 0,1,2,\cdots$
16	Hypergeometric	$H(N,n,p)$	$\binom{N_p}{x}\binom{N_q}{n-x}\binom{N}{n}$	x, integer $0 \leqslant x \leqslant N_p$ $0 \leqslant n-x \leqslant N_q$

Continued

No.	Name	Symbol	Density or Probability Function	Domains
17	Negative binomial	$NB(m,p)$	$\Gamma(m+x)[\Gamma(m)x!]^{-1}p^m q^x$	$x=0,1,2,\cdots$
18	Geometric	$G(p)$	pq^x	$x=0,1,2,\cdots$
19	Logarithmic	—	$K_L q^x/x, K_L\equiv -1/\log p$	$x=1,2,3,\cdots$
20	Multidimensional normal	$N(\boldsymbol{\mu},\Sigma)$	$(2\pi)^{-k/2}\mid\Sigma\mid^{-1/2}$ $\cdot\exp[-(\boldsymbol{x}-\boldsymbol{\mu})\Sigma^{-1}(\boldsymbol{x}-\boldsymbol{\mu})'/2]$, $\boldsymbol{x}=(x_1,\cdots,x_k),\boldsymbol{\mu}=(\mu_1,\cdots,\mu_k)$, $\Sigma=(\sigma_{ij})$	$-\infty<x_1,\cdots,$ $x_k<\infty$
21	Dirichlet	—	$\frac{\Gamma(v_1+\cdots+v_{k+1})}{\Gamma(v_1)\cdots\Gamma(v_{k+1})}x_1^{v_1-1}\cdots x_{k+1}^{v_{k+1}-1}$, $x_{k+1}=1-(x_1+\cdots+x_k)$	$x_1,\cdots,x_k>0$, $x_1+\cdots+x_k<1$
22	Multinomial	—	$n!(x_1!\cdots x_{k+1}!)^{-1}p_1^{x_1}\cdots p_{k+1}^{x_{k+1}}$, $x_{k+1}=n-(x_1+\cdots+x_k)\geqslant 0$	$x_1,\cdots,x_k$ $=0,1,\cdots,n$ $x_1+\cdots+x_k\leqslant n$
23	Multidimensional hypergeometric	$M(n;p_1,\cdots,p_{k+1})$	$\binom{Np_1}{x_1}\cdots\binom{Np_{k+1}}{x_{k+1}}/\binom{N}{n}$ $x_{k+1}=n-(x_1+\cdots+x_k)$	$x_1,\cdots,x_k$ integers $0\leqslant x_i\leqslant Np_i$ $(i=1,\cdots,k+1)$
24	Negative multinomial	—	$\frac{\Gamma(m+x_1+\cdots+x_k)}{\Gamma(m)x_1!\cdots x_k!}p_0^m p_1^{x_1},\cdots,p_k^{x_k}$,	$x_1,\cdots,x_k$ $=0,1,2,\cdots$

A. 2
Distributions of Typical Random Variables

No.	Parameters	Characteristic Function	Mean	Variance
1	$-\infty<\mu<\infty$, $\sigma>0$	$\exp\left(\mathrm{i}\mu t-\frac{\sigma^2t^2}{2}\right)$	μ	σ^2
2	$-\infty<\mu<\infty$, $\sigma>0$	—	$e^{\mu+(\sigma^2/2)}$	$e^{2\mu}(e^{2\sigma^2}-e^{\sigma^2})$
3	$p,\sigma>0$	$(1-\mathrm{i}\sigma t)^{-p}$	σp	$\sigma^2 p$
4	$-\infty<\mu<\infty,\sigma>0$	$e^{\mathrm{i}\mu t}(1-\mathrm{i}\sigma t)^{-1}$	$\mu+\sigma$	σ^2
5	$\sigma>0$	$(1+\sigma^2t^2)^{-1}$	0	$2\sigma^2$

Continued

No.	Parameters	Characteristic Function	Mean	Variance
6	n positive integer	$(1-2it)^{-n/2}$	n	$2n$
7	$p,q>0$	—	$\frac{p}{p+q}$	$\frac{pq}{(p+q)^2(p+q+1)}$
8	m,n positive integers	—	$\frac{n}{n-2}(n>2)$	$\frac{2n^2(m+n-2)}{m(n-2)^2(n-4)}$ $(n>4)$
9	m,n positive integers	—		
10	n positive integer		$0(n>1)$	$n/(n-2)(n>2)$
11	$-\infty<\mu<\infty$, $\sigma>0$	$\exp(i\mu t-\sigma\mid t\mid)$	none	none
12	$c>0$	$\exp[-c\mid t\mid^{1/2}(1+it/\mid t\mid)]$	none	none
13	$-\infty<\alpha<\beta<\infty$	$(e^{i\beta t}-e^{i\alpha t})/[it(\beta-\alpha)]$	$(\alpha+\beta)/2$	$(\beta-\alpha)^2/12$
14	$p+q=1,p,q>0$, n positive integer	$(pe^{it}+q)^n$	np	npq
15	$\lambda>0$	$\exp[-\lambda(1-e^{it})]$	λ	λ
16	$p+q=1,p,q>0$, N,Np,n positive integers $N\geqslant n$	$(Nq)^{[n]}(N^{[n]})^{-1}\cdot$ $F(-n,-Np;Nq-n+1;e^{it})$, $m^{[n]}\equiv m!/(m-n)!$	np	$\frac{npq(N-n)}{N-1}$
17	$p+q=1,p,q>0$, $m>0$	$\frac{p^m}{(1-qe^{it})^m}$	$\frac{mq}{p}$	$\frac{mq}{p^2}$
18	$p+q=1,p,q>0$	$\frac{p}{1-qe^{it}}$	$\frac{q}{p}$	$\frac{q}{p^2}$
19	$p+q=1,p,q>0$	$-K_L\log(1-qe^{it})$	K_Lq/p	$K_Lq(1+K_Lq)/p^2$
20	$-\infty<\mu_1,\cdots,\mu_k<\infty$, $\sum$ symmetric positive definite quadratic form	$\exp\left(i\mu t'-\frac{t\Sigma t'}{2}\right)$, $t=(t_1,\cdots,t_k)$	$E(x_i)=\mu_i$	$V(x_i)=\sigma_{ii}$, $\mathrm{Cov}(x_i,x_j)=\sigma_{ij}$
21	$v_1,\cdots,v_{k+1}>0$	—	$E(x_i)=\frac{v_i}{v_1+\cdots+v_{k+1}}$	$V(x_i)=Cv_i(v_1+\cdots+v_{k+1}-v_i)$, $\mathrm{Cov}(x_i,x_j)=-Cv_iv_j$, $C\equiv(v_i+\cdots+v_{k+1})^{-2}\cdot(v_1+\cdots+v_{k+1}+1)$
22	$p_1+\cdots+p_{k+1}=1$, $p_1,\cdots,p_{k+1}>0$, n positive integer	$(p_1e^{it_1}+\cdots+p_ke^{it_k}+p_{k+1}e^{it_{k+1}})^n$	$E(x_i)=np_i$	$V(x_i)=Cnp_1(1-p_i)$, $\mathrm{Cov}(x_i,x_j)=-Cnp_ip_j$

Continued

No.	Parameters	Characteristic Function	Mean	Variance
23	$p_1+\cdots+p_{k+1}=1$, $p_1,\cdots,p_{k+1}>0$, $N, Np_1,\cdots,Np_k, n$ positive integers	—	$E(x_i)=np_i$	$V(x_i)=Cnp_i(1-p_i)$, $\text{Cov}(x_i,x_j)=-Cnp_ip_j$, $C\equiv\dfrac{N-n}{N-1}$
24	$p_0+p_1+\cdots+p_k=1$, $p_0,p_1,\cdots,p_k>0$, $m>0$	$p_0^m(1-p_i e^{it_1}-\cdots-p_k e^{it_k})^m$	$E(x_i)=\dfrac{mp_i}{p_0}$	$V(x_i)$ $=mp_i(p_0+p_i)/p_0^2$, $\text{Cov}(x_i,x_j)=mp_ip_j/p_0^2$

Appendix B

Tables

B. 1 Table of Binomial Probabilities
B. 2 Table of Random Digits
B. 3 Table of Poisson Probabilities
B. 4 Table of Standard Normal Distribution Function
B. 5 Table of χ^2 Distribution
B. 6 Table of t Distribution

B. 1 Table of Binomial Probabilities

$$P(X=k)=\binom{n}{k}p^k(1-p)^{n-k}$$

n	k	$p=0.1$	$p=0.2$	$p=0.3$	$p=0.4$	$p=0.5$
2	0	0.8100	0.6400	0.4900	0.3600	0.2500
	1	0.1800	0.3200	0.4200	0.4800	0.5000
	2	0.0100	0.0400	0.0900	0.1600	0.2500
3	0	0.7290	0.5120	0.3430	0.2160	0.1250
	1	0.2430	0.3840	0.4410	0.4320	0.3750
	2	0.0270	0.0960	0.1890	0.2880	0.3750
	3	0.0010	0.0080	0.0270	0.0640	0.1250
4	0	0.6561	0.4096	0.2401	0.1296	0.0625
	1	0.2916	0.4096	0.4116	0.3456	0.2500
	2	0.0486	0.1536	0.2646	0.3456	0.3750
	3	0.0036	0.0256	0.0756	0.1536	0.2500
	4	0.0001	0.0016	0.0081	0.0256	0.0625
5	0	0.5905	0.3277	0.1681	0.0778	0.0312
	1	0.3280	0.4096	0.3602	0.2592	0.1562
	2	0.0729	0.2048	0.3087	0.3456	0.3125
	3	0.0081	0.0512	0.1323	0.2304	0.3125
	4	0.0005	0.0064	0.0284	0.0768	0.1562
	5	0.0000	0.0003	0.0024	0.0102	0.0312
6	0	0.5314	0.2621	0.1176	0.0467	0.0156
	1	0.3543	0.3932	0.3025	0.1866	0.0938
	2	0.0984	0.2458	0.3241	0.3110	0.2344
	3	0.0146	0.0819	0.1852	0.2765	0.3125
	4	0.0012	0.0154	0.0595	0.1382	0.2344
	5	0.0001	0.0015	0.0102	0.0369	0.0938
	6	0.0000	0.0001	0.0007	0.0041	0.0156
7	0	0.4783	0.2097	0.0824	0.0280	0.0078
	1	0.3720	0.3670	0.2471	0.1306	0.0547
	2	0.1240	0.2753	0.3176	0.2613	0.1641
	3	0.0230	0.1147	0.2269	0.2903	0.2734
	4	0.0026	0.0287	0.0972	0.1935	0.2734
	5	0.0002	0.0043	0.0250	0.0774	0.1641
	6	0.0000	0.0004	0.0036	0.0172	0.0547
	7	0.0000	0.0000	0.0002	0.0016	0.0078
8	0	0.4305	0.1678	0.0576	0.0168	0.0039
	1	0.3826	0.3355	0.1977	0.0896	0.0312
	2	0.1488	0.2936	0.2965	0.2090	0.1094
	3	0.0331	0.1468	0.2541	0.2787	0.2188

Continued

n	k	$p=0.1$	$p=0.2$	$p=0.3$	$p=0.4$	$p=0.5$
8	4	0.0046	0.0459	0.1361	0.2322	0.2734
	5	0.0004	0.0092	0.0467	0.1239	0.2188
	6	0.0000	0.0011	0.0100	0.0413	0.1094
	7	0.0000	0.0001	0.0012	0.0079	0.0312
	8	0.0000	0.0000	0.0001	0.0007	0.0039
9	0	0.3874	0.1342	0.0404	0.0101	0.0020
	1	0.3874	0.3020	0.1556	0.0605	0.0176
	2	0.1722	0.3020	0.2668	0.1612	0.0703
	3	0.0446	0.1762	0.2668	0.2508	0.1641
	4	0.0074	0.0661	0.1715	0.2508	0.2461
	5	0.0008	0.0165	0.0735	0.1672	0.2461
	6	0.0001	0.0028	0.0210	0.0743	0.1641
	7	0.0000	0.0003	0.0039	0.0212	0.0703
	8	0.0000	0.0000	0.0004	0.0035	0.0176
	9	0.0000	0.0000	0.0000	0.0003	0.0020
10	0	0.3487	0.1074	0.0282	0.0060	0.0010
	1	0.3874	0.2684	0.1211	0.0403	0.0098
	2	0.1937	0.3020	0.2335	0.1209	0.0439
	3	0.0574	0.2013	0.2668	0.2150	0.1172
	4	0.0112	0.0881	0.2001	0.2508	0.2051
	5	0.0015	0.0264	0.1029	0.2007	0.2461
	6	0.0001	0.0055	0.0368	0.1115	0.2051
	7	0.0000	0.0008	0.0090	0.0425	0.1172
	8	0.0000	0.0001	0.0014	0.0106	0.0439
	9	0.0000	0.0000	0.0001	0.0016	0.0098
	10	0.0000	0.0000	0.0000	0.0001	0.0010
15	0	0.2059	0.0352	0.0047	0.0005	0.0000
	1	0.3432	0.1319	0.0305	0.0047	0.0005
	2	0.2669	0.2309	0.0916	0.0219	0.0032
	3	0.1285	0.2501	0.1700	0.0634	0.0139
	4	0.0428	0.1876	0.2186	0.1268	0.0417
	5	0.0105	0.1032	0.2061	0.1859	0.0916
	6	0.0019	0.0430	0.1472	0.2066	0.1527
	7	0.0003	0.0138	0.0811	0.1771	0.1964
	8	0.0000	0.0035	0.0348	0.1181	0.1964
	9	0.0000	0.0007	0.0116	0.0612	0.1527
	10	0.0000	0.0001	0.0030	0.0245	0.0916
	11	0.0000	0.0000	0.0006	0.0074	0.0417

Continued

n	k	$p = 0.1$	$p = 0.2$	$p = 0.3$	$p = 0.4$	$p = 0.5$
15	12	0.0000	0.0000	0.0001	0.0016	0.0139
	13	0.0000	0.0000	0.0000	0.0003	0.0032
	14	0.0000	0.0000	0.0000	0.0000	0.0005
	15	0.0000	0.0000	0.0000	0.0000	0.0000
20	0	0.1216	0.0115	0.0008	0.0000	0.0000
	1	0.2701	0.0576	0.0068	0.0005	0.0000
	2	0.2852	0.1369	0.0278	0.0031	0.0002
	3	0.1901	0.2054	0.0716	0.0123	0.0011
	4	0.0898	0.2182	0.1304	0.0350	0.0046
	5	0.0319	0.1746	0.1789	0.0746	0.0148
	6	0.0089	0.1091	0.1916	0.1244	0.0370
	7	0.0020	0.0545	0.1643	0.1659	0.0739
	8	0.0003	0.0222	0.1144	0.1797	0.1201
	9	0.0001	0.0074	0.0654	0.1597	0.1602
	10	0.0000	0.0020	0.0308	0.1171	0.1762
	11	0.0000	0.0005	0.0120	0.0710	0.1602
	12	0.0000	0.0001	0.0039	0.0355	0.1201
	13	0.0000	0.0000	0.0010	0.0146	0.0739
	14	0.0000	0.0000	0.0002	0.0049	0.0370
	15	0.0000	0.0000	0.0000	0.0013	0.0148
	16	0.0000	0.0000	0.0000	0.0003	0.0046
	17	0.0000	0.0000	0.0000	0.0000	0.0011
	18	0.0000	0.0000	0.0000	0.0000	0.0002
	19	0.0000	0.0000	0.0000	0.0000	0.0000
	20	0.0000	0.0000	0.0000	0.0000	0.0000

B. 2 Table of Random Digits

2671	4690	1550	2262	2597	8034	0785	2978	4409	0237
9111	0250	3275	7519	9740	4577	2064	2086	3398	1348
0391	6035	9230	4999	3332	0608	6113	0391	5789	9926
2475	2144	1886	2079	3004	9686	5669	4367	9306	2595
5336	5845	2095	6446	5694	3641	1085	8705	5416	9066
6808	0423	0155	1652	7897	4335	3567	7109	9690	3739
8525	0577	8904	9451	6726	0876	3818	7607	8854	3566
0398	0741	8787	3043	5063	0617	1770	5048	7721	7032
3623	9636	3638	1406	5731	3978	8068	7238	9715	3363
0739	2644	4917	8866	3632	5399	5175	7422	2476	2607
6713	3041	8133	8749	8835	6745	3597	3476	3816	3455
7775	9315	0432	8327	0861	1515	2297	3375	3713	9174
8599	2122	6842	9202	0810	2936	1514	2090	3067	3574
7955	3759	5254	1126	5553	4713	9605	7909	1658	5490
4766	0070	7260	6033	7997	0109	5993	7592	5436	1727
5165	1670	2534	8811	8231	3721	7947	5719	2640	1394
9111	0513	2751	8256	2931	7783	1281	6531	7259	6993
1667	1084	7889	8963	7018	8617	6381	0723	4926	4551
2145	4587	8585	2412	5431	4667	1942	7238	9613	2212
2739	5528	1481	7528	9368	1823	6979	2547	7268	2467
8769	5480	9160	5354	9700	1362	2774	7980	9157	8788
6531	9435	3422	2474	1475	0159	3414	5224	8399	5820
2937	4134	7120	2206	5084	9473	3958	7320	9878	8609
1581	3285	3727	8924	6204	0797	0882	5945	9375	9153
6268	1045	7076	1436	4165	0143	0293	4190	7171	7932
4293	0523	8625	1961	1039	2856	4889	4358	1492	3804
6936	4213	3212	7229	1230	0019	5998	9206	6753	3762
5334	7641	3258	3769	1362	2771	6124	9813	7915	8960
9373	1158	4418	8826	5665	5896	0358	4717	8232	4859
6968	6428	8950	5346	1741	2348	8143	5377	7695	0685
4229	0587	8794	4009	9691	4579	3302	7673	9629	5246
3087	7785	7097	5701	6639	0723	4819	0900	2713	7650
4891	8829	1642	2155	0796	0466	2946	2970	9143	6590
1055	2968	7911	7479	8199	9735	8271	5339	7058	2964
2983	2345	0568	4125	0894	8302	0506	6761	7706	4310
4026	3129	2968	8053	2797	4022	9838	9611	0975	2437
4075	0260	4256	0337	2355	9371	2954	6021	5783	2827
8488	5450	1327	7358	2034	8060	1788	6913	6123	9405
1976	1749	5742	4098	5887	4567	6064	2777	7830	5668
2793	4701	9466	9554	8294	2160	7486	1557	4769	2781

Continued

0916	6272	6825	7188	9611	1181	2301	5516	5451	6832
5961	1149	7946	1950	2010	0600	5655	0796	0569	4365
3222	4189	1891	8172	8731	4769	2782	1325	4238	9279
1176	7834	4600	9992	9449	5824	5344	1008	6678	1921
2369	8971	2314	4806	5071	8908	8274	4936	3357	4441
0041	4329	9265	0352	4764	9070	7527	7791	1094	2008
0803	8302	6814	2422	6351	0637	0514	0246	1845	8594
9965	7804	3930	8803	0268	1426	3130	3613	3947	8086
0011	2387	3148	7559	4216	2946	2865	6333	1916	2259
1767	9871	3914	5790	5287	7915	8959	1346	5482	9251
2604	3074	0504	3828	7881	0797	1094	4098	4940	7067
6930	4180	3074	0060	0909	3187	8991	0682	2385	2307
6160	9899	9084	5704	5666	3051	0325	4733	5905	9226
4884	1857	2847	2581	4870	1782	2980	0587	8797	5545
7294	2009	9020	0006	4309	3941	5645	6238	5052	4150
3478	4973	1056	3687	3145	5988	4214	5543	9185	9375
1764	1860	4150	2881	9895	2531	7363	8756	3724	9359
3025	0890	6436	3461	1411	0303	7422	2684	6256	3495
1771	3056	6630	4982	2386	2517	4747	5505	8785	8708
0254	1892	9066	4890	8716	2258	2452	3913	6790	6331
8537	9966	8224	9151	1855	8911	4422	1913	2000	1482
1475	0261	4465	4803	8231	6469	9935	4256	0648	7768
5209	5569	8410	3041	4325	7290	3381	5209	5571	9458
5456	5944	6038	3210	7165	0723	4820	1846	0005	3865
5043	6694	4853	8425	8571	1322	1052	1452	2486	1669
1719	0148	6977	1244	6443	5955	7945	1218	9391	6485
7432	2955	3933	8110	8585	1893	9218	7153	7566	6040
4926	4761	7812	7439	6436	3145	5934	7852	9095	9497
0769	0683	3768	1048	8519	2987	0124	3064	1881	3177
0805	3139	8514	5014	3274	6395	0549	3858	0820	6406
0204	7273	4964	5475	2648	6977	1371	6971	4850	6973
0092	1733	2349	2648	6609	5676	6445	3271	8867	3469
3139	4867	3666	9783	5088	4852	4143	7923	3858	0504
2033	7430	4389	7121	9982	0651	9110	9731	6421	4731
3921	0530	3605	8455	4205	7363	3081	3931	9331	1313
4111	9244	8135	9877	9529	9160	4407	9077	5306	0054

Continued

6573	1570	6654	3616	2049	7001	5185	7108	9270	6550
8515	8029	6880	4329	9367	1087	9549	1684	4338	5686
3590	2106	3245	1989	3529	3828	8091	6054	5656	3035
7212	9909	5005	7660	2620	6406	0690	4240	4070	6549
6701	0154	8806	1716	7029	6776	9465	8818	2886	3547
3777	9532	1333	8131	2929	6987	2408	0487	9172	6177
2495	3054	1692	0089	4090	2983	2136	8947	4625	7177
2073	8878	9742	3012	0042	3996	9930	1651	4982	9645
2252	8004	7840	2105	3033	8749	9153	2872	5100	8674
2104	2224	4052	2273	4753	4505	7156	5417	9725	7599
2371	0005	3844	6654	3246	4853	4301	8886	5217	1153
3270	1214	9649	1872	6930	9791	0248	2687	8126	1501
6209	7237	1966	5541	4224	7080	7630	6422	1160	5675
1309	9126	2920	4359	1726	0562	9654	4182	4097	7493
2406	8013	3634	6428	8091	5925	3923	1686	6097	9670
7365	9859	9378	7084	9402	9201	1815	7064	4324	7081
2889	4738	9929	1476	0785	3832	1281	5821	3690	9185
7951	3781	4755	6986	1659	5727	8108	9816	5759	4188
4548	6778	7672	9101	3911	8127	1918	8512	4197	6402
5701	8342	2852	4278	3343	9830	1756	0546	6717	3114
2187	7266	1210	3797	1636	7917	9933	3518	6923	6349
9360	6640	1315	6284	8265	7232	0291	3467	1088	7834
7850	7626	0745	1992	4998	7349	6451	6186	8916	4292
6186	9233	6571	0925	1748	5490	5264	3820	9829	1335

This table is adapted from *Handbook of Statistical Tables*, which is published by Addison-Wesley Publishing Company, 1962, Massachusetts, edited by Donald B. Owen.

B. 3 Table of Poisson Probabilities

$$P(X=k)=\frac{\lambda^k e^{-\lambda}}{k!}$$

k	$\lambda=0.1$	0.2	0.3	0.4	0.5	0.6	0.7	0.8	0.9	1.0
0	0.9048	0.8187	0.7408	0.6703	0.6065	0.5488	0.4966	0.4493	0.4066	0.3679
1	0.0905	0.1637	0.2222	0.2681	0.3033	0.3293	0.3476	0.3595	0.3659	0.3679
2	0.0045	0.0164	0.0333	0.0536	0.0758	0.0988	0.1217	0.1438	0.1647	0.1839
3	0.0002	0.0011	0.0033	0.0072	0.0126	0.0198	0.0284	0.0383	0.0494	0.0613
4	0.0000	0.0001	0.0003	0.0007	0.0016	0.0030	0.0050	0.0077	0.0111	0.0153
5	0.0000	0.0000	0.0000	0.0001	0.0002	0.0004	0.0007	0.0012	0.0020	0.0031
6	0.0000	0.0000	0.0000	0.0000	0.0000	0.0000	0.0001	0.0002	0.0003	0.0005
7	0.0000	0.0000	0.0000	0.0000	0.0000	0.0000	0.0000	0.0000	0.0000	0.0001
8	0.0000	0.0000	0.0000	0.0000	0.0000	0.0000	0.0000	0.0000	0.0000	0.0000

k	$\lambda=1.5$	2	3	4	5	6	7	8	9	10
0	0.2231	0.1353	0.0498	0.0183	0.0067	0.0025	0.0009	0.0003	0.0001	0.0000
1	0.3347	0.2707	0.1494	0.0733	0.0337	0.0149	0.0064	0.0027	0.0011	0.0005
2	0.2510	0.2707	0.2240	0.1465	0.0842	0.0446	0.0223	0.0107	0.0050	0.0023
3	0.1255	0.1804	0.2240	0.1954	0.1404	0.0892	0.0521	0.0286	0.0150	0.0076
4	0.0471	0.0902	0.1680	0.1954	0.1755	0.1339	0.0912	0.0573	0.0337	0.0189
5	0.0141	0.0361	0.1008	0.1563	0.1755	0.1606	0.1277	0.0916	0.0607	0.0378
6	0.0035	0.0120	0.0504	0.1042	0.1462	0.1606	0.1490	0.1221	0.0911	0.0631
7	0.0008	0.0034	0.0216	0.0595	0.1044	0.1377	0.1490	0.1396	0.1171	0.0901
8	0.0001	0.0009	0.0081	0.0298	0.0653	0.1033	0.1304	0.1396	0.1318	0.1126
9	0.0000	0.0002	0.0027	0.0132	0.0363	0.0688	0.1014	0.1241	0.1318	0.1251
10	0.0000	0.0000	0.0008	0.0053	0.0181	0.0413	0.0710	0.0993	0.1186	0.1251
11	0.0000	0.0000	0.0002	0.0019	0.0082	0.0225	0.0452	0.0722	0.0970	0.1137
12	0.0000	0.0000	0.0001	0.0006	0.0034	0.0113	0.0264	0.0481	0.0728	0.0948
13	0.0000	0.0000	0.0000	0.0002	0.0013	0.0052	0.0142	0.0296	0.0504	0.0729
14	0.0000	0.0000	0.0000	0.0001	0.0005	0.0022	0.0071	0.0169	0.0324	0.0521
15	0.0000	0.0000	0.0000	0.0000	0.0002	0.0009	0.0033	0.0090	0.0194	0.0347
16	0.0000	0.0000	0.0000	0.0000	0.0000	0.0003	0.0014	0.0045	0.0109	0.0217
17	0.0000	0.0000	0.0000	0.0000	0.0000	0.0001	0.0006	0.0021	0.0058	0.0128
18	0.0000	0.0000	0.0000	0.0000	0.0000	0.0000	0.0002	0.0009	0.0029	0.0071

Continued

k	$\lambda=1.5$	2	3	4	5	6	7	8	9	10
19	0.0000	0.0000	0.0000	0.0000	0.0000	0.0000	0.0001	0.0004	0.0014	0.0037
20	0.0000	0.0000	0.0000	0.0000	0.0000	0.0000	0.0000	0.0002	0.0006	0.0019
21	0.0000	0.0000	0.0000	0.0000	0.0000	0.0000	0.0000	0.0001	0.0003	0.0009
22	0.0000	0.0000	0.0000	0.0000	0.0000	0.0000	0.0000	0.0000	0.0001	0.0004
23	0.0000	0.0000	0.0000	0.0000	0.0000	0.0000	0.0000	0.0000	0.0000	0.0002
24	0.0000	0.0000	0.0000	0.0000	0.0000	0.0000	0.0000	0.0000	0.0000	0.0001
25	0.0000	0.0000	0.0000	0.0000	0.0000	0.0000	0.0000	0.0000	0.0000	0.0000

B. 4 Table of Standard Normal Distribution Function

$$\Phi(x)=\int_{-\infty}^{x}\frac{1}{(2\pi)^{1/2}}\exp\left(-\frac{1}{2}u^2\right)\mathrm{d}u$$

x	$\Phi(x)$	x	$\Phi(x)$	x	$\Phi(x)$	x	$\Phi(x)$	x	$\Phi(x)$
0.00	0.5000	0.60	0.7257	1.20	0.8849	1.80	0.9641	2.40	0.9918
0.01	0.5040	0.61	0.7291	1.21	0.8869	1.81	0.9649	2.41	0.9920
0.02	0.5080	0.62	0.7324	1.22	0.8888	1.82	0.9656	2.42	0.9922
0.03	0.5120	0.63	0.7357	1.23	0.8907	1.83	0.9664	2.43	0.9925
0.04	0.5160	0.64	0.7389	1.24	0.8925	1.84	0.9671	2.44	0.9927
0.05	0.5199	0.65	0.7422	1.25	0.8944	1.85	0.9678	2.45	0.9929
0.06	0.5239	0.66	0.7454	1.26	0.8962	1.86	0.9686	2.46	0.9931
0.07	0.5279	0.67	0.7486	1.27	0.8980	1.87	0.9693	2.47	0.9932
0.08	0.5319	0.68	0.7517	1.28	0.8997	1.88	0.9699	2.48	0.9936
0.09	0.5359	0.69	0.7549	1.29	0.9015	1.89	0.9706	2.49	0.9938
0.10	0.5398	0.70	0.7580	1.30	0.9032	1.90	0.9713	2.50	0.9938
0.11	0.5438	0.71	0.7611	1.31	0.9049	1.91	0.9719	2.52	0.9941
0.12	0.5478	0.72	0.7642	1.32	0.9066	1.92	0.9726	2.54	0.9945
0.13	0.5517	0.73	0.7673	1.33	0.9082	1.93	0.9732	2.56	0.9948
0.14	0.5557	0.74	0.7704	1.34	0.9099	1.94	0.9738	2.58	0.9951
0.15	0.5596	0.75	0.7734	1.35	0.9115	1.95	0.9744	2.60	0.9953
0.16	0.5636	0.76	0.7764	1.36	0.9131	1.96	0.9750	2.62	0.9956
0.17	0.5675	0.77	0.7794	1.37	0.9147	1.97	0.9756	2.64	0.9959
0.18	0.5714	0.78	0.7823	1.38	0.9162	1.98	0.9761	2.66	0.9961
0.19	0.5753	0.79	0.7852	1.39	0.9177	1.99	0.9767	2.68	0.9963
0.20	0.5793	0.80	0.7881	1.40	0.9192	2.00	0.9773	2.70	0.9965
0.21	0.5832	0.81	0.7910	1.41	0.9207	2.01	0.9778	2.72	0.9967
0.22	0.5871	0.82	0.7939	1.42	0.9222	2.02	0.9783	2.74	0.9969
0.23	0.5910	0.83	0.7967	1.43	0.9236	2.03	0.9788	2.76	0.9971
0.24	0.5948	0.84	0.7995	1.44	0.9251	2.04	0.9793	2.78	0.9973
0.25	0.5987	0.85	0.8023	1.45	0.9265	2.05	0.9798	2.80	0.9974
0.26	0.6026	0.86	0.8051	1.46	0.9279	2.06	0.9803	2.82	0.9976
0.27	0.6064	0.87	0.8079	1.47	0.9292	2.07	0.9808	2.84	0.9977
0.28	0.6103	0.88	0.8106	1.48	0.9306	2.08	0.9812	2.86	0.9979
0.29	0.6141	0.89	0.8133	1.49	0.9319	2.09	0.9817	2.88	0.9980
0.30	0.6179	0.90	0.8159	1.50	0.9332	2.10	0.9821	2.90	0.9981
0.31	0.6217	0.91	0.8186	1.51	0.9345	2.11	0.9826	2.92	0.9983
0.32	0.6255	0.92	0.8212	1.52	0.9357	2.12	0.9830	2.94	0.9984
0.33	0.6293	0.93	0.8238	1.53	0.9370	2.13	0.9834	2.96	0.9985
0.34	0.6331	0.94	0.8264	1.54	0.9382	2.14	0.9838	2.98	0.9986

Continued

x	$\Phi(x)$	x	$\Phi(x)$	x	$\Phi(x)$	x	$\Phi(x)$	x	$\Phi(x)$
0.35	0.6368	0.95	0.8289	1.55	0.9394	2.15	0.9842	3.00	0.9987
0.36	0.6406	0.96	0.8315	1.56	0.9406	2.16	0.9846	3.05	0.9989
0.37	0.6443	0.97	0.8340	1.57	0.9418	2.17	0.9850	3.10	0.9990
0.38	0.6480	0.98	0.8365	1.58	0.9429	2.18	0.9854	3.15	0.9992
0.39	0.6517	0.99	0.8389	1.59	0.9441	2.19	0.9857	3.20	0.9993
0.40	0.6554	1.00	0.8413	1.60	0.9452	2.20	0.9861	3.25	0.9994
0.41	0.6591	1.01	0.8437	1.61	0.9463	2.21	0.9864	3.30	0.9995
0.42	0.6628	1.02	0.8461	1.62	0.9474	2.22	0.9868	3.35	0.9996
0.43	0.6664	1.03	0.8485	1.63	0.9485	2.23	0.9871	3.40	0.9997
0.44	0.6700	1.04	0.8508	1.64	0.9495	2.24	0.9875	3.45	0.9997
0.45	0.6736	1.05	0.8531	1.65	0.9505	2.25	0.9878	3.50	0.9998
0.46	0.6772	1.06	0.8554	1.66	0.9515	2.26	0.9881	3.55	0.9998
0.47	0.6808	1.07	0.8577	1.67	0.9525	2.27	0.9884	3.60	0.9998
0.48	0.6844	1.08	0.8599	1.68	0.9535	2.28	0.9887	3.65	0.9999
0.49	0.6879	1.09	0.8621	1.69	0.9545	2.29	0.9890	3.70	0.9999
0.50	0.6915	1.10	0.8643	1.70	0.9554	2.30	0.9893	3.75	0.9999
0.51	0.6950	1.11	0.8665	1.71	0.9564	2.31	0.9896	3.80	0.9999
0.52	0.6985	1.12	0.8686	1.72	0.9573	2.32	0.9898	3.85	0.9999
0.53	0.7019	1.13	0.8708	1.73	0.9582	2.33	0.9901	3.90	1.0000
0.54	0.7054	1.14	0.8729	1.74	0.9591	2.34	0.9904	3.95	1.0000
0.55	0.7088	1.15	0.8749	1.75	0.9599	2.35	0.9906	4.00	1.0000
0.56	0.7213	1.16	0.8770	1.76	0.9608	2.36	0.9909		
0.57	0.7157	1.17	0.8790	1.77	0.9616	2.37	0.9911		
0.58	0.7190	1.18	0.8810	1.78	0.9625	2.38	0.9913		
0.59	0.7224	1.19	0.8830	1.79	0.9633	2.39	0.9916		

This table is adapted from *Handbook of Statistical Tables*, which is published by Addison-Wesley Publishing Company, 1962, Massachusetts, edited by Donald B. Owen.

B. 5 Table of χ^2 Distribution

If X has a χ^2 distribution with n degree of freedom, this table gives the value of x such that $P(X \leqslant x) = p$.

n	p								
	0. 005	0. 01	0. 025	0. 05	0. 10	0. 20	0. 25	0. 30	0. 40
1	0. 0000	0. 0002	0. 0010	0. 0039	0. 0158	0. 0642	0. 1015	0. 1484	0. 2750
2	0. 0100	0. 0201	0. 0506	0. 1026	0. 2107	0. 4463	0. 5754	0. 7133	1. 022
3	0. 0717	0. 1148	0. 2158	0. 3518	0. 5844	1. 005	1. 213	1. 424	1. 869
4	0. 2070	0. 2971	0. 4844	0. 7107	1. 064	1. 649	1. 923	2. 195	2. 753
5	0. 4177	0. 5543	0. 8312	1. 145	1. 610	2. 343	2. 675	3. 000	3. 655
6	0. 6757	0. 8721	1. 237	1. 635	2. 204	3. 070	3. 455	3. 828	4. 570
7	0. 9893	1. 239	1. 690	. 2167	2. 833	3. 822	4. 255	4. 671	5. 493
8	1. 344	1. 647	2. 180	2. 732	3. 490	4. 594	5. 071	5. 527	6. 423
9	1. 735	2. 088	2. 700	3. 325	4. 168	5. 380	5. 899	6. 393	7. 357
10	2. 156	2. 558	3. 247	3. 940	4. 865	6. 179	6. 737	7. 267	8. 295
11	2. 603	3. 053	3. 816	4. 575	5. 578	6. 989	7. 584	8. 148	9. 237
12	3. 074	3. 571	4. 404	5. 226	6. 304	7. 807	8. 438	9. 034	10. 18
13	3. 565	4. 107	5. 009	5. 892	7. 042	8. 634	9. 299	9. 926	11. 13
14	4. 075	4. 660	5. 629	6. 571	7. 790	9. 467	10. 17	10. 82	12. 08
15	4. 601	5. 229	6. 262	7. 261	8. 547	10. 31	11. 04	11. 72	13. 03
16	5. 142	5. 812	6. 908	7. 962	9. 312	11. 15	11. 91	12. 62	13. 98
17	5. 697	6. 408	7. 564	8. 672	10. 09	12. 00	12. 79	13. 53	14. 94
18	6. 265	7. 015	8. 231	9. 390	10. 86	12. 86	13. 68	14. 43	15. 89
19	6. 844	7. 633	8. 907	10. 12	11. 65	13. 72	14. 56	15. 35	16. 85
20	7. 434	8. 260	9. 591	10. 85	12. 44	14. 58	15. 45	16. 27	17. 81
21	8. 034	8. 897	10. 28	11. 59	13. 24	15. 44	16. 34	17. 18	18. 77
22	8. 643	9. 542	10. 98	12. 34	14. 04	16. 31	17. 24	18. 10	19. 73
23	9. 260	10. 20	11. 69	13. 09	14. 85	17. 19	18. 14	19. 02	20. 69
24	9. 886	10. 86	12. 40	13. 85	15. 66	18. 06	19. 04	19. 94	21. 65
25	10. 52	11. 52	13. 12	14. 61	16. 47	18. 94	19. 94	20. 87	22. 62
30	13. 79	14. 95	16. 79	18. 49	20. 60	23. 36	24. 48	25. 51	27. 44
40	20. 71	22. 16	24. 43	26. 51	29. 05	32. 34	33. 66	34. 87	36. 16
50	27. 99	29. 71	32. 36	34. 76	37. 69	41. 45	42. 94	44. 31	46. 86
60	35. 53	37. 48	40. 48	43. 19	46. 46	50. 64	52. 29	53. 81	56. 62
70	43. 27	45. 44	48. 76	51. 74	55. 33	59. 90	61. 70	63. 35	66. 40
80	51. 17	53. 54	57. 15	60. 39	64. 28	69. 21	71. 14	72. 92	76. 19
90	59. 20	61. 75	65. 65	69. 13	73. 29	78. 56	80. 62	82. 51	85. 99
100	67. 33	70. 06	74. 22	77. 93	82. 86	87. 95	90. 13	92. 13	95. 81

This table is adapted from *Biometrika Tables for Statisticians* (Vol. 1, 3rd ed.), which is published by Cambridge University Press, 1966, edited by E. S. Pearson and H. O. Hartley; and from "A New Table of Percentage Points of the Chi-square distribution," *Biometrika*, Vol.51(1964), pp. 231-239, by H. L. Harter, Aerospace Research Laboratories.

Continued

0.50	0.60	0.70	0.75	0.80	0.90	0.95	0.975	0.99	0.995
0.4549	0.7083	1.074	1.323	1.642	2.706	3.841	5.024	6.635	7.879
1.386	1.833	2.408	2.773	3.219	4.605	5.991	7.378	9.210	10.60
2.336	2.946	3.665	4.108	4.642	6.251	7.815	9.348	11.34	12.84
3.357	4.045	4.878	5.385	5.989	7.779	9.488	11.14	13.28	14.86
4.351	5.132	6.064	6.626	7.289	9.236	11.07	12.83	15.09	16.75
5.348	6.211	7.231	7.841	8.558	10.64	12.59	14.45	16.81	18.55
6.346	7.283	8.383	9.037	9.803	12.02	14.07	16.01	18.48	20.28
7.344	8.351	9.524	10.22	11.03	13.36	15.51	17.53	20.09	21.95
8.343	9.414	10.66	11.39	12.24	14.68	16.92	19.02	21.67	23.59
9.342	10.47	11.78	12.55	13.44	15.99	18.31	20.48	23.21	25.19
10.34	11.53	12.90	13.70	14.63	17.27	19.68	21.92	24.72	26.76
11.34	12.58	14.01	14.85	15.81	18.55	21.03	23.34	26.22	28.30
12.34	13.64	15.12	15.98	16.98	19.81	22.36	24.74	27.69	29.82
13.34	14.69	16.22	17.12	18.15	21.06	23.68	26.12	29.14	31.32
14.34	15.73	17.32	18.25	19.31	22.31	25.00	27.49	30.58	32.80
15.34	16.78	18.42	19.37	20.47	23.54	26.30	28.05	32.00	34.27
16.34	17.82	19.51	20.49	21.61	24.77	27.59	30.19	33.41	35.72
17.34	18.87	20.60	21.60	22.76	25.99	28.87	31.53	34.81	37.16
18.34	19.91	21.69	22.72	23.90	27.20	30.14	32.85	36.19	38.58
19.34	20.95	22.77	23.83	25.04	28.41	31.41	34.17	37.57	40.00
20.34	21.99	23.86	24.93	26.17	29.62	32.67	35.48	38.93	41.40
21.34	23.03	24.94	26.04	27.30	30.81	33.92	36.78	40.29	42.80
22.34	24.07	26.02	27.14	28.43	32.01	35.17	38.08	41.64	44.18
23.34	25.11	27.10	28.24	29.55	33.20	36.42	39.36	42.98	45.56
24.34	26.14	28.17	29.34	30.68	34.38	37.65	40.65	44.31	46.93
29.34	31.32	33.53	34.80	36.25	40.26	43.07	46.98	50.89	53.67
39.34	41.62	44.16	45.62	47.27	51.81	55.76	59.34	63.69	66.77
49.33	51.89	54.72	56.33	58.16	63.17	67.51	71.42	76.15	79.49
59.33	62.13	65.23	66.98	68.97	74.40	79.08	83.30	88.38	91.95
69.33	72.36	75.69	77.58	79.71	85.53	90.53	95.02	100.4	104.2
79.33	82.57	86.12	88.13	90.41	96.58	101.9	106.6	112.3	116.3
89.33	92.76	96.52	98.65	101.1	107.6	113.1	118.1	124.1	128.3
99.33	102.9	106.9	109.1	111.7	118.5	124.3	129.6	135.8	140.2

B.6 Table of t Distribution

If X has a t distribution with n degree of freedom, the table gives the value of x such that $P(X \leqslant x) = p$.

n	$p=0.55$	0.60	0.65	0.70	0.75	0.80	0.85	0.90	0.95	0.975	0.99	0.995
1	0.158	0.325	0.510	0.727	1.000	1.376	1.963	3.078	6.314	12.706	31.821	63.657
2	0.142	0.289	0.445	0.617	0.816	1.061	1.386	1.886	2.920	4.303	6.965	9.925
3	0.137	0.277	0.424	0.584	0.765	0.978	1.250	1.638	2.353	3.182	4.541	5.841
4	0.134	0.271	0.414	0.569	0.741	0.941	1.190	1.533	2.132	2.776	3.747	4.604
5	0.132	0.267	0.408	0.559	0.727	0.920	1.156	1.476	2.015	2.571	3.365	4.032
6	0.131	0.265	0.404	0.553	0.718	0.906	1.134	1.440	1.943	2.447	3.143	3.707
7	0.130	0.263	0.402	0.549	0.711	0.896	1.119	1.415	1.895	2.365	2.998	3.499
8	0.130	0.262	0.399	0.546	0.706	0.889	1.108	1.397	1.860	2.306	2.896	3.355
9	0.129	0.261	0.398	0.543	0.703	0.883	1.100	1.383	1.833	2.262	2.821	3.250
10	0.129	0.260	0.397	0.542	0.700	0.879	1.093	1.372	1.812	2.228	2.764	3.169
11	0.129	0.260	0.396	0.540	0.697	0.876	1.088	1.363	1.796	2.201	2.718	3.106
12	0.128	0.259	0.395	0.539	0.695	0.873	1.083	1.356	1.782	2.179	2.681	3.055
13	0.128	0.259	0.394	0.538	0.694	0.870	1.079	1.350	1.771	2.160	2.650	3.012
14	0.128	0.258	0.393	0.537	0.692	0.868	1.076	1.345	1.761	2.145	2.624	2.977
15	0.128	0.258	0.393	0.536	0.691	0.866	1.074	1.341	1.753	2.131	2.602	2.947
16	0.128	0.258	0.392	0.535	0.690	0.865	1.071	1.337	1.746	2.120	2.583	2.921
17	0.128	0.257	0.392	0.534	0.689	0.863	1.069	1.333	1.740	2.110	2.567	2.898
18	0.127	0.257	0.392	0.534	0.688	0.862	1.067	1.330	1.734	2.101	2.552	2.878
19	0.127	0.257	0.391	0.533	0.688	0.861	1.066	1.328	1.729	2.093	2.539	2.861
20	0.127	0.257	0.391	0.533	0.687	0.860	1.064	1.325	1.725	2.086	2.528	2.845
21	0.127	0.257	0.391	0.532	0.686	0.859	1.063	1.323	1.721	2.080	2.518	2.831
22	0.127	0.256	0.390	0.532	0.686	0.858	1.061	1.321	1.717	2.074	2.508	2.819
23	0.127	0.256	0.390	0.532	0.685	0.858	1.060	1.319	1.714	2.069	2.500	2.807
24	0.127	0.256	0.390	0.531	0.685	0.857	1.059	1.318	1.711	2.064	2.492	2.797
25	0.127	0.256	0.390	0.531	0.684	0.856	1.058	1.316	1.708	2.060	2.485	2.787
26	0.127	0.256	0.390	0.531	0.684	0.856	1.058	1.315	1.706	2.056	2.479	2.779
27	0.127	0.256	0.389	0.531	0.684	0.855	1.057	1.314	1.703	2.052	2.473	2.771
28	0.127	0.256	0.389	0.530	0.683	0.855	1.056	1.313	1.701	2.048	2.467	2.763
29	0.127	0.256	0.389	0.530	0.683	0.854	1.055	1.311	1.699	2.045	2.462	2.756
30	0.127	0.256	0.389	0.530	0.683	0.854	1.055	1.310	1.697	2.042	2.457	2.750
40	0.126	0.255	0.388	0.529	0.681	0.851	1.050	1.303	1.684	2.021	2.423	2.704
60	0.126	0.254	0.387	0.527	0.679	0.848	1.046	1.296	1.671	2.000	2.390	2.660
120	0.126	0.254	0.386	0.526	0.677	0.845	1.041	1.289	1.658	1.980	2.358	2.617
∞	0.126	0.253	0.385	0.524	0.674	0.842	1.036	1.282	1.645	1.960	2.326	2.576

This table is taken from Table Ⅲ of Fisher & Yates: *Statistical Tables for Biological, Agricultural and Medical Research*, published by Longman Group Ltd., London (previously published by Oliver and Boyd Ltd., Edinbrugh).